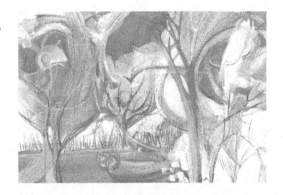

More and *Different*

notes from a thoughtful curmudgeon

Philip W. Anderson

Princeton University, USA

More and *Different*

notes from a thoughtful curmudgeon

 **World Scientific**

NEW JERSEY · LONDON · SINGAPORE · BEIJING · SHANGHAI · HONG KONG · TAIPEI · CHENNAI

Published by

World Scientific Publishing Co. Pte. Ltd.

5 Toh Tuck Link, Singapore 596224

USA office: 27 Warren Street, Suite 401-402, Hackensack, NJ 07601

UK office: 57 Shelton Street, Covent Garden, London WC2H 9HE

Library of Congress Cataloging-in-Publication Data
Anderson, P. W. (Philip W.), 1923–
  More and different : notes from a thoughtful curmudgeon / by Philip W. Anderson.
    p. c
  ISBN-13 978-981-4350-12-9 (hard cover : alk. paper)
  ISBN-13 978-981-4350-13-6 (pbk : alk. paper)
  1. Science. I. Titl
  Q171.A527 2011
  500--dc23

                    2011017192

British Library Cataloguing-in-Publication Data
A catalogue record for this book is available from the British Library.

Printed in Singapore.

Preface

About 1990 I began to imagine writing as an essential part of my retirement plans. I had recognized that running a research group past the age of 75 was unlikely to be subsidized by the agencies, nor was I eager to go on writing proposals, even though most of the burden was assumed by the excellent staff team Princeton physics has managed to maintain. In my writing I was encouraged, first of all, by Gloria Lubkin with her "Reference Frame" column in *Physics Today*, for which she chose me as one of an initial handful of contributors; the book which follows contains a few of these columns. I had all along responded to the occasional request to do a book review for *Science, Nature, Physics World,* and *Physics Today*, but reviewing too became a quasiregular assignment when my friend Andrew Robinson took over science books for the *Times (London) Higher Education Supplement*, and again several THES reviews are here. My dream of giving up physics for writing, however, never materialized; physics continues to dominate my professional life, and I probably produced many more technical than expository words. It turns out, contrary to my expectations, that funding is by no means necessary for creativity, in fact the relationship may be inverse. If travel subsidies are to be had (and an internet server), a more or less full life, at least in theoretical physics, can continue without direct government subsidy. So the product I've mined for the present volume is thinner than it should be.

I hear of the occasional book or article nowadays which denies the cliches about age and math or physics ability, which say that no one does anything original after the age of 35, and the like. I'm sorry I never wrote an essay on the subject, because in my experience the cliché is wrong, wrong, wrong. I think it is more like this — everyone has an age at which he stops thinking originally, but that age is enormously variable. I know good physicists whose useful lifespan is in the forties or younger — the name I have for them is "young fogies" — but I know plenty of others who haven't reached it at 80 or more. One contributing factor to the misappprehension is that the time for recognition of a revolutionary advance is measured in decades, so those of the later years often outlive their authors. Anyhow, forgive my going on about a pet subject — to my surprise physics has remained exciting and wholly absorbing as I age, and while I love to write, that occupation has always been secondary.

Selection — there is hardly a subject within science that doesn't fascinate me (I read *Science* magazine for pleasure, and perhaps it was my father's *Science* subscription that hooked me in the first place). Not to mention a lot of other subjects — I'd love to write about early medieval history if given the chance, for instance — but fortunately I am seldom allowed to sound off in print on any of my other hobbies. What is here is mostly what someone or other has asked me to write or speak about.

I am informed, by those from the publishing business who should know, that collections of reviews don't sell. Therefore I've tried to lard this book with other writings, some almost as brief as book reviews but others more substantial, pieces based for instance on public lectures I've given. But almost every book review I've done contains some thoughts of my own on the subject, so I do not apologize for them. There are a lot I didn't include, but I wanted to keep the book at a manageable size.

I must acknowledge the indefatigable services of Hongjun Li in helping to assemble the contents of this book from the incredible hodgepodge of my files, and putting up with my many foibles along the way. The book would not have existed without her patience and persistence.

Last of all — at least until 2008 I was the beneficiary, for most of my nontechnical writing, of the services of one of the world's best editors: Joyce Gothwaite Anderson, my wife of 63 years (so far). If anyone finds an unnecessary adverb it's my fault, not hers.

Contents

IV. *Science Tactics and Strategy*

V. *Genius*

VI. *Science Wars*

1. Personal Reminiscences

Introduction

This and the piece titled "A Mile of Dirty Lead Wire" are chapters of an unfinished book on the history of superconductivity which I worked on desultorily in spare moments in the '70s and early '80s. Clearly they predate "high T_c" proper and postdate the Standard Model. I include them for the bits of personal history they contain — they will serve as a personal introduction — and for some insights on various characters. I bogged down on the attempt to explain the physics in popularly acceptable form — I have enormous appreciation for those who can do so, I can't.

"BCS" and Me

1. A Personal Interlude

My buddy in the Antenna Group at NRL in Anacostia, D.C., Don Antaya, was drafted and made a Lieutenant (jg); I, who wore glasses, a Chief Specialist (X), but we went on working where we were except for a somewhat dubious interval of "boot" training at which we were taught to "fire" empty '38s. Don's girl from home in Mansfield, Mass., Clytemnestra — we called her "Clit" in those innocent days — was a merry, dynamic red-blonde who broke up when she saw me in my new dress whites: she thought I looked like a street sweeper in an old movie! Some of us failed to fulfill the image the uniform designer had in mind.

Not enough credit has been given to the U.S. Navy for its sensible treatment of science and scientists in those crucial years between the war and the building of the present structure of civilian management and funding of science. The arrangement I enjoyed, and its abrupt termination six weeks after VJ day, were good examples. By October 5, 1945, I was attending graduate courses in my (hopefully) less ridiculous Harvard bow tie and sports jacket.

Very immature socially, and provincial to boot, I'd been bundled through undergraduate school in three fast years to get off to war at 19. So the first two years of graduate school were for me a period of general intellectual and social growth that most of my contemporaries had had in college or high school, and while I kept on top of my courses I was not, like Freeman Dyson or the group of students under Fermi at Chicago or Oppy at Berkeley, immersed in the great and exciting things which were happening in physics at that time, or even aware that I might contribute to them. Two Nobel prizes at least were being earned at Harvard in those two or three years. The 29-year-old full professor, Julian Schwinger, was one of the major figures in the quantum electrodynamics revolution. The graduate student consensus was that one *must* attend his advanced course, which I did, vaguely realizing that it was full of wholly new ways of looking at physics. At the same time Nico Bloembergen, a Dutch postdoc, was collaborating with Purcell and Pound to discover and exploit nuclear magnetic resonance. (Purcell and Felix Bloch shared that prize; Nico has only just gotten it, nominally for his (also superb) later work.) The man who was to become my professor, John Van Vleck, was the theoretical grey eminence of that effort, and the next generation of Purcell students were classmates, so I absorbed that stuff partly by osmosis in my usual lazy fashion, hanging around the lab with Al Sachs, for instance, while he wound magnets.

Mostly, however, I gravitated to a group which was fairly unusual for graduate students in containing men from an assortment of subjects. We had two mathematicians (Chandler Davis, whom the HUAC later sent to Danbury prison, and Tom Lehrer of the songs who was well on his way to achieve the greatest eminence of all); a guitar-playing English major; an engineer; an anthropologist; a biologist; and several chemical physicists. We played bridge, did acrostic puzzles, sang, ate out, drank when we could afford it, went to the beach, read (and, in Chan's case, wrote) science fiction, and in general had a great time. The kind of all-consuming ambition to reach the intellectual core of things that Jim Watson talks about in *The Double Helix*, or Freeman Dyson hints at among the Fermi group and the Los Alamos alumni, and that was undoubtedly present in the big group of Schwinger-followers at Harvard, was far from our minds, or at least far from mine.

I had rather casually observed that there was a lot of action going on in the new field of radio-frequency coherent spectroscopy (the NMR work already mentioned, plus at least the chemical physics being done by several of my friends such as Dave Robinson under Bright Wilson) and that Van Vleck seemed to be in the center of it; all of my instincts rejected the competitive, intellectually snobbish, even somewhat sycophantic group around Schwinger, perhaps for the negative reason that I could see that everyone else was having more what I deemed to be fun than they were. So in early '47, after barely passing my qualifying oral, but being permitted to do a theoretical thesis (my classical mechanics was ever weak) I chose to work with Van in that area.

In hindsight, it was a beautifully chosen topic, at least for a good student. Wartime electronics had made available all kinds of microwave electronics gear, and immediately prewar and during the war a number

of people, Van among them, had pointed out a very rich variety of fundamental molecular spectroscopy which could be done with this gear.

Atoms and molecules, in isolation, are characterized by sequences of
"energy levels" in quantum mechanics, pairs of these being connected by
"spectral lines": more or less sharp frequencies where the atoms absorb
energy according to Planck's relation $hv = E_1 - E_2$. These sharp lines are
familiar in ordinary light, for instance from the sun, but to see them with
coherent, man-made radiation from electron tubes was exciting, especially given that the pre-laser precision of optical spectroscopy was laugh-
able compared to microwaves. This made precise measurement of width
and shape easy. Oceans of data were pouring in from at least four major
experimental centers. The positions of most of the new lines were full of
molecular information, but finding it was a matter of rather routine calculation. The theory of their shapes and breadths was a different matter. A
lot of concepts had been left around in the literature by early masters such
as Lorentz, Wigner, the German physicist Kuhn who had worked with
Landau, and Viki Weisskopf, then at MIT. But only a couple of papers
really addressed the problem from first principles by relating it to the real,
calculable forces between molecules. Like all good problems, it could be
attacked at any level. I could take a tiny piece and relate it to existing theory; I could become the guru of a whole field without really solving much
of anything, by being useful in correlating experimental data; or I could
give the existing methods the kind of major boost which usually happens
if you've got a *lot* of quantitative questions instead of a few qualitative
ones to ask.

For quite a while, I did none of the above. During the spring and summer I carried the two or three papers I was supposed to study back and
forth to Urbana and to New York several times, perhaps in the hope that
the solution might rub off from them. The major outcome of this period
in my life was marriage. By the fall, Joyce was finishing a semester of
graduate school and teaching back in Urbana, and I returned with the
intention of starting serious work on my problem. At first, even with a lot
of library work, nothing seemed to be happening. At this time I believe
Van called me in to see if I was alive, and gave me a little side homework problem to see if I really understood *anything* about molecules and
intermolecular forces. (I did, by then.) But I reached the end of the
semester — which also brought the first job recruiters around — without
really seeing where the problem and I fit together.

I have always admired Jim Watson's book *The Double Helix*, and
particularly the description which occupies the early parts of it, of the

process of getting into a research problem. Jim starts out at the age of 22 or so, totally ignorant of stereochemistry, x-ray diffraction, and most of the other things he needs to know, and thrashes about making a lot of embarrassing false starts, having fantasies of success and fame based on ridiculously wrong guesses; but what he almost hides by his frankness about all these dreams of glory is that he has learned, at the end, an enormous amount, and is totally surefooted when the real solution comes along. It is a perfect description of how doing research really feels — and one feels pretty hopeless after a year or so of false starts. I am afraid that most of the graybeard scientists (of all ages), who condemn Jim for speaking of these dreams and fears, are either hypocritical or have forgotten their youth (if they were really creative, that is).

That first job recruiter was almost a disaster. One was, in those days, honestly expected to earn a doctorate in three years at Harvard, one and a half for courses and one and a half for research. During one of my more optimistic days I accepted the Bell Labs recruiter's invitation to come give a talk about my problem at Bell, since I nominally had only one semester to go. Even though my training in solid state physics was fragmentary, I knew that Bell was something of a special place; in fact, one might have argued that at that time it had almost half of the productive solid state theorists in the world, since Bill Shockley and Jim Fisk had, almost miraculously, anticipated the revolution in electronics, which Bell was about to cause by inventing the transistor, and had recruited an incredibly powerful solid state research group.

Before this group I went with nothing — I now realize — but a scholarly dissertation on what everyone else had done before me. I vaguely remember a series of interviews with theorists and experimentalists which must have been all right — I know that what I always look for in an interview is "natural curiosity", and I surely had that — but of course the result was negative. This was a bit of a disaster since I had set my heart — strongly influenced by my mother, who was determined that I should not follow what she saw to be the frustrating, impecunious academic career of my father — on working in industry or possibly government (not to mention the need for solvency, since by now a child was on the way).

Rather suddenly — I remember groping in the right direction in the sleeper coming back from Bell — I began to see the answers to my problem. I still have a sheaf of informal notes of January–February '48 — which I filed under "Dear V YOB" following on to an earlier set saying "Dear Van" which was the answer to the homework problem — the acronym is "Van, you old bat", literally — and these contain the key ideas. Van left for one of his many honorific visiting professorships for the spring semester '48, and in that semester Joyce came to join me, my daughter was born, and I solved my problem. I did the third alternative suggested earlier — I clobbered it, developing a new method which in that field was eventually to be apothesized by at least two other authors in review papers, and is still used. The implications of what I did still reverberate in the methodology of modern physics, but that is another story, and it was not until much later that I understood how revolutionary some of it was. I had effectively combined some of Schwinger's new mathematical tools for dealing with abstract quantum-mechanical problems with Van's sense of the real physics of experimental systems of many particles, which had not in fact been done before in anything like that way. In my awe of my elders, I credited such gurus as Van, Weisskopf and Schwinger himself with really having understood the basic thrust of what I did, so I didn't make much of a fuss about methodology but concentrated on just solving the problem at hand. I've never quite made up my mind whether this combination of modesty with a distaste for the purely formal and theoretical was a good or bad thing at the time; I've retained the latter characteristic anyway.

My next recruiting trip, to the industrial labs at Westinghouse and G. E. and the government lab at Brookhaven, was more comfortable than the one to Bell but basically not more successful. G. E. was directed by Van's student Harvey Brooks, who later became his successor as dean at Harvard and the prototypical establishment-Washington science guru, but no offer was forthcoming although I met some very congenial scientists with whom I thought I got along. At Brookhaven the great Sam Goudsmit, an idol of mine after his witty summer course at Harvard and his vital role in the early quantum theory, asked what problem I wanted to work on. I said glibly "something else, but it's up to you; I did the one I had, and I see many things outside my field where my ideas and my brain might work" and Sam made what may have been one of his few mistakes in judgment — he later told me — and turned me down on the basis that I should surely see something definite to do. At Westinghouse was a theorist I admired from his work in my field of gas spectra, Ted Holstein, and in a separate building a big room with eight empty desks and a group leader (let him be nameless — he later graduated to a figurehead post in one of the defense department granting agencies) who wanted to fill the room with people in order to try to understand what was inside the little box of new-fangled transistors which Bell sent to Westinghouse according to some patent licensing arrangement. I was not impressed by that — even I knew that the men I'd met at Bell had published a complete explanation of how they worked, and must be far, far ahead. At the other three places I had been assured of some autonomy at least, if not complete scientific freedom, and I didn't want to settle for less; I also hadn't seen any eight-desk rooms.

Through the summer and fall we rewrote and retyped several versions of my massive 350-page thesis, tended Susan, scraped by on $70 a month, and worried about a job. It was a terrible year for that: the country may have forgotten that postwar economic dip just before the Korean war and the 50's take-off, but I haven't. One college recruiter alone showed any interest in *me,* with an offer of an assistant professorship — also financially viable — at Pullman State College in Eastern Washington. We agonized about Westinghouse, chose Pullman, and were preparing for the long winter drive West over the Rockies when Van finally asked me what I was going to do. I told him and he seemed as unimpressed as I with the alternatives. By now my best contemporaries were settled with prestigious postdocs at such glamorous places as Columbia, Copenhagen, or Birmingham (not for me: we had no money, we had Susan; it wasn't until ten years later that the salary structure accommodated to the married postdoc) or even, like Tom Kuhn, the Society of Fellows right there at Harvard. One or two, especially the few married ones, had somehow achieved assistant professorships at good universities.

After Van's death someone told me that he had been very proud of my thesis. He did invite Viki Weisskopf and Schwinger to join him on the examining committee, but that to me meant only a cause for trauma. I suppose that was the point of this interview, but again there was no word of praise or comment: only a dismissal of these possibilities and: "where do you *want* to go!" I said Bell, with Shockley and Kittel and Bardeen, and after some amusing maneuvers which Joyce wrote a little skit about called "Bellephony" that is where I went. Financially, all three offers were identical: $450 per month, which felt like easy street.

Four of us arrived at Bell within roughly a month of each other: the Swiss theorist Gregory Wannier, adding yet another jewel to the diadem of mature, active solid-state theorists Bell had collected; Jack Galt, an experimentalist back from a postdoctoral year in England; Bernd Matthias who will be a part of this story, and myself. Some of the complex relationships involved in — and resulting from — this coincidence are available to a speculative reconstruction, although you must realize that I, never very perceptive, had little idea as to what was going on around me.

Every viable institution has a "repressor" mechanism to prevent cancerous growth. Bell's is called "nosecount". Bernd and I, at least, of the four, were told we were hired "above nosecount" in that recession year. This implied extraordinary maneuvers by Shockley, in response to extraordinary pressure from Van and, I suspect, from Charlie Kittel and the Bell recruiters at Harvard. (The reason for those maneuvers in Bernd's case was more obvious: he made technologically interesting materials, and in fact came with the only process known for making crystalline barium titanate, a ferroelectric[a], very interesting to Bill at that time.) These maneuvers left me saddled with a couple of debits which I didn't really know about until much later. First, I was coming in strictly as a replacement for the man who had been working with Bill Shockley on his ideas about ferroelectrics, John Richardson. He was a large, craggy, sweet-tempered, brilliant man who later had an excellent career in various West Coast industrial labs, but who turned Bill off with his diffident but uncompromisingly formal and thermodynamic way of doing physics. John never seemed to resent me in the slightest — we actually did a couple of papers together, one with Gregory Wannier, which was symbolic of the easy, communicative atmosphere we all felt as a small knot of theoretical physicists in a big, technology-oriented milieu. Jack Galt and Bernd joined the group of jolly young bachelors of which John was already a member· I believe Jack at least moved into an apartment with John and a couple of others — and in this group the easing out of Richardson may have left a bad taste.

Bernd was something else. He was originally German, of a wealthy family with a Jewish mother who fled Germany and put him and his sister through a succession of schools in Switzerland. He had always — so one hears — used his wits, his charm, and his genuine if conveniently available personal warmth to achieve extraordinary things through his friends of both sexes and all ages, as well as to shortcircuit many of the conventional requirements of life. A dark shock of hair, brilliant eyes, a pale face, a slim, wiry, build; there was always a kind of electricity in the complete focus of his attention on the individual he was talking to. His

---

[a]A "ferroelectric" is a crystal which has a spontaneous dielectric polarization which can be reversed by external forces. The analogy is to "ferromagnetism", which is spontaneous *magnetic* polarization like iron bar magnets have. In principle, there is an electric field near a ferroelectric, although almost always this is neutralized by free charges from the air or on electrodes applied to the crystal. It has nothing to do with iron, which is seldom a component or the crystal. Ferroelectricity is useful nowadays for rather esoteric reasons in conjunction with laser technology, but the simple-minded applications we had in mind then (for instance, for computer memory) didn't work out.

German accent lent his English mellifluous undertones; he never wrote English well but spoke it with his own unique colloquial fluency and was widely read in German, selectively so in English, and knew several other languages. He exuded sophistication and machismo (a word not yet current, but perfect for his effect both on women and his contemporaries). He must have been mostly self-taught, since nothing on earth could make him learn what he didn't want to. He had a very sensitive feel for chemistry, for unusual properties of materials and how to measure them simply, and a contempt for received wisdom which was the exact opposite of my (then) diffidence in the face of authority. This of course partly reflected conditioning. I had been trained by Van (in whose works I have yet to find a misstatement), Schwinger, and their like, and even the experimentalists I had known worked in clean, well-ordered, well-lighted subjects where answers came out yes or no and surprises were rare; while Bernd grew up in the continental school of separate institutes, where the clean part of physics, and theory, were hermetically sealed away from the kind of materials-oriented things he and his professor did, in which the intellectual level has never been elevated. In that atmosphere it is not unusual to find professors of great authority and very limited competence managing large and well-financed institutes. This happens in the U.S. but less frequently, and I had not encountered the phenomenon. Bernd had learned little about the nature of theoretical physics, and his attitude toward authority was strictly ad personam; we were headed for confrontation.

My second hidden debit was in Bill Shockley's head. I had come to work with him on his ideas on ferroelectricity; he had fired the previous occupant of that slot, and he was determined he would fire me if I didn't do the job he wanted.

I don't need to introduce Bill to a general audience, but what I may need to explain is that he is one of the most brilliant men I have ever met, and that includes about all of the great theoretical physicists of my time. His quick, sure approach to all kinds of problems was a marvel. It is not widely known that as an intelligence exercise he and Jim Fisk made an entirely independent design for the Fermi atomic pile using no classified data — which was good enough to cause patent problems after the war. They also designed the atomic bomb from scratch on the same basis.

It is not at all uncommon for such overwhelming ability to be accompanied by what would be arrogance and overconfidence in one's own judgment in a lesser man. This was true of Pauli, for instance, or Landau, each of whom is known to have delayed or discouraged certain valid work on the basis of snap personal judgments. Bill made some similar technical misjudgments in the early days of semiconductor devices, which would normally now be forgotten in the face of his enormous achievements not only at Bell but in the early history of Silicon Valley. I feel Bill's better-known misjudgment of the realities of the IQ controversy, while it contains to some extent the kind of impatience with prejudice and apparent muddleheadedness which is like that of Pauli or Landau, also stems from a much deeper trait of personality, which seemed to be evident to many of us even all those years ago in his heyday at Bell. He had a lack of empathy, and a determination to see relationships with those around him as competitive or testing situations, which led more or less naturally to his later views. He is not, or at least wasn't, at all politically rightist or conservative in the usual sense. (He played a liberal, helpful role in the case of the famous battery-additive controversy involving the Bureau of Standards and E.U. Condon, for instance.)

His scientific work was characterized by quickness and clarity. I often have said of him that he could take the first few steps on a problem faster than anyone I know, but that he would not necessarily see as deeply into it as someone else might, and, in fact, none of his achievements that I know of have the depth of insight of a London, a Landau, or an Onsager (to mention some of my personal idols, i.e., put my cards on the table).

Bill had made his usual quick judgment about the theory of ferroelectricity, at least in Bernd's crystals of barium titanate, and it was my responsibility to prove him right, within limits. Even Bill would have accepted, I'm sure, a clear demonstration that he was wrong, but in fact I came very quickly to feel that Bill's concept was pretty close to the mark and I worked rather hard to give it what I thought was a somewhat sounder basis. Meanwhile I learned, not least from Bill himself, great quantities of solid state physics: what is a local field, how to do "Ewald" sums, crystallography, etc., etc. The kind of education I received is almost not available any more, now that solid state physics (which we now call condensed matter physics, because nowadays we're interested not only in solids but in liquids, gels, polymers, etc.) has divided itself up into specialties, each with its distinctive techniques and concepts, and the ideas of one specialty are often totally foreign — or even contradictory — to those of another. I was learning magnetism from Charlie Kittel, kinetic theory of gases and phase transitions from Gregory, crystal chemistry and crystallography from Bernd, Betty Wood and Alan Holden, statistical mechanics from John Richardson and Conyers Herring, and reading voraciously about a lot of other subjects.

Frankly, I thought I was doing pretty well, even with Bill's pet project for me. Looking back these many years later, in fact, I was doing very well indeed, in the sense that I had seen much more deeply into the problem than he ever had, and in that the three or four papers I wrote about various aspects of it were each forerunners of what, later, were to be major themes of research, either my own or others'. But I received little encouragement and what I did publish was incomplete or simply inserted into my collaborators' papers.

Bill seems, in fact, to have felt that I was a disaster. Fortunately we did not, at that time, have the hard and fast rules of personnel management that we now use at Bell Labs. When it came time to give me my small, very late raise, in the spring of 1950, the news was broken by Bill's long-suffering co-department head, Stan Morgan, and Stan carefully allowed me to ignore the fact that in the rapidly inflating economic climate of 1950, with universities expanding rapidly and industry increasingly aware of the new nuclear and electronic technologies spawned by the war, my raise was an open invitation to pack up and go to greener pastures. Later I was told that only bitter arguments from Stan and Kittel had saved my job.

While I labored mightily — for me — to make the detailed numerical calculations which Bill's idea required, my own theoretical studies convinced me that these calculations were rather meaningless except as a rough guide to one's thinking, and to some extent I lost interest. Therefore, when Charlie Kittel made me aware of some things happening in magnetism which Gregory and I could do something about very simply and immediately, I was happy to take time off and write a couple of papers which were so well-received that I received a coveted invitation to present an "invited paper" at the spring '51 meeting of the American Physical Society in Oak Ridge. Meanwhile, the same kind of somewhat unimaginative numerical calculation on $BaTiO_3$ which Bill had urged on me was completed by his old associate, John C. Slater at MIT, and in a rather stiff little luncheon meeting with Slater and Shockley my attempt to urge on Slater my own cautions and the more phenomenological ideas which Richardson and I had come up with was treated with condescension at best, which was gall in Bill's throat.

Bernd also fell afoul of Bill in that ferroelectric year. Bell Labs has always tended to produce a certain type of scientist of great reputation but little honor in his own country: characters who, through diligent application of the resources and subordinates which Bell puts at their disposal in a very limited field over many years, create worldwide reputations which glow exceedingly overseas, especially in the kind of isolated institute which was Bernd's origin. These peoples' limitations are often very apparent to their reasonably alert colleagues at Bell. Bell is all too prone to the "but what have you done for us lately!" attitude, emphasizing flexibility and "growth" at some sacrifice in depth. But normally the evaluation is correct, as it was in the case of a scientist named Warren P. Mason, who had taken as his own the field of piezoelectricity which is closely related to our joint concerns about ferroelectricity. Warren Mason had an unerring eye for two things: not quite worthless patents and what is now called the M. P. U. (for minimum publishable unit). As a result he led the labs for years in patents granted and had enormous numbers of papers, most of these containing minimal experimental data fitted in a somewhat casual fashion to some currently fashionable theoretical concept. Some good work came out of his group; but most of it was done by a wispy, shy little man named Herbert J. McSkimin, who, after Mason's retirement, also did some very nice things on his own; I knew him at the time only because we carpooled for a year.

Bernd, still dazzled by Mason's European reputation, was glad to collaborate with him, during the months before he actually arrived at Bell, on a theoretical paper about barium titanate, putting forth what they called the "rattling titanium" model. Since this contradicted not only Bill's concept, but also could be shown to disagree with several aspects of the experimental data, Bill, Richardson and I used it as a straw figure to illustrate the superiority of our theories. Although Bernd accepted the failure of this first foray into theory with reasonably good grace, it left him well-disposed towards neither me in particular nor theorists in general.

Undaunted, he proceeded to do superbly what he really could do, namely, innovative, energetic, and imaginative preparation of new materials, and simple but alert tests of their properties. With his characteristic combination of insight and luck, he chose a technical assistant of Lithuanian immigrant stock, with little formal education but enormous innate ability, Joe Remeika. Together they discovered, within the year, processes for growing beautiful, large crystals of barium titanate, as well as a group of similar compounds many of which also were ferroelectric. So far as I know, Bill Shockley never retained the loyalty of any subordinate more than a few years, and Bernd never lost it. What's more, also unlike Shockley, he never lost the support of any one of his organizational superiors.

Bernd always managed to give the feeling — and in many cases I am sure it was reality — that the relationship was personal, that at least for your time together he related to you as a person, with concerns and values outside the work relationship. This directness is surprisingly rare, especially when aimed at one's management superiors, and it works, even though in fact Bernd was one of those people whom a friend once characterized as "on duty 18 hours a day every day". The second ability was to manage an inclusiveness in his enterprises: what he did was always his *team* showing up a hostile outside world, and both bosses and subordinates felt included in a marvelous quest for exciting new results. Those who were *in* his team were always cozened into cooperation, never competed with; it was sometimes a dreadful shock to move outside the team. So the first, and the most loyal (and justifiably so) of the Matthias team was Joe; the second, probably the most level-headed and skeptical of all, was Stan Morgan.

Collaboration with Warren P. Mason was a real anomaly in Bernd's career, never repeated. Normally Bernd had a very sharp eye for quality and a disquieting ability to make his kind of close personal relationship with an extraordinary variety of exceptional people — later on, people like Wernher von Braun and Linus Pauling. Particularly, at that time European refugees were in many of the most prominent positions in American physics, and I gather that Bernd could be even more charming in German or Italian than he was in English. He came to be friendly with several of the key figures in the great 1925–35 decade of quantum physics: Fermi, Bloch, Pauli, and particularly Eugene Wigner at Princeton. To a young theorist like myself these people were beyond my ken: I had slaved over Pauli's Handbuch article about quantum mechanics, and Wigner's group theory, both in the original German, and the unfamiliar language undoubtedly added (like Sanskrit or Latin) to the hieratic flavor of their words. The knowledge thus wrung out of Wigner's book had been all-important to the formal success of my thesis. He was, in addition, the much-admired professor of several of the generation I saw as honored teachers: John Bardeen and Conyers Herring, at Bell, and Fred Seitz, who had written the only reasonably comprehensive modern text of solid state physics (later to be President of the National Academy).

Bernd, then, had persuaded Eugene Wigner into believing — as I am sure he still does — that Bernd was an unrecognized genius, deeply maligned by the narrow minded theoretical whiz kids like Bill Shockley and myself at Bell, and full of exciting generalizations about the thrilling phenomenon or ferroelectricity. This conviction led directly to a confrontation which in my own deepest feelings will never be anything but painful, much as I may intellectually realize that it contains mainly low comedy.

Let me first explain some things about Eugene. He is Hungarian, one of five extraordinary contemporaries from that small nation — Szilard, Teller, von Neumann, von Karman and Wigner — each of whom has written his signature all over one or more of the major developments of our high technology age: von Neumann with computers, Wigner with solid-state physics, Wigner, Szilard and Teller on The Bomb, von Karman on the whole science of aerodynamics and rocketry. It is not characteristic of any of these gentlemen to accept the opinions of others with easy grace, and like Teller — perhaps even more than Teller — Wigner has maintained an attitude of total suspicion of left wing, or even liberal, actions and intentions, often — as in the case of his advocacy of bomb shelters, and his testimony against Oppenheimer — in the face of nearly unanimous opposition from colleagues. At the time this was not public knowledge, and I did not expect Wigner to be any more or less open-minded than other scientists in my experience. I was also unprepared for another Austro-Hungarian foible of Eugene's: his absolute punctiliousness on matters of courtesy, a code of self-deprecation and indirection which could leave a Confucian gentleman feeling that he had behaved like a first-class boor. (Three years later I had the pleasure of seeing Wigner with a Japanese scholar of the old school, Professor Yamanouchi of Tokyo, and this observation was apparently confirmed.) If Wigner does not agree with a speaker, his question is always phrased — in his characteristic high-pitched, accented, precise voice — with much respect for the speaker and for his capacity and that of the rest of the audience for comprehension beyond Eugene's own. It often seemed that the Wignerian manner was designed to extract maximum pain from the speaker's embarrassment. It was only *much* later than the time I speak of that it was at all likely that the speaker was right, and Wigner wrong.

The "rattling titanium" was also very nearly Bernd's last foray into detailed molecular mechanisms. A quick study, he very soon realized that we theorists are almost always able to make specific statements once we know *what* a given substance — like $BaTiO_3$ — *is*, i.e., whether insulating, magnetic, metallic, ferroelectric. But to take a given combination of elements and predict even the most trivial fact about it *a priori* — like whether it is a stable compound at all, or what is its crystal structure (i.e., how the atoms go together) — is almost completely beyond us even now. We know why: all of the differences in energy which determine this kind of thing, the differences between different stable forms of a given substance, are very tiny compared to the total chemical energy which binds the atoms together, and it is impossible — at least has been until very recently — to calculate that sufficiently accurately to see the differences.

Failing theory, this kind of question is open to chemical intuition, analogy, and most important of all, luck and hard work. All of these Bernd had, and from the moment he realized this situation, it might be said he had neatly spotted the theorists' Achilles' heel and was eager to exploit it. He could breathlessly announce that he had the secret to finding ferroelectrics, or superconductors, or whatever, and gullible theorists would fall over each other to explain whatever alchemical formulation he chose to give to his results. In the case of ferroelectrics, it was the "oxygen octahedron" (or, as Bernd chose to leave it in his native German, "octa(h)eder"). This was the configuration which was common to the several compounds he had found by rather obvious analog reasoning from $BaTiO_3$. The central European accent added much to the mythic significance of this "discovery", and he seems to have bewitched Eugene Wigner, who assigned a graduate student to the task of explaining this mysterious result.

This foray of Bernd's infuriated Bill Shockley, who put it in terms I could not but agree with: "Bernd has got at Eugene and convinced him of a bunch of nonsense — why don't you go down there and straighten him out?" If it were anyone but Bill, I would not suspect him of anticipating the outcome of sending an unknown subordinate on such an errand. In the natural course of events, I was listened to with the utmost politeness, and obsequiously assured that he, Eugene, was very sure I must be right, and that his most good friend Bill Shockley surely had understood the problem more deeply than he, Eugene Wigner, possibly could; and equally naturally Eugene paid absolutely no attention, the student continued to work, and on the few occasions — much later — when I did get opportunities to speak before the Princeton physics department, I was particularly sharply questioned in the inimitable Wigner style. It was made clear to all that little or nothing that I said was to be believed. All things pass, of course; 25 years later (after his retirement) I do not think Eugene seriously opposed my appointment at Princeton, although I still do not believe he accepts my ideas about superconductivity and it is with some difficulty that he recalls that I too have a Nobel Prize. As for me, it wasn't until 20 years later that I realized that the villain of the story was probably Bill Shockley and not Bernd or Eugene.

2. First Glimmers in the Theory of Superconductivity

John Bardeen had always been intrigued by the problem of superconductivity. Readers of Don Marquis may remember a piece in which Mehitabel meets an old trouper cat who claims to be the reincarnation of Shakespeare. After a number of drinks, the Shakespeare personality became very maudlin about being driven by necessity to write all those potboiling dramas, when all he had really wanted was to compose delicate sonnet sequences to his ladyloves. I always felt this was the way that John felt about the transistor: it was all very well, but what he *really* wanted was to solve the problem of superconductivity. On his third try, he did.

John is a remarkable person of whom I would have said he was unique except for the close resemblance to him of his two sons, Bill and Jim — both first-rate theoretical physicists. Dark-haired, medium height, with somewhat unblinking, spherical dark eyes, his most distinctive characteristic is quiet: to describe him as a man of few words, and those in a low voice, is a gross overstatement of his communicativeness.

Joyce still remembers the first large party we attended at his house. He greeted us, cordially smiling, at the door; took the ladies' coats, served drinks, helped Jane usher us into the dining room, all with no words that any of us can recall but with warm, smiling hospitality. (Jane, it should be said, is if anything a bit more than normally verbal, and extraordinarily warm, so that with the two of them there is no social discomfort at all.) After dinner, as we all sat on the porch, after sitting awhile listening carefully to what we all had to say, he rose, brought a chair into the middle of our conversational group, and without a word set on it a small radio which he turned on. Most of the ladies assumed that he had finally succumbed to complete boredom with their chatter, but of course he was simply showing us the very first transistor radio, as his contribution to the conversation.

I did talk — almost all on my side — with him from time to time about various scientific matters, and at least one paper of his with Bill Shockley impressed me very much and remains part of my personal physics education. But mostly he was not someone one could learn from: if he could solve something, he did; if he couldn't, he didn't say much. I have always liked to bat my ideas around with someone when possible, especially before they amount to anything, and John was too off-putting a figure for that: at that time I talked mostly to Gregory Wannier or Charlie Kittel.

Suddenly the news arrived that something very important had happened in superconductivity, and a delegation of experimentalists from Rutgers came up and waited upon John with their new result. In the early '40s, John had made his first attempt at a theory of superconductivity, and the Rutgers group felt it might be related to their results. (It wasn't, but he soon produced a second.)

***

The great wartime atomic plants, especially the isotope separation facility at Oak Ridge, had converted some of their capacity to the production of pure isotopes for scientific and medical purposes. The nuclear charge of each element, in units of the electron charge e, is called its atomic number, and determines its chemical properties almost completely. Most elements consist of a mixture of nuclei of the same charge but slightly different masses, so that the atomic weight (in neutron or proton masses) is the average of the masses A of the different "isotopes". Since their chemistry is so similar, the different isotopes are difficult to separate, and it was only the military purpose of separating U_{235}, the more fissionable isotope of uranium with $A = 235$, from U_{238} that inspired us to develop facilities for large-scale isotope separation.

The Rutgers group had purchased enough of a heavy isotope of mercury, Hg, to be able to measure the dependence, if any, of the superconducting transition temperature T_c on the isotopic mass, and had found a surprisingly strong one as these things go: T_c was unmistakably less for the heavier isotope, and within the rather large errors of the measurement the dependence was like the inverse square root: $M^{-1/2}$.

Even on the face of it this result was very significant. There was no doubt in anyone's mind that superconductivity was a property of the free electrons in the metal, not of the atoms as a whole or of the ion cores: if nothing else, this is immediately implied by the London equation as well as by the thermodynamics, where the thermal entropy which is affected is that of the electrons alone. The ions' mass had to enter in some indirect way, by their effect on the electrons.

Second, the form of the dependence was immediately suggestive. All of the failed attempts at explaining superconductivity — for instance, John's own, or the equally unsuccessful attempt of the great physicist Heisenberg — had had to come to terms with the strange fact that all of the transition temperatures are extraordinarily small: the highest at the time about 15 K, the present record near 23 K, but the great majority less than 10 K. One of the theoretical physicists' sharpest tools is a sense of magnitude and scale. In most problems, the very scale on which phenomena take place is a strong hint as to their origin. In most systems, there are only a few magnitudes of the right sort involved. In the ordinary electron gas without any interaction with the outside world, there are only two such scales, and for not particularly mysterious reasons they are nearly the same so there is really only one: either the typical size of the electrostatic repulsion energy between two electrons, e^2/r_s, or the typical Fermi energy $\bar{h}^2/mr_s^2$ where r_s is the average distance between nearby electrons, or equally the typical wavelength of the electron waves. These energies are both about 10 electron volts or, in our preferred energy units of degrees Kelvin, rather large: about 100,000 K. The typical T_c is about one-ten-thousandth of this, where almost all the electrons are trapped forever in very deeply bound states and only a few at the "Fermi surface" are still agitating themselves. There is no obvious "small parameter" which will account for something rather spectacular occurring at this point.

Part of the dilemma is immediately solved when we discover that the nuclear mass is involved. The atom cores bring in a different, much smaller energy scale, namely the energy associated with the vibration frequency of the atoms.

The atoms are much heavier than the electrons, but the forces between them, and between them and the electrons, are about the same; therefore, just as we know that a heavy string, with the same tension, will vibrate more slowly than a light one, giving off a lower-pitched sound, the atoms move more slowly than the electrons, at frequencies which have a wide spectrum but can be characterized by a typical frequency called the Debye frequency ν_D. This leads by Einstein's relation to an energy scale $h\nu_D$ which is typically about 100–1000 times smaller than E_F. The characteristic ratio of the two is, theoretically, the square root of the nuclear mass (about 100 protons, typically) divided by the electron mass, which is indeed about 300.

This energy, converted into a temperature, is called the Debye temperature and is typically of the order of 100 K or so, or a bit higher. It was perhaps the first empirical triumph of the quantum theory that Einstein, and more accurately Debye, in 1910–12 predicted that below this temperature the atomic vibrations could no longer participate in the thermal motion, so the heat capacity of solids would rapidly decrease below T_{Debye}.

***

Thus the energy of a typical atomic vibration depends on the nuclear mass, is much smaller than typical electronic energies, and depends on the nuclear mass as the inverse square root $M_N^{-1/2}$. Although this energy is still too big, at least we have a congenial scale against which to measure T_c: it is a small fraction, but not an infinitesimal one, of T_D the Debye temperature.

All of this was obvious, to John Bardeen at least, immediately upon hearing the Rutgers result. Within a few days he was able to propose a more complete theory, which he hoped would account for the whole phenomenon. This theory went off to the *Physical Review* for publication as soon as he could be sure that the experimentalists' paper had been submitted and was to be published.

Alas, he learned immediately that in the same issue was to appear an article by Herbert Fröhlich of Liverpool, announcing a theory composed *before* he knew of the experiment, and *predicting* the result!

Fröhlich was by no means an unknown. Another of Hitler's gifts to England, he was not unusual in having settled down in a "Redbrick" university and flourished there. Even stronger was the group with Rudolf Peierls at Birmingham, for instance, while the U.S. had by now contributed its share of McCarthy-era refugees to British Redbricks, Bohm in London and Gerry Brown, the young nuclear physicist, with Peierls. There were even a few refugees from the stifling conservatism of the Oxbridge world, particularly Mott and his group at Bristol. This brief period after the war, while Oxbridge's monopoly of the best people and best students seemed irrevocably broken, and before the brain drain began, was a glory of British science. The greatest strength of American science over the 35 years I have observed it has been an institutional diversity, the survival of great industrial, governmental, and non-elite university institutions; and whenever other countries have experienced diversity, even briefly, one sees a golden age.

I was familiar with Fröhlich's little book on dielectrics from my ferroelectric work, which I liked very much because it was exploring the same kind of new territory I had looked into in my thesis. His theory of superconductivity, although (I need not leave you in suspense) it was ultimately, with Bardeen's, unsuccessful, was very important and seminal, also. To this day, many Russian papers in the subject begin with an obligatory reference to the "Frelik Gamiltonian" (the best transliteration they seem to manage, never having equated English H in the Russian).

It is worth stopping to explain why, and briefly introducing Fröhlich himself. One must recognize that almost immediately after the breakthroughs of Schrödinger and Heisenberg in producing the quantum theory, and Dirac's synthesis of the two approaches into a more general formal structure, theoretical physics began, at first imperceptibly, to go off in two directions. Stimulated at first by the desire to make the quantum theory compatible with relativity, and to really understand fully the interactions of matter with radiation, an increasingly sophisticated mathematical apparatus began to be built up which came to be called "quantum field theory": the quantum theory not just of electrons and nuclei but of the electromagnetic field, at first, and then all the complicated particles (for instance, like the positron and the neutrino) and fields (such as the mesons) which began to appear in the search for the secrets of the nucleus.

A second set of problems, those of ordinary atoms, molecules and solids, required the development of the quantum theory along quite different lines: the development of methods of dealing with crystal and molecular symmetry, which involved the use of the branch of mathematics called group theory, and of dealing with large numbers of atoms simultaneously. The former was the specialty of Eugene Wigner and his students, the latter of my mentor Van. Under such leaders as these two, Darwin (the grandson) and Nevill Mott in England, John Slater at MIT, and quite a number of others, these special skills evolved in a community which called themselves variously "solid-state", "atomic" or "chemical", physicists, or possibly they called their field statistical mechanics, as did Darwin and Lars Onsager at Yale.

For the solution of the problem of superconductivity, it was to turn out that both sets of skills would be needed, and a new field of physics, embodying a combination of the two, would have to appear. While some of those who pioneered this field, notably Landau, Feynman and Peierls, always worked on both types of problems, more of us were of a new breed, working entirely on the solid-state type of problem but with the new and more sophisticated tools of quantum field theory. We called ourselves "many-body theoretical physicists" because we specialized in large systems, but, unlike statistical mechanics, we were interested mainly in quantum properties of low-temperature states, not in the properties dominated by the thermal motion of the atoms and electrons.

Fröhlich's role in history was to be the first to pose some crucial problems of solid-state physics clearly in field theoretical form. Some, like myself, and later Bohm and Pines, borrowed various techniques from the field theorists; others, like Landau, never really made any distinction; but Fröhlich was the first to really wave a flag and call attention to the fact that in very demonstrable and obvious ways solid state problems *were the same as* key problems in field theory.

Quantum field theory is a magnet for unconscionably brilliant, arrogant, elite-oriented young men. It is saved from narcissistic stagnation only by the fortunate fact that its practitioners are often so very bright that they are also open-minded. It seems somehow to bother these bright young men to accept that their beautiful problems almost all have equally difficult homologues in the mundane antics of electrons in niobium, silicon, cerium, gallium arsenide or some such material. Nonetheless, sometimes they do; and from time to time I rejoice — giving myself far more credit for the fact than I deserve, of course — that in my lifetime the two branches of theoretical physics have resisted their natural centripetal tendencies and maintained, and perhaps even increased, a web of communication which now seems almost unbreakable. But in the beginning, as is nowadays — except for our Russian colleagues — almost forgotten, there was almost no one but the slightly raffish figure of Herbert Fröhlich.

Fröhlich's first contribution was to pose what he called the "polaron problem".

***

This is a model for the motion of an electron injected into an ionic crystal like ordinary table salt, NaCl. The sodium atoms in such a crystal are charged positively, i.e., are Na^+ (sodium plus) "ions", while the chlorine also forms ions, negative (Cl^-) of course. The electron attracts the Na^+, being itself negative, and repels the Cl^-, so that as it moves it is surrounded by a cloud of ionic motion which co-moves with the electron. This motion is called a "polarization", because it represents a net positive charge motion towards the electron, and negative away from it. The "polar" of "polaron" comes from this; the "on" implies that the electron has in some sense become a new kind of quantum particle, carrying its cloud of polarization with it and therefore having a different effective mass and charge from the "bare electron".

***

This problem is remarkably similar to a number of problems that come up in quantum field theory. The fundamental particles of physics, in quantum theory, are always accompanied by a cloud of disturbance in the vacuum, since in the quantum theory the vacuum is not really empty but full of all kinds of "zero-point fluctuations". Thus solid state physics has a strange, but intellectually satisfying, parallelism with particle physics: we tend to replace the very tangible regular lattice of ions which make up the solid by a kind of featureless, vacuum-like medium, where the particle theorists turn the apparently empty, featureless vacuum into a medium swarming with vibrating life; but the two of us treat our media in much the same way.

In the course of doing this Fröhlich was the first to point out that the vibrations of the ions of a solid, the so-called "lattice vibrations" with energies in the ballpark of the Debye temperature, could be treated as a quantum field on a very parallel basis with the electromagnetic "photon" or meson fields which particle theorists were then discussing. To emphasize this field-like character of the vibrations, we came to call them "phonons": quanta of the sound waves which propagate in any solid. (This term, used first for sound quanta in liquid helium, goes back to Landau.)

After posing the polaron problem, and giving a first, useful but rough solution for it. Fröhlich next turned his attention to metals. Again, the electrons in the metal interact with the vibrations of the atoms, but the vibrations which the electrons affect are the ordinary sound waves, not the "polar" vibrations of ions in opposite directions which are important in the polaron problem; and there are great numbers of electrons in the metal, with the only important ones being those near the "Fermi energy". The point, however, is that one can write down a "model Hamiltonian": a formal, very simple description including the quantized electron mass, the quantized ion lattice waves or phonons, and their interaction. As I said, this Fröhlich did and it is still an oversimplified, but formally valid, starting point.

This Fröhlich did, without prompting from any experimental data; and then, of course, he attempted to solve his creation, at least as accurately as he had the polaron problem. The way he did this was more or less exactly what is done in the polaron problem, namely, to surround each electron in the metal with the appropriate cloud of ionic motion, moving along with the electron as it zips past the ions and the other electrons.

Just as for the electrons in an ionic crystal, the result is to change the properties of the electrons. To express this change we now use another word due to Landau: we call them "quasiparticles" which is not particularly apt because in no sense are they not whole or not real, but they are "changed" or "effective" or "clothed" as opposed to "bare" particles. But at that time Landau hadn't invented this word (if it was his).

A particularly serious catastrophe can happen if the properties of the electrons are changed too much. For instance, one might find that the effective mass had changed so much as to become negative, so that as the momentum and velocity of the electron is increased, its energy gets smaller. But the whole calculation is based on the idea of the "Fermi momentum", a momentum below which almost all states are occupied because they are lower in energy (using the simple energy equation of first-year physics, $E = mv^2/2 = p^2/2m$). Thus when m gets negative, there must be some massive rearrangement of the state of the metal. This kind of "catastrophe" was exactly what Fröhlich — and, in his almost identical paper, Bardeen — found might happen in certain metals, and they suggested that this catastrophe was the source of superconductivity.

***

It is not hard to see how this might come about. You will recall that Pauli's exclusion principle eliminates all overall states in which two electrons occupy the same state at the same time. As the electron moves past the ions, it will give them little pulses of acceleration, and vice versa, leaving behind an altered state of vibration of the elastic lattice of ions. This state can be described by adding up states containing either one more or one less quantum of the vibrational field ("phonon"). If such a scattering of the electron, with absorption or emission of phonon, is to really take place, the energy of the particles must all add up, so the state the electron goes to will have its energy changed by something like the Debye energy $h\nu_D$.

Electrons deep down inside the Fermi surface are trapped by the exclusion principle and can't undergo such scatterings, because the state they would want to occupy is already full. Electrons more than $h\nu_D$ above the Fermi surface can scatter all they like. It is a general rule of physics, not just of quantum mechanics, that allowing a particle to adjust ("perturb" is the word we use) its state always lowers its energy, so this says that the particles above the Fermi surface can lower their energy more easily, and Fröhlich and Bardeen supposed that they could do so until they got right down below the Fermi surface and stood the distribution on its head.

***

This catastrophe could happen only if the phonons scattered the electrons very strongly. John Bardeen, in particular, suggested that you could estimate how strong this scattering is from the size of the electrical resistance in the metal. At ordinary temperatures, almost all the resistance of most reasonably pure metals comes from the thermal vibrations of the ions, so of course one can estimate how strongly the electrons feel these vibrations by looking at how good a conductor it is. The correlation is very good: metals with high superconducting transition temperatures like lead and mercury are bad conductors, while good conductors, like silver and copper, are not superconductors: a paradox which long had been recognized as such but not understood.

Of course the theory must explain the new fact from which Bardeen, at least, started: the isotope effect. It does, very nicely. The electrons move much more rapidly than, the ions, and therefore do not make them resonate very well. The faster the natural frequency of motion of the ions, the less ineffective is the electron, so the natural scale of energy of the Fröhlich-Bardeen theory is $h\nu_D$, the Debye energy.

This energy is precisely proportional to the correct inverse power of the mass of the ion. This is caused by exactly the effect which makes a heavier guitar string vibrate at a lower frequency. When one gets all through with the arithmetic, all the energies which are purely electronic drop out and the transition temperature turns out to be just a number depending on the strength of coupling — if the coupling, and hence the phonon resistance, is big enough — times $h\nu_D$.

***

Bardeen, in his paper, flourished a table of numbers which showed the strong correlation of lattice phonon-scattering resistance and T_c; but at first neither he nor Fröhlich pointed out that all T_c's then known were small — or even tiny — fractions of the Debye energy. There is no clear reason in the theory for this and in the end this was to be the first symptom of failure.

Just at this time, in mid-1950, Fröhlich came over on a tour of the U.S., beginning with visits to Rutgers and Bell Labs. I was, of course, eager to see him to talk about dielectric problems and his book: I was only beginning to make progress with my ideas on ferroelectrics, but he was very encouraging indeed. He had — has — a bronzed, aquiline, sardonic face, and a strong Rhineland German accent. In looks and manner he was not all that dissimilar to Bernd Matthias, with whom he always hit it off well, especially later, when Fröhlich represented the opposition and could be used to *épater les bourgeoises*. He encouraged a fine, at that time unusually long and bouffant, head of black hair, later to turn a spectacular white and make him a very striking figure.

I don't want to oversell my memory of this visit, except for my part of it: I was barely aware of what was going on in superconductivity. But I have the impression that it was only at this time that, gradually, John Bardeen and Fröhlich revealed, like poker players turning over their cards, their two theories to each other; certainly it was the first opportunity they had had to discuss their similarities and differences. I even remember that at the big party Charlie Kittel gave for Fröhlich, in his new house in Featherbed Lane in the "New Vernon Riding Country", Bardeen and Fröhlich spent most of the time in earnest discussion in a corner, presumably settling their differences, priorities, etc. As far as I know they parted amicably at that time: both always gave credit for essentially simultaneous ideas, and Bardeen was always careful to point out that only he had had the experimental hint to work from. It was a bitter pill for him to swallow, surely; in so far as he is capable of exhibiting excitement — a slight twinkle in the eye, a slight tremble of the hands — he had surely been excited for the previous few weeks. Fröhlich was more clearly elated. Surely this trip, which became something of a triumphal progress, must have been the high point of his career, a peak from which it must have been very hard to climb back down. He did have to climb down because, of course, it was *not* the right theory, and seven more years of fumbling and speculation still had to be gone through before we began to know the answer.

Three years later I saw Fröhlich in Japan, at the first of the three great postwar confabulations of the world of theoretical physics before modernday austerity began to set in (Kyoto, 1953; Seattle, 1956; Trieste, 1968; we'll hear more about some of these later). Eager young Japanese had decorated his theory with elaborate detail, but his own talk was unenthusiastic about it: a celebration of what it *couldn't* explain, in fact a markedly open and honest — as these things go — confession of its lack of success.

In fact, he now proposed an entirely new concept, still based on lattice waves but very different. The new mechanism, unfortunately, would only work in a strictly one-dimensional metal: a metal composed of a single line of atoms. Real superconductors are, of course, three-dimensional and therefore for a number of years Fröhlich's new idea was just a mathematical curiosity. He wanted to form a rigid, wavelike structure in the gas of electrons, which could slide frictionlessly along past the ions carrying the electrons with it. Oddly enough, such chain-like metals are now known, and they indeed have the "density waves" predicted by Fröhlich; but his idea that they would superconduct turned out to be quite wrong, if — like most of his work in those days — immensely stimulating, original and prophetic.

Nine years later, in spring '62, I visited him in Liverpool. By now he had married an attractive young wife and his white mane was well-developed. He was surrounded by an admiring group of young people, and we enjoyed a ritual visit to his favorite café where he held forth in quite continental style. But the Redbrick effusion was pretty much over in England. Fröhlich himself took up various speculations in the fringe areas of biology and physics, presumably hoping to relive his earlier brilliant isolation; but nothing amounted to much. As for the BCS theory (by now six years old), he felt it "Once Cooper invented the pairs, which gives you the exponential for T_c, what else could it be!" Sour grapes, of course, not totally negative, but still very influentially so. It is a well-known phenomenon to historians and sociologists of science, this reluctance to finally accept a real breakthrough unless it's one's own, so much so that someone once said "no one is ever *convinced*; they just die off." (This is not really true, even in this case: really, most scientists are surprisingly flexible.)

That is my memory of him: a determinedly cosmopolitan figure, perhaps a little pathetic at that point, looking for new glories and in the process inadvertently losing the really well-deserved acclaim he had coming to him for his old ones. I cannot help but speculate a little, on why it was these slightly condescending, subtly misguided putdowns of the BCS theory almost always came in a Central European accent. Fröhlich, Wigner, Uhlenbeck, Bloch, Wentzel, John Blatt, even Landau (a little), Bernd Matthias himself: was it the down-home, pragmatic Midwestern *style* of such people as John Bardeen and some of the other chief actors, which put them off? Was it a battle of the old quantum theory generation against the new? Or was Bernd Matthias simply a very effective opponent who used his best weapon, the German-speaking refugee Mafia, to great effect! And why did *he* oppose BCS? We'll see.

I was at that meeting in Japan, to listen to Fröhlich and his admirers talking at cross-purposes, not because I belonged — nor, probably, was wanted — at such a high-level international gabfest, but because a series of events had led to my being a Fulbright lecturer at Tokyo that fall, and I was therefore tolerated as a hanger-on. This, like its successor conferences, had the ambitious goal of uniting the whole world of theoretical physics (cosmology, astrophysics, elementary particles, nuclei, solid state, atoms, plasma physics, etc.) in some kind of unified format. The Japanese were bitterly disappointed that few of the then-existing Nobel prize-winners accepted their invitation: the Diracs, Heisenbergs, Schrödingers stayed home, but they should have been proud of their prescience in acquiring roughly 15 future Nobelists, one by accident — myself. Bernd had broken the ice at Bell with a two-year leave of absence (won a painful few months at a time) to go to Chicago and learn cryogenic (low temperature, i.e., liquid helium) techniques with John Hulm at Fermi's Institutes building, so I in my usual thick-skinned manner confronted my boss, Stan Morgan, with the Japanese fait accompli, and the Labs arranged a temporary leave, without pay except for the meeting part of it. Perhaps by this time some frissons at the coming flood of competition were over-taking the Labs' staid structure. John Bardeen had already left for a double professorship at Illinois, each half (in two separate departments) of which came to nearly his entire Bell salary. John's departure speech was a mild-sounding but actually rather bitter diatribe against what he called Bell's attack of "transistoritis": by which I believe he meant the great drain of research resources into the exploitation of the transistor breakthrough, now being built up mostly under Bill Shockley, for whom his dislike was more and more apparent. Charlie Kittel was soon to depart for Berkeley, to pick up the group of brilliant students left stranded there by the famous California Loyalty Oath, which had scattered Oppenheimer's successors at the great Berkeley department to a variety of jobs. Industrial science — in the guise of Bell Labs — was not for the first time or the last having to cope with the consequences of having built up a whole new field outside of academia, stimulating competition for its practitioners not only from other industries but from the flourishing multiversities, growing fat and happy on the new Federal research support system.

Overseas travel — in fact, in those days even travel past the Mississippi — at conservative, engineer-oriented Bell Labs, with its two-week vaca-tions, had long been the reward of long and faithful service. Every 10 or 15 years (or never) you traveled, always first class as befitted a jun-ior executive of the Bell System, usually in a delegation, to one of the big international meetings in your field. Wives often came along and a week of your vacation would then be used up in some watering spot. The delegation would make an obligatory visit to at least one appropriate sister lab — Philips Eindhoven, in Holland, was a favorite — and several sensibly chosen universities, moving from one to the other in appropri-ate style; and would prepare a little report on the whole junket for less fortunate colleagues.

Academics, with their long vacations, sabbaticals, and wide opportu-nities for consulting, speechifying, summer jobs, and fellowships, always had traveled far more freely than we industrialists. With the advent of the grant system this traveling soon came to seem frenetic, if in many cases very far from first class. For years most of the grants were military, and even non-military grants could ante up a virtually free seat on the MATS (Military Air Transport) network to anywhere in the world. To foster

European recovery, summer schools were financed where one lived in the rough at Les Houches in the French Alps, Varenna in the Italian Lakes, or St. Andrews in Scotland, while a number of attractive places on this continent set themselves up in similar imitation of the Michigan summer schools of the prewar era at which American theoretical physics was born.

I didn't wake up to the peripatetic frenzy going on among my university colleagues until 1955 or so, but my friends at Bell were less obtuse and I suppose the Japanese adventure was deeply envied by several, as I know it was by Bernd. I was not eager to travel alone, and Lord knows Joyce, with a toddler at home and a girlhood in Chicago traumatized by burglars, did not welcome my leaving home for long periods nor was she eager to come along. But Japan was different, and we duly packed up Susan and sailed from Seattle on the rattly old Hikawa Maru. Susan was an enormous success — she upstaged Sir Nevill Mott and Hideki Yukawa on every newspaper front page, and — much to our embarrassment — even displaced Eugene Wigner from a shared bathroom at the hotel in Nara. After a day of misery, Eugene tapped almost inaudibly at our door and whispered that he would be *very*, very grateful if from time to time we would unlatch the door on his side.

The Fröhlich–Bardeen theory was indeed dying. It foundered in strict logic on two specific failures, but in fact I think everyone had a sense that when a real theory appeared it would illuminate the entire landscape rather than throw a flash of light here and there. This is the nature of such breakthroughs as the quantum theory of metals following from Bloch's bands and Pauli's statistics.

But one could hang — and many did — a distaste for the Fröhlich–Bardeen theory on two things: energies and the London equation. The former point is straightforward: the physicist is always asking himself the vital question about what "sets the scale", i.e., what are the factors which decide what the general range of dimensional quantities such as energy in a phenomenon will be. Any physical theory has as its answer for the sizes of some physical quantity the product of two things: first, some kind of number such as 1, 2, e, π or something much more complicated, which is dimensionless (i.e., just a number) and comes from detailed computation; and second, a dimensional quantity more or less setting the units or "scale" on which the dimensionless factors play. This latter depends on the basic physics, i.e., on the answer to the question: "Have we really identified what phenomenon and which term of an equation for that phenomenon is involved!"

The answer to that question for Fröhlich–Bardeen is obviously "the phonon energy" $hv_D = k\theta_D$. The electron energies are shifted by some such amount at the Fermi surface — more, perhaps, or less, but this is the only scale in the problem, as appeared all too clearly in Fröhlich's way of writing the problem down (the famous "Fröhlich Hamiltonian").

The actual T_c's, however, range from somewhat less than 1/5 of the Debye (phonon) energy hv_D (itself $k \times 100$–200 K) to much less than that — down, at that time, to numbers of order 10^{-2} K, which was one-thousandth of the energy scale. Clearly only a conspiracy of Nature would produce such a biased distribution of dimensionless numbers, and in our lexicon "conspiracy of Nature" means "a better theory", or at least it should do.

Worse than that was the result of any attempt to derive London's equation from the theory. John Bardeen seemed to be able to make it come close, or at least claimed so. He insisted that this theory should have a very large diamagnetism, because it was in some sense a dynamic version of an unsuccessful one he had prepared in 1944, which did. But even Fröhlich was adamant that London's equation really is not just a large diamagnetism, and no plausible derivation of the true electromagnetic properties ever seemed to result. This may be about the time at which John added to his reputation for inscrutability by his answer to Brian Pippard at some public discussion in Cambridge, when asked why superconductivity followed from this model. He thought long and hard, and then the right hand started to move in a circle. "Because it's a diamagnetic effect," he said, and added no more.

Let me add a historical footnote. 1953 was the height of the McCarthy era and of the Stalinist terror in Russia. Naturally, therefore, there were no Russians in Japan, and we were almost unaware of the great strides in the phenomenological description (which have been described elsewhere) due to V. L. Ginzburg and L. D. Landau in 1951.

Bernd Matthias had returned to Bell and immediately set about putting Bell Labs on the cryogenic map. He employed, trained, and enslaved to his charm a bright gnome named Ernie Corenzwit to run the helium cryostat reliably and with Ernie and Joe Remeika, at first, set about making a remarkable series of discoveries of new superconducting alloys and compounds. This was an even more wide open field than ferroelectrics, and the relevant experiment even easier, involving a vacuum flask of liquid helium, a pump, two copper coils, and a little winch with a lot of string for lowering ampoules of samples into the helium-filled coil to test it for Meissner effect. This was a period in which we developed a fairly warm relationship — I was in no sense a competitor, and now deeply engaged in various works about magnetism, so was in no position to challenge his various rationalizations of the admittedly remarkable intuitive feel he soon acquired for likely materials. He even decided that my presence, chatting, on a lab stool was good luck, since his first really unexpected compound, $CoSi_2$, showed up one day when I was there. (Unexpected because Co is a magnetic metal and Si a semiconductor: one of the first metallic compounds of purely non-superconducting elements to show superconductivity. No one could claim, in this case, that an impure or unreacted sample was the cause of his observations, crude as they were.)

Bernd's envy of my trip to Japan should have been fully assuaged when he was invited, and went, to the prestigious 1954 Solvay Congress, focused on electrons in metals.[b] Bernd's many new compounds gave a bright spot to the superconductivity picture, where the failure of Bardeen–Fröhlich was made fully evident.

Bernd Matthias was not the only staff member at Bell working in superconductivity in those few years. One of Oppenheimer's brighter students, Harold W. Lewis, had refused the University of California Loyalty Oath and been rescued by Oppenheimer at the Institute for Advanced Study for a couple of years. He took it into his head that he would enjoy a stint of solid state physics, and rather to our surprise no particular obstacle was seen by the Labs, partly I suppose because Hal's actual security clearance never was touched. To us he was considerable of a radical hero at the time, but we also were proud of the Bell Labs for its very minor response to McCarthyite hysteria. A very pro forma "security questionnaire" was asked of us, but those who did not fill it out (there were three I know of) never felt any great pressure for compliance.

For those years 1953–56 Hal puzzled away at the theory of superconductivity, which was pleasant for us, since he and the visitors he invited

---

[b]I too had been invited to the Solvay, (for what reason I know not, since my work had no relevance to that subject at all. It may have been Van, it may have been a reflex from my previous year's presence at Kyoto, or it may well have been H. A. Kramers, a very great Dutch physicist who had rather out of the blue, called up from the Institute for Advanced Study at Princeton one day and spent an afternoon discussing my work on antiferromagnetism. It was a revelation to me to discover in writing a memoir recently that it was in 1923, the year I was born, that my professor Van had been befriended by the young Kramers and made thereby his first serious contact with the great world of European theoretical physics. In any case, Bell Labs made it quite clear that I was not to be financed to two major international meetings in two years, and, unaware of the signal honor extended to me, I turned it down with hardly a second thought.)

kept us abreast of the subject, but frustrating for him since he got nearly nowhere despite his undoubted brilliance. He mostly whiled away the time carrying out some rather ingeniously instrumented and highly competent experimental projects: an oil viscometer, a ballistic cardiograph, and an attempted measurement of the Hall effect in a superconductor (I have heard recently that he may have been less competent or more unlucky than he seemed, since the effect has since been measured and was probably big enough to see). Hal's chief cross was, however, Bernd, since Hal could under no circumstances brook the obviously egregious nonsense which Bernd kept spouting about the mystic, alchemical formulas he used in his continually successful search for new superconductors. Into this search Bernd was also steadily attracting more collaborators, such as Betty Wood, who helped him with crystallographic identification of his samples, and later on Ted Geballe, a low temperature specialist who eventually became his department head, moving himself gradually from his semiconductor work into superconductivity.

It is hard to explain, especially to non-scientists, the strength of such a clash of personalities. Hal, slight and dark, had a marked personal resemblance to Oppenheimer which many of us suspected was enhanced intentionally by imitation of mannerism, speech and style: the emphasis on precision, rationality, mathematical rigor, intellectual breadth and force.

He must have seen himself as confronted with the darkest forces of superstition, mysticism and black magic. To find at the same time that these satanic forces were surely much more attuned to the purely experimental bias of the Bell Labs ethos than Hal's intellectual austerity ever could be, must have been very frustrating indeed. To many outsiders, especially in industry, we look like a badly housed branch of the Ivy League, but in fact, even in research the management is much more admiring of inspired tinkerers with an eye to the technological main chance than of deep thinkers. All of these frustrations I understand well from personal experience.

Bernd was in fact a better inorganic chemist than most major figures of his generation, and no physicist at all, in spite of what he called himself. Chemists quite properly, since they are forced by their subject to deal with the real complexity of all possible compounds of the 100+ chemical elements, are delighted to find empirical, heuristic, qualitative guides to their thinking, which lead them, entirely independent of any intellectual justification, into an interesting set of compounds, and when they find such a group will happily characterize them and move on. From the pure empiricism of Paul Ehrlich's search for the "magic bullet" comes most of the force even of the most modern chemistry.

Thus it was that when the University of Wisconsin called him, Hal Lewis happily answered and, one presumes, soon forgot the traumas of puzzling over superconductivity (but never those of dealing with Bernd). In fact, in the ensuing years he somehow became one of the most respected members of the "Jason" group of scientific advisers to the Defense Department, along with Bernd as a matter of fact, and eventually its Chairman as the group nearly disintegrated under Vietnam pressures — by all accounts, Hal was a thoroughgoing hawk on Vietnam, faithful to the DOD long after Bernd (with whom his relationship was never easy) and several others had left it.

3. BCS: Pairs at Last

The summer of 1956 was a very busy one for me, and in fact for the whole of the theoretical group at Bell Labs. Starting with the departure in 1951 of John Bardeen for the green fields of Illinois and in 1953 Charlie Kittel, off to Berkeley with Art Kip of MIT to replace Berkeley's Oath-caused losses, it seemed we were rapidly to lose every theorist of note we had. Shockley was busy building transistor empires and then his own companies, and Hal Lewis and Gregory Wannier were both testing various alternatives (the one soon to be off to Wisconsin, the other to Geneva and then Oregon). There was a great ferment; starting salaries were rising raster than our own could be revised, and George Feher was hired from Berkeley in 1953 at a rate only a last minute raise kept below my own on my return from Japan. In 1955 Bob Shulman made a public act of Gregory Wannier's private strategy of keeping a list of everyone's salaries (at Bell always a carefully guarded management secret!) by the simple scheme of letting anyone look at the whole list if he added his own, and it turned out all of us were making within 10% of the same salary, independent of experience, age or agreed ability. That very year a group of "heroes" like Claude Shannon and Conyers Herring received special giant raises in salary. In the midst of this ferment, caused by the explosive growth of a science and a technology we had ourselves spawned, we theorists somehow felt ourselves particularly adrift. Each summer some of our university colleagues, such as Walter Kohn of Carnegie, Quin Luttinger of Michigan, or David Pines of Illinois, would visit at consulting fees well in excess — it seemed — of our salaries, talking of their new and healthy ONR contracts, the essentially free worldwide travel that made possible — even MATS (Military Air Transport) seemed glamorous to us — the exciting summer schools, Leisure of the Theory Classes as Jeremy Bernstein was to call them, in places like Varenna and Les Houches in the Alps, their postdocs, their students, and their total freedom. We felt put upon to say the least.

Hal Lewis left us one legacy: he distilled, in his rational way, the nature of our uneasiness as theorists at Bell, and persuaded Conyers Herring, Peter Wolff and myself to propose a theory department of very innovative structure. The resulting department was a great success, as was almost everything — except the theory of superconductivity — to which Hal put his hand. With the new department came the right — almost the obligation — to flood the Labs with visitors from the outside world of theoretical physics, both to give talks and to come for longer stays. Our attempt was to increase our visibility, for recruiting purposes especially but also because a number of our few previous visitors — especially Walter Kohn and Quin Luttinger — had done collaborative and consulting work which very much pleased our experimental masters. At least four or five of these visitors will play a key role in this story: Leon Cooper, David Pines, John Blatt, J. Robert Schrieffer, and Gregor Wentzel.[c]

Leon was actually at the Institute for Advanced Study during the year 1955–56. He did not ever appear on our payroll, as far as I remember, but he did come by and give a couple of talks, one memorable in the light of later events.

I have never quite been sure whether Leon is an inspired dilettante or a great man. He has written very few papers, but these few are all respectable, and one or two besides BCS quite influential. He very much enjoys the rather sophisticated, elegant international world which his charm, his Nobel prize, and his considerable business success have made available to him (in fact, I gathered from chatting with a woman from that world at a Cambridge dinner party that he was not above advertising his Nobel prize some years before he actually got it). He does not spend much time slogging around in the workaday world of departmental affairs, publication problems, etc., showing up from time to time at the pleasanter meetings with invariable charm, elegance, cheerfulness, and a certain diffidence — impossible not to like very much, but — probably quite incorrectly — seeming a bit trivial.

---

[c]The roster of our visitors in those two or three summers reads like an almost unbelievable Who's Who in theoretical physics of that time: Erich Vogt, E. Montroll, J. C. Ward, Kerson Huang, Keith Brueckner, E. Abrahams, Bob Brout, and no doubt others I have forgotten must be added to those I have mentioned.

The talk at Bell was about a strange little calculation in which he purported to show that the standard state of a normal metal — the so-called "Fermi sea" —was not stable in a very peculiar way: that one could lower the overall energy by taking electrons out of the states near the surface of the Fermi sea in pairs, and putting the pair of electrons together into a peculiar correlated state which we only later came to describe as a "bound state".

The calculation seemed tremendously oversimplified. The electrons were assumed to be able to proceed through the metal as though it was absolutely free of encumbrances, in spite of the many other electrons and ions in the way, *except* for the interaction with its partner in the pair; and of that interaction, only a tiny special part was kept. Now, after the great 1957 papers of Landau and Migdal on the state of the Fermi sea are well understood and accepted, the motivation for these steps is clearer; but then it was completely fuzzy, at least to those of us not really well up on the very rapid developments taking place in what was coming to be called "many-body theory". But the calculation had some very suggestive and positive things going for it which I will talk about shortly.

Now if a system can make a change in its wave-function which lowers its energy, it is the most fundamental principle of physics that it will do so. A stone at the bottom of a hill is stabler than it is at the top; if it can roll down, or if some slight jiggle will set it free to do so, it will. Water stays at and goes to the bottom of the well, keeping its surface as level as possible. The instability of higher-energy states towards those of lower energy is instinctive knowledge in all of us and especially in physicists. (I should have thought, that is, until 1 encountered some of the eminent theoretical physicists of the type more inclined towards mathematical abstraction, who often seemed more prepared to quibble with this kind of simple argument from classical stability — in fact hardly to have seen simple stability arguments — than with all the rest of the apparatus of the BCS theory which was to develop.) This reduction in energy, if true, certainly signaled a very significant instability of the commonly accepted state of a metal, an instability which could lead to a phenomenon as bizarre as superconductivity. But what could honestly be objected to was the manifest fact which Cooper more than anyone else was quite aware of, that he didn't know what to do next. He could make one bound pair (we will now call them) and then another and then another — when did it stop? What could it do for him? What was the resulting state!

Balanced against these two difficulties were two very positive aspects. One, Leon himself had not really so much to do with, except for suggesting a way to use what was already known. It was really John Bardeen — who here first returns to the stage, after his failure with the Fröhlich–Bardeen theory who, with David Pines, identified and formalized a way of bringing the lattice vibrations (phonons) into the problem which, this time, turned out to be near enough to correct to be at least temporarily useful during the first crucial years. John and David pointed out that in addition to the modification of the energy of single electrons which motivated the Fröhlich theory, there followed from the lattice vibrations an *attraction* between *pairs* of electrons, which had certain special properties.

The reason for this attraction may be seen from a number of simple analogies. For example, when two acrobats stand at the same place on a safety net or trampoline, they stand in a deeper depression than either one alone would make, and either one, to get away, will have to run up a little hill out of the depression. The elastic motion of the atoms of a metal crystal is not so exaggerated as that, nor does the electron motion take place in two dimensions, rather in three; but the principle is exactly the same: the two particles deform the material around them in the *same* direction, hence they attract. Of course, at the same time the electrons repel each other because they have the same electronic charge — like pith balls in the beginning physics experiment — but that force can be weaker or stronger, and also has a very different dynamic behavior.

A second string to the acrobat analogy: if the two acrobats run fast in intersecting paths the trampoline may bounce back and make the second acrobat *less* rather than *more* satisfied to be nearby, if they cross not at the same time. Thus there can be special dynamical effects. John and David brought out this dynamical character, showing that when two electrons have nearly the same energy (which, you will remember, translates quantum mechanically into frequency) the resulting slowly-varying electron density leads to an attractive interaction; if not, it is repulsive. The relevant energy difference or frequency is the Debye energy, the characteristic energy difference which played such a role in the Fröhlich–Bardeen theory; and it is again true that, using this interaction, we can hope to make everything scale at the same rate with isotopic mass as the Debye energy and obtain the isotope effect.

What Leon did with this Bardeen–Pines interaction (which in itself is not a particularly brilliant discovery, since in principle all of the particle interactions of physics result from the intervention of an underlying field, and the techniques for such a calculation were even then straightforward, if the results were not completely obvious) was to make a brilliant oversimplification which allowed its essential nature to be comprehended immediately. This process of "model-building", essentially that of discarding all but essentials and focusing on a model simple enough to do the job but not too hard to see all the way through, is possibly the least understood — and often the most dangerous — of all the functions of a theoretical physicist. I suppose all laymen and most of my scientific colleagues see our function as essentially one of calculating the unknown, B, from the most accurate approximation possible to the known, A, by some combination of techniques and hard work. In fact, I can barely use an electronic calculator — almost all of my experimental colleagues being far more skilled at that than I — and I very seldom produce actual numerical results — and, if so, make some graduate student or other junior colleague do the actual work.

Actually, in almost every case where I have been really successful it has been by dint of discarding almost all of the apparently relevant features of reality in order to create a "model" which has the two almost incompatible features:

(1) enough *simplicity* to be solvable, or at least understandable;
(2) enough *complexity* left to be interesting, in the sense that the remaining complexity actually contains some essential features which mimic the actual behavior of the real world, preferably in one of its as yet unexplained aspects.

I said dangerous, and the sense in which this is true is that one is laying a trap for the majority of one's colleagues, who are too literal-minded to understand either the necessity or the reality of the model-building process. A really well-built model can often stand a great deal of weight if used judiciously, but it can never hold up against being taken completely literally. We will see how this happened to Leon's.

***

Leon's oversimplification of the Bardeen–Pines interaction was to suppose that any two electrons within an energy $\bar{h}\omega_D$ (the mean Debye phonon energy) of the Fermi surface felt an attraction to each other, and all others felt nothing. Aside from that, almost all of the rest of his extremely short paper was pure hope: mostly the hope that it was justifiable to ignore all the rest of the electrons except the paired partner, and also the hope that the pairs might somehow condense together into a Bose superfluid: in the end a prescient suggestion, but not, as we shall see, original.

But from his oversimplified interaction he gained one terribly suggestive quantitative result: the binding energy of the pair of electrons, a new
energy scale, which we now call Δ, and along with this a new length scale,
the radius of the bound electron pairs in space.

Leon's model contained two relevant quantities, the cutoff energy
which I have above called $\bar{h}\omega_D$, and the strength V of the attractive interaction, which could be reduced to a dimensionless parameter $N(0)V$ by
multiplying it by the density of electron states $N(0)$ near the Fermi level,
measured per metallic electron. The former, as in all phonon theories, sets
the overall energy scale, the latter tells one the strength of the coupling
of electrons to each other (these being field-theoretical ways of putting
things only just then becoming second nature to us many-body theorists).
The marvelous equation for Δ was simply

$$\Delta = \bar{h}\omega_D e^{-1/(N(0)V)}$$

and for the accompanying length

$$\xi = \bar{h}V_F/\Delta = V_F/\omega_D e^{+1/(N(0)V)}$$

(It is interesting, looking back, to see that much of the notation ($N(0)V$,
and ξ) came after this seminal paper, though the concepts and quantitative arguments were there.)

These equations were marvelous because they remedied at a stroke
just the worst quantitative deficiencies of Fröhlich–Bardeen. $N(0)V$ is
a parameter which experimental data and the Bardeen–Pines theory
roughly provides for us, and it could be expected to vary in a range
considerably less than 1 (say 1/10 to 1/2). The exponential quantity thus
is, first, quite small (never bigger than 1/7, and down to 1/400), and
second, very widely variable, so that the experimentally very wide varia-
tion of characteristic temperatures below a rather low upper limit seems
reasonable indeed. Nonetheless we retain the isotope effect. In addition,
the realization that both Landau's and Pippard's different phenomenolo-
gies required a different, and quite long, characteristic length from the
London λ was growing on the community, so ξ was a welcome quantity to be given (it too could vary widely, but usually exceeds λ by a
fair amount in the simple pure metals then of interest). This was the
"exponential" that Fröhlich spoke about and which almost at once struck
many of us as very positive about this suggestion.

Finally, very briefly, Cooper made his most prescient and vital remark:
that the actual superconducting transition could represent, not the

breakup of some kind of superfluid condensation of the pairs, but the disintegration of the pairs themselves due to thermal agitation. It is this feature which was indeed to be the clear distinction between BCS and its opponents and predecessors, and which in the end led to most of its extraordinarily detailed experimental confirmations.

The talk I heard must have been in summer '56: I remember walking along a sunlit corridor somewhere in Bell Labs, and talking it over with Hal, possibly Bob Schrieffer, and others, and agreeing that it was a tantalizing speculation; but much of what appears in the letter (published in late '56, submitted September 18 from Illinois with acknowledgments already of Bardeen and Pines' comments) either was not there yet or completely failed to reach my consciousness. The facts of the publication suggest the former. I clearly missed completely the opportunity — offered freely, generously and fairly to all comers by Leon — to be in at the birth of the theory of superconductivity. My excuse may be better than most — I was undergoing the painful, slow gestation of what may be my own biggest contribution to theoretical physics. It seems, looking back, that we all must have been writing papers from morning till night: many of the most important results of many-body theory appeared somewhere in the world during the year 1957, or at most between '56 and '59.

It is important too to keep a certain perspective on Leon's pairs. Fröhlich's remarks were not that unusual, and Landau has been heard to claim that if *he* had just seen Leon's paper soon enough he would have done everything in a few weeks. Perhaps; but the "pairs forming bosons" remark had been around for 20 years, and widely publicized already in these hectic '56 months, as we shall soon see — and no one else produced BCS as a consequence.

The other claimants to the electron pair idea both had unusual histories.

Way back in the 1930s a chemist named Ogg, from Columbia, published an unusual sequence of observations on liquid ammonia with sodium dissolved in it. This is a fascinating system in which there are free electrons released by the dissolved sodium ions, a liquified gas mimicking a metal in many ways including luster and conductivity. But Ogg claimed that as the ammonia solidified — at relatively very high temperatures of about 150 K — he saw evidence of superconductivity. The evidence was never repeated, and after stoutly maintaining its genuineness and then retracting it Ogg, tragically, committed suicide. But before doing so he advanced the hypothesis that his results were caused by the electrons being attracted together in pairs, which then, I suppose, Bose condensed like liquid helium to give superconductivity.

It is now by no means so implausible to anyone familiar with other examples of electron pairing to accept that pairing could exist, although I believe experimental evidence on magnetism of the electrons rules it out in this case. But the pairing leading to superconductivity is theoretically unlikely; the resulting particles would carry great holes in the ammonia fluid with them which are much too massive objects to Bose condense. Of course, there is also no evidence since Ogg for that to occur. Nonetheless, this gives Ogg — or his ghost — 20 years' clear priority on the mention of the words "electron pairs".

Much more serious and realistic were the labors of a group of theorists centered in Sydney, Australia. This group produced a remarkable group of papers which were mostly about liquid helium in rotating containers or "buckets", and we called them the "Australian Bucket Brigade". They consisted of quite well-known physicists, M. Schafroth, Stuart Butler, John Blatt and M. J. Buckingham, of whom Butler remained a well-known nuclear theorist (he died in 1982); Buckingham did only this mostly, though he had an eminent chemist brother I knew in Cambridge. Schafroth and Blatt were central European (this again!). Schafroth, I believe, was a brilliant man, the core of the group; and I think he was really on the right kind of track — no more confused, essentially, than all the rest of us, and capable of adjusting his thinking to reality; but he died tragically sometime in 1958(!) in a small plane accident.

John Blatt was another matter. He wrote, with Viki Weisskopf at MIT, the major textbook of nuclear physics of the time; and along with all the other great names of theoretical physics who graced our new department in the late '50s, we had John for a few months on two occasions. He had decided that he didn't believe the new theories of Keith Brueckner and associates (such as Hans Bethe) of what went on in the nucleus, and he was determined to give them — actually, to give the set of nuclear forces on which they were based — a crucial test. This test involved an explicit "brute force" calculation of the nucleus tritium (two neutrons and one proton making up the mass 3 isotope of hydrogen). This involved heavy use of our new electronic computer, so heavy that he discarded what we would now call the "user software" and demanded to be permitted to program, in the interest of extra speed, directly on the machine in "machine language", to the despair and considerable inconvenience of our computing department. He was basically on his way at the time from MIT to Australia. All of this rather typifies John, flying quixotically in the face of the wave of the future, rather dogmatically shrill in his Viennese accent, slight, dark, intense — often something basically right about what he says, but never as right as he thinks.

Although no one had really put all the pieces together (unless perhaps Lars Onsager, who could never explain it clearly enough) the theory of liquid helium was actually in pretty good shape, what with various key ideas by Feynman and by Penrose and Onsager, along with Landau's two-fluid model and various other contributions. It was the Feynman vortex picture which was relevant to rotating buckets, and the Aussies didn't really have that right, just nearly so — really, close enough. Historically, their work has sunk without a trace.

Not so correct their work on superconductivity. When, later, I began to work in this field, I found Buckingham's paper to be a key one as far as posing the relevant questions was concerned; and Schafroth and Blatt had also begun to develop a theory in which *pairs* of electrons bound together to make an effective set of bosons played the key role. They followed Landau — without recognizing his equivalent idea — in believing that a Bose gas of charged particles would be a superconductor.

***

Electrons are what is called "Fermions" or particles obeying Fermi-Dirac statistics and having a spin quantum number of ½ so that they are necessarily magnetic: the "spin" implies a rotating charge cloud. When a pair binds together they become a boson or bose particle — the helium atom, for instance, is a boson, containing bound pairs of three sets of Fermions, i.e., neutrons, protons and electrons. Bosons have integer spins — none at all, for instance, is common, so they can be non-magnetic. When a pair of electrons bind together in a superconductor, the resulting pair has zero spin, at least in all cases we know of so far.

Einstein, and independently a young Indian physicist named Bose, first wrote down the statistical theory of a large system of Bosons. I would remind you that electrons — and in fact all Fermions — can only exist one to a quantum state, which is responsible for the enormous "Fermi energy" of the metal, since all of the states of low momentum and low velocity get filled up, leaving the last filled states to be those near a "Fermi surface" in a hypothetical space in which the coordinates are the three components of momentum (which I will probably carelessly refer to as "momentum space"). Another momentum space concept is the "Fermi sphere" of the filled states: not really a sphere except in very simple, isotropic metals, but anyhow a solid figure bounded by the "Fermi surface". This is the "Fermi statistics". Actually, the surface smears out a bit as the temperature rises, but not much at normal temperatures.

Classical statistics is often called "Maxwell statistics" and does not apply to quantum particles except at very high temperatures. You see, quantum particles differ in a deep *philosophical* way from classical particles of the sorts we feel we know intuitively, in that they are *absolutely indistinguishable* one from another — each is simply a quantum of one particular kind of field, and just as we can't distinguish one bit of electric or magnetic field from another, we can't tell one electron from another *in principle* (not just because we're unable to paint it red or blue or paint a number on it).

This makes the operation of interchanging two particles occupying the same state a special one in quantum mechanics, since the new state can't be distinguished from the old one. There are two possibilities: we can either forbid double occupation entirely, i.e., assume that the interchange makes them interfere destructively. This leads to the Fermion type of particle. On the other hand we can allow such multiple occupation which mathematically, it turns out, requires us even to encourage it, i.e., gives us *constructive* interference: this is the "Bose" statistics in which there is a special bias in favor of having many particles in the same quantum state.

Einstein very early realized that a gas of Bose particles (as for instance He atoms) would, if the temperature were lowered far enough, undergo a kind of catastrophe in which the very lowest quantum state would become irresistibly attractive and swallow up more and more of the particles. Thus from the late '20s we had a model for a peculiar quantum transition in this case. It wasn't, however, until Feynman's work of the '50s that the identification of the superfluid transition of helium with this one was really accepted, though London's and especially Landau's work had prepared us for this conclusion, and the really correct way to understand it waited for Penrose, Onsager, and some of Landau's collaborators to make the proper mathematical statement quite clear. I wish I could tell this story more fully, because it too has its fascinating (and occasionally tragic) personal aspects.

***

A successor meeting to the great 1953 Tokyo congress of theoretical physics was arranged for the fall of 1956 in Seattle, Washington. One of its exciting features was the very first sizable post-Stalin delegation from the Soviet Union, led by a great mathematical physicist named Bogoliubov (translates as "lover of God"). He was a sight — bright orange new shoes, a heavy rumpled suit of some nondescript material, a wooden bowl style haircut and a habit of walking like a plump mechanical toy; one of the few really great non-Jewish Russian mathematicians, and regrettably reliable politically, he has become a relatively familiar figure in the west — and a much more urbane one — with the passage of years. Of him more later.

Air travel was only just becoming commonplace. I lumbered back and forth across the U.S. in 10 hours in a Stratocruiser. Feynman, whom I met for the first time through the ubiquitous David Pines, gave a beautiful tour de force of a talk about his work on superfluid liquid helium, full of deep insight, spontaneity, and carrying utter conviction, which he concluded with a few words on superconductivity: "I have assumed that this phenomenon must appear if I calculated carefully enough, and with sufficient accuracy, the magnetic susceptibility of an interacting electron gas. I have calculated for months, and thought as hard as I can, and I have to admit it: I am totally at a dead end. I can only conclude that it cannot be done!" (or words to that effect). After the beauty of his understanding of helium, and the utter conviction that one was listening to the words of one of the greatest living physicists, there was silence and total acceptance in that lecture hall, crowded with the best of world theoretical physics.

Before the applause had even begun, a figure leaped up onto the stage — John Blatt. In his high, dogmatic, accented voice he declaimed "But it can! We have the answer, and it is electron pairs!" He went on to explain very briefly that he and Schafroth had invented a method in which they made up bound pairs of electrons, which were therefore Bosons, and allowed them to condense to make a superfluid Bose-condensed state which was superconductivity; and he pointed out that they were presenting this work in contributed papers at another session.

I do not remember — perhaps the proceedings of the meeting, which I don't own, would say — whether any of the key questions were asked in public at that time. Elsewhere at the meeting, Blatt and his friends gave a set of contributed papers, unfortunately for them mixing up the liquid helium work — which was much less convincing than Feynman's, and in the end much less fruitful for the future than Onsager's three-year-old papers — with whatever seminal ideas on superconductivity they really had. But I can tell you what the questions should have been in *fact*: do you understand T_c and the energetics? Do you understand the "energy gap" (just now being experimentally verified)? Do you understand Landau's and Pippard's second length scale? Do you understand persistent currents? The answer to all of these was "no", if they were answered directly at all.

In essence, the question could be stated succinctly: "do you have a *program* or a *theory*?" Most theoretical physicists follow K. Popper in feeling that the criterion of a *theory* is whether it allows one to calculate something: i.e., "can it be falsified?" (i.e., tested quantitatively).

The Schafroth-Blatt theory is, unfortunately, permanently commemorated in a book published by Benjamin Press, of which David Pines (Bardeen's close friend and colleague) is the series editor and at that time a major stockholder. It turned out to be possible, by means of the most excruciating effort, to formally transform their method into a version of the BCS theory, although this could be done only if you knew what answer you wanted ahead of time. Almost immediately after the book was submitted, Schafroth was killed in a plane accident, and it was clearly not suitable to turn it down under the circumstances. It did not contain any actually demonstrably false statements by the time it was published; but it is now forgotten.

Shortly after Leon's letter appeared, to return to the main narrative, two things happened: Bob Schrieffer seems to have determined to either show Leon it wasn't right or else; and John Bardeen was awarded the Nobel prize along with Walter Brattain and Bill Shockley, for the transistor. It is not easy to work effectively in the midst of the rather intense ballyhoo of the Nobel prize, and in particular both the University of Illinois, not previously so honored I believe, and Bell Labs felt that they had a right to fête John; and of course there was the Nobel trip to take. I happened to attend the Illinois party at Urbana because I was home visiting my parents. The party was a warm, delightful occasion at which the fact that the work actually occurred at (bleep). Laboratories was not mentioned very much. John seemed relaxed then, but on his way to Sweden he stopped at the Bell Labs for a series of festivities and seemed by no means to be so. As old bridge partners and friends of both Brattain and John, Joyce and I were permitted to put John and Jane up (though not invited to the strictly business official party) and we felt for the first time a few breaths of the Bardeen temper (he was heard, for instance, to swear at a shoe) which is normally confined to the golf course (and *always* to inanimate objects). From this I deduce that by about December 1 (a few days previously) Bob and/or John must have guessed the form of the BCS ground state, and (since this discovery was the thing that John wanted above all else) the prospect of a trip full of talkative people in full dress, accompanied by Shockley whom he hated and Brattain, who, though a warm friend, could be difficult in his own way, while Leon and Bob were at home solving *his* problem, did not appeal very much to John.[d]

As it turned out he got back in time. While Bob and Leon had produced a ground state wave function and made great progress in some

---

[d]Note added in proof — Bob had the first epiphany very shortly after John returned, not before he left.

calculations, it wasn't until John was back that he walked in one morning with the method of doing statistics, i.e., getting the temperature dependence and T_c. (This and the Cooper letter of September are the only solid facts about the actual collaboration which have been admitted to outsiders. All the rest of what I say here is pure speculation. It is easy knowing the strengths and weaknesses of the collaborators to guess at this and that, but they have determined to share in absolute equality and I think everyone feels that is probably fair and surely admirable.) Anyway, in February 1957 a Phys. Rev. letter was submitted, containing a remarkable number of results (given that the time available can have been only two months of which at least two weeks were "wasted" going to Stockholm) all of which seemed to be in excellent agreement with various aspects of experiments. This letter was followed by a full paper, submitted in July of 1957 and finally published in the December 1 issue of *Phys. Rev.* This full paper is extraordinarily complete in providing recipes for calculating almost all the interesting and characteristic phenomena of superconductors.

It is based wholly on Leon Cooper's original oversimplification of the forces between electrons, hence has the essential virtue of a "model" calculation: the metal is replaced by a very few parameters: the velocity of electrons at the Fermi surface, V_F; the Debye energy $\bar{h}\omega_D$; and the coupling constant $N(0)V$ for electrons below the Fermi surface. In terms of these all of the conventional measurements can be calculated: electromagnetic response at all frequencies including Landau's penetration depth, Pippard's "coherence length", and the newly discussed "energy gap" which was already being measured by microwaves for aluminum, and soon was to be observed in the infrared; the whole gamut of thermal behavior; responses to ultrasound and to magnetic resonance applied to the nuclei; and, in fact, many, many further phenomena which were to be discovered such as electron tunneling measurements of the energy gap.

It was an extraordinary — and, to the unprejudiced, an instantly convincing achievement. An example of the kind of unexpected bonuses which appeared so quickly was the beautiful experimental demonstration of the so-called "coherence factors" in the response to ultrasound as opposed to the relaxation of magnetic resonance. Charlie Slichter, a colleague at Illinois, and his student Chuck Hebel (soon to come to Bell) were just in the process of an ingenious measurement of magnetic relaxation rates of aluminum nuclear moments. To their surprise these peaked just inside the superconducting state, only to drop hurriedly at lower temperatures, which agreed precisely with an unexpected prediction of BCS. This contrasts violently with the absorption of sound, which *drops* abruptly in the same region. This kind of thing was why Fröhlich was so wrong in saying the content of BCS was meager; every detail of any measurement of this conventional kind is predicted, many of these predictions being quite complex and unexpected; the mathematical structure is quite unique to this theory.

We now understand almost everything about superconductivity in a very deep way, mostly due to the beautiful later work of Bob Schrieffer, in distant collaboration with a series of very able Russians of the Landau school.

I realize that I have left out Bob. He, too, was at Bell Labs during that momentous summer of 1956, visiting as a summer student working for Conyers Herring. He must have been highly recommended by John Bardeen to have been there at all. I remember his coming out to swim in — or at least to sit around — our new plastic swimming pool, and a slightly amused discomfiture when we found his father owned orange groves in Florida and an olympic-size concrete pool. The father invariably provided Bob with his annual slightly-used Cadillac. Bob was — and is — what my wife's sorority sisters would have classified as "smooth" — good looking, blond, reasonably athletic, very confident, clearly from a moneyed family: oddly enough, rather a type among the very few WASP theoretical physicists of my generation. Protective coloration, perhaps — actually, later events have shown that he never quite developed the tough self-confidence which such a background and his incredible level of achievement would justify. Deeply self-critical and careful in his own work and career, he sometimes seems to overvalue others, to suffer fools too gladly, and to evade controversy even where that might clear the air.

Among the many heroes of this story Bob is perhaps the least likely and most important. London was after all a visibly complex personality: a philosophy student, lover of poetry, and deeply cultivated man; Landau the founder of a school and favorite subject of lovingly preserved anecdotes, Bernd Matthias a personality no one seemed to be able to ignore. I can't remember ever hearing a Schrieffer anecdote or an unkind word about anyone from Bob's lips. Even his letters of recommendation for junior associates are famously overoptimistic. All of his complexities are very well concealed. Yet behind those layers of protective coloration the most important steps in our scientific revolution took place.

Bob, fortunately, was of a graduate student generation which received the fruits of the first flowering of quantum field theory. For reasons now obscure, among them Fermi's presence at Chicago and a quick-thinking administration at the University of Illinois, the Midwest was the world center of theoretical physics, and several brilliant young people were at Illinois — on their way, mostly, to one or another coast within a few years. Bob was always easy with field theory concepts as well as with solid state physics, then as now a fairly unusual combination; and what sent him to Illinois to work with John Bardeen I'll never know.

***

From this point on this excerpt becomes excessively technical for a volume of this sort. As I said in the introductory note, this was one of the unfinished chapters of what was to be a personal history of the whole field of superconductivity, written in the late '60s apparently, when the excitement of what we had done was fresh, at least to me, and long before the recent three decades of turmoil caused by the discovery of "non-BCS" superconductivity in myriad forms — organics, heavy electrons and He3 all about 1970, and the cuprates and relatives in 1986+. To it I will append a few words related to the breaking-off point.

First I'd like to mention the enormous consequences of the BCS breakthrough to the understanding of the quantum theory and thereby to our whole view of the universe. The BCS wave-function gradually came to be accepted as the exemplar of a whole New Thing: spontaneous symmetry-breaking. It was the key step — or at least one of them — in the intellectual freeing of physicists' minds from the strong deterministic thinking of Newton and Descartes. The real world *emerges* from the fundamental laws, but not in any obvious way or maintaining the categories and concepts in those laws. The many Nobel Prizes — up to the most recent one, for Nambu — which have followed on from BCS are still only a weak reflection of the intellectual revolution it caused.

Second, I'd like to comment on the brief character sketch of Schrieffer at the end, in view of later developments. I used the phrase "protective coloration" and in doing so I seem to have said a mouthful. From Illinois he went to a prestigious professorship at Penn, where in fact we worked in parallel and cooperatively on the consolidation of the BCS theory (see "when the fat lady sang", elsewhere in this book), and he did other excellent stuff, among other things collaborating with Alan Heeger on the early stages of Heeger's Nobel-earning work. (There were tensions with the Bell group in that field — none of Bob's (or my) doing — which were the first inklings of the unfortunate sociology now prevalent in much of materials physics.) Exxon, fat with profits from the first oil crisis, decided to build up a Bell-style research capability in Northern NJ and hired Bob in some kind of "Chief Scientist" mode, I think retaining his Penn connection — a lab which survived a couple of decades, more or less, and did good work in some fields. It was from a colleague and mutual friend who shared administrative responsibility with him at that time that I first learned that Bob had had a number of depressive episodes during that period, and that he was diagnosed and was under treatment as a bipolar personality — incidentally, not at all an uncommon problem among theoretical physicists. He went on to be Director of the NSF's Institute for Theoretical Physics in Santa Barbara for a five-year term at least, reasonably successfully. At that time, the late '80s, his disability was fairly generally known and under control.

It was inevitable that we would clash when the enormous furor about the high-T_c Cu compounds broke. He, of course, was the survivor of BCS and the author of the book on superconductivity — the natural "owner" of the field. I had a record only inferior to his in that field — but I had the same kind of claim to "own" the field of magnetic oxides, of which this was clearly a subgroup; and I had published the first paper in which the physical model behind the phenomena was set out. Indeed, when the Nobel committee wanted advice as to whether to award the discovery promptly, they consulted the two of us, who happened to be together at Varenna on Lake Como, and we agreed (perhaps for the last time) enthusiastically, that it should be.

Well, clash we did, the first inklings coming out at that same summer school in Varenna. Neither details nor blame matter much now — there was more than enough hubris inflating both of our wrong theories at that time. But our estrangement meant that I failed to notice his increasing depression and decreasing scientific output; when, in 2003, he suddenly went off his medicine and began traveling the world seeming his original ebullient self, but hawking a truly crackpot theory, all of his friends were apprehensive and correctly so — a few months later he caused a tragic auto accident, and of his later career the less said the better.

A Mile of Dirty Lead Wire

A Fable for the Scientifically Literate

As the (probably apocryphal) story goes, one dark afternoon in the late winter of 1910–11 a student was completing the apparatus for measuring the resistance of a frozen wire of mercury metal in Prof. Kamerlingh Onnes' new laboratory in Leyden, using the low temperatures Onnes had just achieved by liquefying helium (about four degrees above the absolute zero of temperature, of which more later). Finally he complained to the Professor that he thought there was a short in the apparatus. For some reason he could measure no electrical resistance in the mercury below a certain temperature, and he was unable to find the mistake he had made. "Never mind, young man, go on home," said Kamerlingh Onnes, "I'll look over it myself tonight." By morning Onnes had well under way a manuscript describing *his* remarkable discovery that the resistance of this metal totally and suddenly vanishes at a temperature a bit above the boiling point of helium, about 5.9 K.[a] This fact is a shock to a physicist's normal apprehension of reality which at first he may casually classify among the many as yet unexplained small mysteries one encounters in the course of the accurate observation of nature (things like the rise of water to the top of a redwood tree, the shape of snowflakes, the persistence of fog, the navigation of birds). But then it gradually grows on him because of its total contradiction of what, previously, had seemed to be simple, universal, obvious laws of nature.

Normally, electrical conduction is what the physicist calls a "transport process": appropriately, the carrying from here to there of something,

---

[a]The story is indeed apocryphal; the assistant was a long-time associate and business partner, and his role was acknowledged by Onnes.

liquid, electricity, heat, magnetism, matter, etc. The analogy implied by the word transport is literally expressed when we use the phrase "flow" of electricity in a wire, like the "flow" of a liquid in a hose or pipe. To make a river flow to the sea, to make the very atmosphere move, requires force or pressure — water won't flow without the application of either the force of gravity or the pressure of a pump. The silent motion of the moon around the earth, the planets around the sun, is "dissipative" — it costs energy, which is randomized and converted into heat in the body of the planet and sun or into random motion of the oceans — all the energy of the ocean tides (and of the earth and atmospheric tides, equally real if not as obvious) must come from that motion and suck energy from it, so that the moon is very gradually falling toward the earth.

There are only two terrestrial exceptions to this universal rule that all motion and all flow dissipates energy, and they are both phenomena which only occur at extremely low temperature: superconductivity, and the superfluidity of liquid helium itself, which occurs below 2.18 K. (The latter was only discovered piecemeal in the '20s and '30s, by Keesom at Leyden, and in the '30s by Kapitza, Allen and others. The occasional arbitrariness of the Nobel awards, ever-present but particularly evident in this area, bypassed Keesom and Allen and got to Kapitza around 40 years after he coined the word "superfluidity".)

The phenomenon of electrical resistance which, in our every day world, always gives energy to and heats the wire which carries a current, is the back pressure of electrical voltage which opposes the flow of current, as a pressure head builds up when water flows. In the superconductor, as it is cooled to a temperature specific to each metal which becomes superconducting (and almost all do), this pressure disappears totally — so totally that I once calculated that a current in a ring of what was a very poor superconductor (as they go) would nonetheless continue to whirl around for 10^{99} years if undisturbed; the 10th power of the age of the universe in years (again in these days of pocket calculators, I do not feel I still have to avoid exponential notation and descend to explaining to the reader the number of zeroes in 10^{99}.) By contrast, in ordinary wires the removal of the voltage will stop the current in less than a microsecond (10^{-6} sec.).

One of the most familiar photographs in the history of science is the official group portrait of the first Solvay Congress in October, 1911. It hangs in the lobby of the Hotel Metropol in Brussels, where the early Solvay Congresses were held; and as I write I am looking at it in my copy of the official Solvay history. Mme. Curie leans back on her hand, she, Poincaré and Jean Perrin are thoughtfully examining some book lying on the conference table, while the young Einstein, James Jeans, a number of bearded elders, and a few youthful scientists in the background look on; Ernest Rutherford, not yet the "father of the nucleus", handsome and manly, stares straight into the camera. Ernest Solvay, a white-bearded Belgian industrialist, is prominent in the foreground; he conceived and personally and totally financed this first high-level international physics congress, which has served as a model for many similar enterprises. It is thought he wished to present his own theories, but was dissuaded. The Solvay Congress still exists, with its reputation for formality, lavish hospitality, and a surprisingly intimate connection with the Belgian royal family — Einstein became a musical friend of the then queen during his several visits to Belgium, and at a recent congress (my own first) the king requested a private session in the Palace's charming Baroque theatre, and fed us in return a dinner at which he intelligently and closely cross-questioned our mainland Chinese delegation.

One of the figures in that photograph, a smiling, plump, balding, mustachioed Dutchman standing next to Einstein, is Kamerlingh Onnes, who received the Nobel prize in 1913 "for the liquefaction of helium". He spoke very briefly on "electrical resistance", and although his official communication does not mention it, he here announced his discovery of some months before, of what seems already to have been called "super-conductivity". The discussions at this and the (roughly) triennial congresses which succeeded it are the history of the excruciating, profound, and incredibly beautiful struggle (which had just begun) to drag from nature the secret of the quantum theory of matter. When this finally had occurred, by the 1927–33 congresses, virtually all of the painful and confusing matters which had been discussed in 1911, 1913, 1921 and 1924 were resolved, or resolvable.

Or at least, all but one — the problem of superconductivity, announced at the first congress, stumped not only these august minds, but the entire world of physics, for no less than 46 years. The 10th Congress in 1954, convened on the highly relevant subject of "electrons in metals", reveals almost as little about this one enigma as the first where it was barely discussed, being at that time too mysterious and bizarre for even the right questions to be formulated. In 1954, as we will see, it was a problem just beginning to be cracked.

Kamerlingh Onnes' liquefaction of helium, in the institute later named for him at the ancient university of Leyden, opened up a new range of low temperatures which was to have an extraordinary and unexpected influence on the development of physics.

Temperature is one of the numerical scales of physics which is not open-ended: there is not a highest temperature, but there is a lowest one, the absolute zero (or zero on the Kelvin scale: 0 K.) The temperature represents the amount of energy of random "thermal" motion of the atoms, electrons and molecules of matter: a hot body feels hot to your hand because its atoms are sharing their motions with your hand's atoms by bouncing them around a little harder. There can't be negative motion: at best, zero, which is 0 K. We live at about 300 K (0 centigrade, the freezing point of water is 273 K.) (I see no point in quoting meaninglessly large negative temperatures on our ordinary scales.) The only usable scale for the range I discuss is the Kelvin one. Dry ice, at a toasty 200 K, is quite cold enough to be as dangerous as cold gets; but helium liquefies at 4.2 K and can be pumped down (cooled by evaporation just like water in the desert) to about 1. This is cold enough that most normal atomic motion has stopped, and (the reason I said that it was unexpectedly interesting) it is in this region where ordinary motion stops, that quantum phenomena and quantum motion become all-important and that the most rigorous early tests of the ideas of those minds in Brussels were to be made.

Kamerlingh Onnes is known to have hoped, for a few months or a year, to have found something of great practical importance: a way to make large magnetic fields (such as are regularly used in large electrical machinery) cheaply, for no cost in electrical power. This hope evaporated when he found that a few hundreds of gauss destroyed the effect in most of the metals he tried. This was disappointing: simple permanent magnets made of iron do not have to be expensively cooled, and give much bigger fields: a few thousand is easy, and with the help of not too big an electric current 20,000 gauss is not hard. (For comparison, the field of the earth is about 1 gauss and that of a toy magnet a few thousand.) An amusing illustration of the complex interplay of theory, technology and invention, some of which will be told later, is that it wasn't until 1961, after the solution of the theory in 1957, that it was discovered that the old limitation was a false one: the right kinds of superconductors will carry enormous currents at very high magnetic fields, at last count up to a half million gauss. But the main discoverers of this fact — my friends and colleagues at Bell Labs, Bernd Matthias and Gene Kunzler — were in the one case antipathetic, in the other indifferent, to theory. Much later Cornelis Gorter, a 1930's student at and later director of

Leyden, remarked that there was nothing really preventing people from making this discovery at that early period but that it was clear that neither technology — even measurement techniques — nor the psychological atmosphere, was yet ready. Kamerlingh Onnes' dream is now reality, at least in scientific institutions around the world and also in pilot industrial applications — if you ever see a fusion power plant, for instance, it will use superconducting magnets; and there really exists an experimental superconducting train in Japan suspended on superconducting magnetic bearings. Magnetic fields are caused by circulating electric currents. Every flow of electricity makes a magnetic field whose lines of force surround it, but the field is much enhanced by winding the current-carrying wire in a coil to multiply the effect; each turn of the coil adding an equal amount to the field. A current in a circle makes a field which loops up through the current loop and back down the outside. What is a "field"? This is a concept we will have to struggle with later, so it would be well to have the idea clearly in mind. A mathematician would describe it as a "continuous function in space", by which he would mean that the field (call it H) can be described as having some numerical value at any point r in the part of space we are interested in, $H(r)$. We imagine a compass needle suspended at the point r, and the force aligning it is proportional to $H(r)$, its direction being the direction of H. But we can really just think: a number (or a set of numbers) at each point. 19th century physicists confused themselves endlessly trying to make more of it.

Inside the atom Bohr's old picture of circulating electrons in orbits (familiarized by the "atoms" of old atomic energy posters) immediately provides a mechanism of circulating currents for atomic magnetism. This mechanism (to return to our 1911 photograph) was formalized by Pierre Curie and Paul Langevin, the latter a ladies' man reputed to be Mme. Curie's lover (a handsome fellow still in 1911). Their friend, Pierre Weiss, postulated a force lining up the atomic magnets all in the same direction and achieved a crude explanation of the existence of ferromagnetism — the permanent static magnetic field of iron — well before that 1911 Solvay. The *mechanism* for Weiss' "molecular field" was unknown: the best one could do was 1000 times too small; but a mechanism existed, no fundamental laws were broken (Weiss' mechanism does work in a few modern examples) and it should have been clear (was it? I don't know) that ferromagnetism — even though, on the atomic level, it also involves permanently flowing currents — was not the terrible mystery that superconductivity was. Or perhaps it was simply that human hands had held magnets since ancient Greece (the name comes from the city of Magnesia) and human minds had already comprehended atomic magnets.

The theory of magnetism is typical of what happened in that incredible breakout of theoretical physics starting in 1925. Goudsmit and Uhlenbeck added spin to the electronic orbital current, and Pauli the exclusion principle; by 1928 Heisenberg had shown that Weiss' "molecular field" was a consequence of these two plus quantum mechanics: a force he called "exchange". The principles of the theory of ferromagnetism were complete; and in 1932 Dirac, in the crowning achievement, explained the electron spin and magnetic current as an inevitable consequence of special relativity and quantum mechanics. Thus almost every concept of that period plays a vital role: relativity, quantum theory, statistics, statistical mechanics — and a workable fundamental theory took six or seven years until all the i's were dotted.

To *really understand* ferromagnetism? — that is a question. We now would say that they had left out half the story, and that is the most interesting half. Not until 1959 did anyone really feel able to calculate whether a given substance would or would not be ferromagnetic and why. But all the components and the orders of magnitude were clear; at the time they did not think that the "details" were a serious intellectual problem. I happen to think they are, for reasons we will come to eventually.

The circulating currents in a superconducting loop are quite different from those in a ferromagnet: they flow from atom to atom all the way through the metal, they allow "sources" and "sinks" — places where you can take some out and put some in, so that there is a genuine flow rather than a circulation. It is like the difference between water trapped in the cells of your body, and water flowing in a river, starting at a spring ("source") and ending in the ocean ("sink"). Even if the water in each of your cells circulated around to the right, there would be no net flow, although from the outside it would look as though a current was flowing under your skin.

One may, if one likes, of course, start a current flowing in a superconducting wire with an external battery or generator and then attach the two ends together to make one or a number of loops; if you do, the current keeps flowing and a magnetic field is maintained just as it would be by a conventional electromagnet. This is called a "persistent current". This process is just what is done in the "persistent current" mode of modern superconducting magnets.

In some real sense, the mystery is transferred from the innards of the electron — where, one gradually came to feel, anything could happen — to a very macroscopic, human level. H. B. G. Casimir, another Leyden physicist of the '30s and later director of the research laboratories of the Philips Corporation, the doyen of European industrial research, put it very clearly: "The remarkable thing is that the electrons somehow manage to maintain a kind of order, whatever it may be, along a mile of dirty lead wire."

In the Fall of 1980 we had a colloquium at Princeton from Jiri Matisoo, a clean-cut youngish man with prematurely grey hair, dressed as we would expect a rising research executive at IBM's central research laboratories at Yorktown Heights to be. He talked matter-of-factly about the business of his large-ish department, which is "The Josephson Computer Technology". His slides were beautifully prepared micrographs of microelectronic gadgetry of the sort one finds inside the modern computer, but with a difference. It was, for once, the graduate students and postdoctoral fellows in the back, not the senior staff in the front, who kept up something of a buzz of interest and questioned him closely at the end. After all, it is they who do the hard work of modern research, which is mostly long unglamorous hours at the computer terminal programming the data handling techniques or the computation, or "massaging" the data as it comes out with various computerized techniques for assessing its reliability or significance.

"How big is the box in which you will have your supersize computer of the future, with 280 kilobits of nanosecond-access memory, etc.?" "About a 5-inch cube", Jiri responded. "And how much power will you have to handle?" "It hasn't really been worth estimating — negligible. And liquid helium is a marvelous coolant. It will easily dissipate the minute amounts of power necessary. Of course, a conventional computer the same size — much slower, of course — would require a doubling of the air-conditioning capability of a fairly large building."

This is only the most massive and spectacular use of the superconducting "Josephson Effect". As a tool for precise scientific measurement this effect is so fabulously sensitive, where applicable, that nearly always the problem is to turn down its sensitivity to what is needed. Mapping currents in the deep interior of the brain and magnetic prospecting for prehistoric artifacts are two serious jobs it has done. The Josephson effect — what it is — we shall see later. Josephson himself?

Joyce and I had only a few hours in the whole week to explore the medieval island town of Lindau on Lake Constance, a town whose patron and major landowner, the Graf Bernadotte of the Swedish royal family, triennially hosts a gathering of physics Nobelists. Shortly beyond the gate, I found myself alone and backed up to find Joyce admiring an oriental rug hung at the back of the corner shop. Soon there were three of us — the owner of the shop, ardently urging us in his rather weird English to come in and see his wares. He did indeed have a marvelous stock, and he and Joyce, a nice lady who attracts friendly relationships, were purring over the rugs in spite of our quick realization that we were out of our depth financially. Upstairs he knelt on a particularly glorious red one and said, "On this I fly." Joyce: "You mean it's so beautiful you can lose yourself and feel like flying — I see that!" Dealer: "No, I fly." He gestured at a wall poster of the Maharishi Yogi and his TM saints, all in flight posture as it is claimed by that sect. Prominent in the left-hand column was Brian Josephson, not in loincloth like some others but unmistakably "flying". "See, there is even one of your Nobel Prize winners there." (He had, it seems, spotted me by now.) "He flies." "Ah, yes — and once upon a time he worked with Phil!" she said, turning to me. Short of a free rug, the shop was ours from that point.

Brian Josephson's contact with reality never seemed strong. He is brilliant; Jewish, of course; raised by a mother who consented to be interviewed on television on the subject of "How I made my son a child prodigy"; and tentative, diffident and insecure in a way I have seen in several brilliant Englishmen who have had to go through the public or grammar school experience, which for such a boy must be brutal indeed. Somehow the British system seems not to destroy such men, as I believe ours often does in its rejection of any kind of outstanding achievement, but in fact, the ones who come out most superbly in England are the all-rounders, who are active, handsome, games-playing talkers as well as — or instead of — creative thinkers. It is probably significant that England's greatest theoretical physicist, and probably her greatest physicist of this century, P. A. M. Dirac, was never knighted, as contrasted with many of not nearly his stature in the world. In Britain the pure thinker is not a figure of honor. In the U.S., on the other hand, except in the minority Jewish and oriental cultures which provide such a disproportionate fraction of our greatest scientists, one has the sense that a high proportion of scientists are a bit out of it and somewhat psychologically warped by peer-group rejection, but very few survive in the deeply scarred condition of Brian or my other friends of similar physical, mental and emotional bent.

Brian always searched out the esoteric and exciting. The idea of broken symmetry intrigued him. Perhaps he nurtured a hope that when quantum mechanics became macroscopic, it could do something really bizarre in terms of weird aspects of consciousness, alternative universes, etc. We will see later that naive ways of thinking of it can lead in that direction. He had a bent for formalism but at the same time a very concrete, engineering mind — and so he had chosen the Cavendish and an experimental thesis, rather than the natural place where one finds such people, in the DAMTP (Dirac's department), "Applied Maths" where the panjandrums of the Higher Knowledge collect. Perhaps that would have been better for him — one can be as imaginative as one likes with the cosmos, black holes, the Big Bang, etc., and no one realizes that you are really shaking the foundations of everything we think we understand, or at least no one minds because it is all too far away. Also, it happens only rarely, and usually long after the fact, that a theoretical speculation in that kind of field is confirmed by experiment and becomes a practical success as well, so that overwhelming success relatively rarely strikes before the original and speculative mind which achieved it has reached some kind of mature adjustment to itself and to the world around it. I have a very definite feeling that honors and responsibility came much too fast to Brian Josephson for his own good. As it was, his hold on reality was always somewhat tenuous. He loped with a one-sided gait along the streets of Cambridge, sometimes talking to himself, and often crossing over to the opposite side to avoid an encounter with an acquaintance or friend, especially if the friend was accompanied by wife or stranger. Some hoaxer told him the water in the U.S. was not drinkable, and U.S. Customs was baffled on his first visit by a suitcase full of mysterious bottles of a clear fluid. After the great discovery and as the various honors piled up leading to the inevitable Nobel prize, he became more and more nervous, bothered by hallucinations and disturbed enough to spend some time in a nursing home. His time in Stockholm seems to have been difficult for him (relieved somewhat by the kindness of Neil Ashcroft and his wife who accompanied him and helped with social burdens); nor was his talk up to the usual standards of formality or completeness (but precise enough and very worth thinking about — he was never a fool in any way.) Then, of a sudden, the tension snapped — he stopped formal work on physics, and even the interest in serious brain science which accompanied his semi-occult leanings, and took up wholly with TM and the sponsoring of mediums and poltergeists; acquired a pretty and competent wife, a much easier and more relaxed manner, and a child.

Quite clearly — Transcendental Meditation cleared up some very serious problems for him in a most amazing way. It worked for him — who is to deny it? But something sharp and critical in that marvelous intellect is gone. I should have said that in the years 1963–69 between his discovery and his most serious problems, there came at least two, and perhaps more, papers of the very highest quality and interest on entirely different physics subjects, as well as a stream of helpful and original — often unpublished — discussions and letters to the various people who were carrying on with the Josephson effect itself, clarifying the effect and many related matters. It is fair to say that these discussions, clarified and extended by myself and others acting as apostles for him, were the real final solution to the puzzle of Casimir's "mile of dirty lead wire". Broken symmetry and its consequences solved so much — is it any wonder he reached for more, and melted his wings?

That Solvay photograph, in its peculiar juxtaposition, throws into strong relief one of the most important, and least generally understood, facets of the intellectual history of modern science. To begin with, let me emphasize that the intellectual revolution wrought by the quantum theory did in actual truth lay a sound foundation which we still use. This incorporates many of the great triumphs of 19th century physics, from the structure now called "classical", and Einstein's other great revolution, relativity. Not one picojot of it all has since been altered, in the sense that Einstein and Dirac really fundamentally altered Faraday and Newton. The structure of the relativistic quantum theory we now use contradicts nothing Dirac, Heisenberg, Schrödinger and their contemporaries such as Born and Jordan wrote down between 1926 and 1933, when it was basically all completed.

In the 1890's people first began to study the true composition of matter: the elementary building blocks, atoms, nuclei, and electrons, and their relationships. Then in 1925 appeared the correct theoretical structure for completely understanding matter, and for two decades thereafter there was an extraordinary burst of creative energy, elaborating this structure both in the sense of discovering the unknown and of understanding the known: discovering new particles like the neutron, the neutrino, the positron, the mesons, and new structures like the interior of the nucleus, as well as understanding old mysteries like the nature of the chemical bond and the structure and properties of solid bodies.

Yet throughout this period of extraordinary success there remained a certain number of deep, nagging difficulties in physics. Readers of Freeman Dyson's book, *Disturbing the Universe*, will already know of one set of these: the "divergences" of quantum electrodynamics which make all calculations come out with infinite answers (and for that matter, even worse problems in any other fully elaborated theory one could attempt which really came to grips with both quantum mechanics and relativity). The apparently illogical and arbitrary "spectrum" (a physicist's analog word which means just the array of quantitative data such as mass, charge, spin, etc.) of the elementary particles as they were revealed to us was the second. Finally, known long before any of these subtle, nearly philosophical problems, an apparently unrelated failure of all of the marvelous structure we had for understanding matter: superconductivity, and its twin superfluidity, remained, of all the major phenomena of macroscopic bodies, the only really mysterious ones for the 30 years after 1927.

Dyson has written, at least in part, of the beginnings of the solution of the first set of problems. What I want to do here is to tell some of the story of the last, in the form partly of a brief intellectual history, and partly of a personal reminiscence. The story of the second, the revelation of the ultimate internal structures of the particles of our world and of its forces, is certainly neither complete nor is it a story for me in particular to tell; but there is a relationship between that and my own subject of superconductivity. The first steps on that road had to wait on the solution of Kamerlingh Onnes' apparently totally separate mystery, because that solution showed us a new way to think about the world. Throughout the papers of Salam and Weinberg representing the first steps on the solution of these problems (for which the 1979 Nobel prize for physics was given), one finds the mysterious phrases "broken symmetry", "Goldstone boson", "Higgs mechanism", shorthands for ways of thinking, all of which appeared in the world only because Nature and the phenomenon of superconductivity showed the way. These are only a few, and perhaps not the greatest, of the imaginative leaps this theory contains: but they started things going.

Part of what I am trying to say here is that there can be a successful and meaningful scientific revolution in which it is neither the accepted laws of nature which change, nor is there some new subject for investigation such as the nucleus or radio astronomy: what changes are techniques, attitudes, interpretations, and the philosophical set of the minds involved. I definitely mean not to preach intellectual relativity in the negative sense fashionable among some philosophers of science: our advances are very real, not a redefinition of the problem. Those 20 men and 1 woman in the 1911 photograph knew as well as I do that Kamerlingh Onnes' superconductivity was a real problem, which they could not yet solve; and most of them would now admit we solved it. We have not redefined the quantum theory; we carry it to its logical conclusions.

It is perhaps significant that almost all of us who carried this work forward were relatively young, not of that generation which built quantum mechanics and had to reconcile its strange dictates with their classical sense of space-time. Mostly, we learned it second or third hand, as an established discipline whose rules and techniques we came to feel as intuitive and natural, not as a peculiar displacement of the classical: we found and find it almost painful to do 19th century physics. The great Bohr–Einstein philosophical debates which fascinate the historians and the philosophers are to us a bit wrong-headed — both sides missed all the relevant points.

It is as if, in order to carry the quantum theory to its logical conclusion, a new breed of people had to be created: the theory was so great that it could stand the strain of going farther, but its creators could not. If Albert Einstein could not be happy with the quantum theory, who could expect any mortal to do better?

But it is time to get back to the subject. I hope in the following to do some of three things: to give some sense of the people of this part of science, how they interact and who they are; to give at least the flavor of the science of superconductivity, what it is, how it was solved, and where it leads; and finally, to start on the road of the intellectual revolution discussed above.

Scientific and Personal Reminiscences
of Ryogo Kubo*

Whatever lucky stars I may have had the fortune to walk under, surely one of the most important was the one which led Ryogo Kubo and me to meet.

In the early '50s Kubo was one of the first Japanese postwar visitors to this country in the field of solid state theory, when he spent two years at the Franck Institute in Chicago. I usually attended the Thanksgiving meeting in Chicago, and I may have met him as early as fall '50. Certainly he attended my talk at the next fall meeting, Thanksgiving '51, since he referred to it in his paper on variational spin wave theory of antiferro-magnestism at the fall '52 Maryland meeting on magnetism, which was published in the January 1953 *Reviews of Modern Physics*. By that time we knew each other fairly well and had discussed not only that mutual interest, but also line broadening and exchange narrowing, which was the subject of my talk at the same meeting. I recently found a very short list of people to whom I sent reprints and preprints at that time, and he was on it. He also visited Bell Labs early in his stay, but my memory of that visit is quite vague.

What is squarely fixed in my memory is that we were both at the American Physical Society March Meeting in Pittsburgh in 1951, at which time we both attended an invited session where one or more of the speakers made very heavy weather of the calculations of irreversible processes, with references to "master equations", Boltzmann's H-theorem, and the like. We had been calculating essentially dissipative responses (spectral line shapes and intensities) by correlation functions, and we

*Originally published in the *Bulletin of the Physical Society of Japan*, Vol. 50, No. 11, 1995, pp. 896–898. © 1995 The Physical Society of Japan.

discussed the fact that our methods avoided all that nonsense. (Kubo, unfortunately, did not remember that first discussion when I asked him recently.) Although Einstein (whose spectral coefficients I had used in my thesis) and Onsager had long since pioneered the idea of a connection between fluctuations and dissipation, and Callen and Welton were developing a general formalism (unknown to us), the idea of turning this into a practical tool was one of Kubo's great contributions which was to develop over the next few years.

In any case it was also at the Pittsburgh meeting that he and Seiji Kaya (his dean at Tokyo University) first approached me with the idea of going to Japan. I was completely unprepared for the offer. Joyce and I had just bought an old house in the country, so we were financially extended and involved in extensive remodeling and repairs. We were in no position to leave for a year in '52–'53, which was the proposition. I said as much to Kubo. But he and Kaya persisted, and finally offered a part-year Fulbright lectureship at Todai (Tokyo U.) for the year '53–'54, in conjunction with the Kyoto International Theoretical Physics Congress. (The Fulbright program was extended to Japan on signature of the Peace Treaty in '52.) In total ignorance as to what was involved, and in a spirit of adventure more than anything else, we accepted.

It was, with hindsight, a remarkable appointment. Kubo, although in his early thirties, seems to have been given the complete confidence of his seniors, a very un-Japanese attitude; moreover he chose for the first postwar visitor to Todai, the most prestigious Japanese university, a 28-year-old from industry of no extraordinary reputation at home or abroad even in the restricted field of solid-state physics. Many of my contemporaries, for instance, had overseas postdoctoral experience. Kubo was reported to have later said privately that he "discovered" me, and I believe that is, in a real sense, true although it is also true that, in a less literal sense, I "discovered" him as well.

Kubo made another trip to Bell and visited our home, where Joyce fed him a meal in complete ignorance of Japanese customs and eating habits, which must have been a problem for him. He absolutely cracked up at our attempts to speak Japanese using the outdated books available to us — real Japanese teaching had been taken over by the military, and their excellent new texts were classified. It was on that visit that we began to appreciate Kubo's dry humor as well as his warmth, and his total willingness to ignore cultural barriers.

The Kyoto international meeting was a fantastic opportunity for me to meet the elite of international physics, and many of my scientific friendships began there: for example, Mott, Frank, Lowdin, Yang, Gorter and Onsager among a host of others from all countries and fields.

That meeting was a triumph for Japanese physics, and the group of relatively young men who organized it deserve great admiration. In low-energy physics, a committee of 15 or so is listed in the program but one knows that Kubo, with his colleagues Kotani and Yamonouchi from Tokyo, was a major driving force. To illustrate the quality of the assemblage, of about fifty foreign visitors no less than a third were later to receive the Nobel prize, and half again as many were individuals such as Bhabha, Slater, and Peierls, who could well have. These were not obvious choices: most of them were in the midst of the work which would earn the prize, if not even earlier in their careers, like Yang, Townes, or myself. Most of the talks were at the cutting edge of developing subjects like superconductivity and liquid helium (Feynman was in the midst of his work on this, as was Bardeen on superconductivity); Onsager chose to give the talk which originated "Fermiology"; Mulliken's contribution was strikingly original; and the same level was maintained in other subjects, covered. No meeting since the early Solvay Congresses has managed to cover so much of the material of theoretical physics at such a level.

It was followed by two successors, one a conference in Seattle in 1956 which approached the atmosphere of the first, and was noteworthy as the first meeting in the West with Russian representation. But then physics grew up and physicists grew apart, and the third attempt in 1968 in Trieste, while praiseworthy, had more of a historical and formal flavor.

Kubo rescued Joyce, Susan and myself from the Japanese YMCA building in which the Fulbrights were housed on arrival, and deposited us in the brand new Marunouchi hotel which was to be our lifestyle during the month of the conference. After managing the essentials, the very first thing with which Kubo greeted me was the Kubo-Tomita generalization of my theory of exchange narrowing — this was his contribution to the international congress. He explained it eagerly and I was delighted with the scheme. In those few days in Tokyo before the conference moved to Kyoto we also settled on a house, with the aid both of the Fulbright office and of Kubo, a few blocks from his house and communicating with Todai by the same streetcar.

On our return to everyday life after the month-long elegance and excitement of the conference, and its associated meetings and tours, Kubo's helpful assistance really became essential. There were literally no local English-speakers except for our ever-helpful and delightful landlords, the Tajimas (he actually spoke French only), and Kubo. On our first morning Chizuko Kubo fed us breakfast at their home and Kubo showed me the way to work. He often dropped over after work accepting a drink tentatively with his characteristic "well ...," the familiar "war ...," to our ears, and talked about physics, the affairs of the world and even their war experiences in Tokyo under the fire-bombing. Of course, there were formal occasions too, most notably when he, Kotani and Yamanouchi took the three of us (my 5-year-old daughter Susan as well) out to a geisha restaurant. This was a breaking of the Japanese custom more unusual even than we realized.

At work we had adjoining offices, and he arranged my teaching schedule: a weekly seminar on line-broadening and relaxation problems, and a course on the theory of magnetism the notes for which his students transcribed (I still have copies of the "Little Red Book" which resulted). Both were attended by a number of people, some more senior, from other universities: Toru Moriya from Osaka, for instance and Kei Yosida. I had never taught before, and most of the students had never, before the Kyoto meeting, heard English speakers lecture. The level of communications was not high. But actually I was very pleased at the seminars, where every week one of the students studied and reported, usually with some success, on an important paper or topic. Kubo's assistance with every aspect of this, and his constant help with communication, made it all possible.

Ryogo Kubo and I worked very much in tandem for that six months (or a bit less, since in March we travelled for some weeks before leaving). Our adjoining offices in the physics building were large and roomy but with floors of some composition of a stickiness renewed periodically by a wet mopping (I remember for a time covering mine with newspapers). For heat we had gas appliances with soft rubber tubing, and at the height of what turned out to be an unusually cold winter the front of the building blossomed in stove pipes for soft coal stoves. One day I visited Kubo to find his room smelling of gas and him nearly unconscious: one of his gas heaters had gone out, the tubing probably stepped on by a visitor. I dragged him out unharmed.

We were both intrigued by the possibilities we saw in the correlation-function methods we had pioneered, that the calculation of response functions from the fluctuations in the equilibrium state avoided all the complications of Boltzmann's equation and the formal difficulties of irreversibility.

Kubo was more mature, and my goals were much more restricted than his. I was interested in solving specific problems of line-broadening or magnetic relaxation. My major work of the period was a long article on the narrowing-out of fine structure in a specific line, published in *J. Phys. Soc. Jpn.* I was still a technician, a problem-solver interested only in specific problems, not in building the general structure of the subject: I had even a certain contempt for general formalism. Kubo saw that there were general possibilities in the correlation function formalism I had pioneered, and he pursued them rather than the specific answers I was after. During that winter he produced the imaginary-time formalism for correlation functions at finite temperature, and invented the boundary condition which has retained his name. I was immature enough to feel it was basically a clever mathematical curiousity. Within a year Matsubara had applied the boundary condition to many-body green functions, producing the many body formalism which then became standardized via the Landau group's elegant use of it with diagrammatic perturbation theory. To me it is a very interesting historical fact that Schwinger's Green function, and correlation functions, which were the basis of my methods, came via Japan to Moscow, where they met Feynman's diagrams going the other way, and made many-body theory! Shortly thereafter, but after I left, Kubo came up with the "Kubo formula" for conductivity, as a paradigmatic example of fluctuation-dissipation methods, and showed its relationship to the standard Boltzmann approach. Mel Lax was producing similar results at the same period, but never with quite the elegance and simplicity of Kubo's work.

It is interesting to look back and realize after all these years, what the experience of working with Kubo in Japan meant to my career. Above all, his confidence in me gave me confidence in myself. I was expected to perform as a leader and teacher, at the age of 29 and 30, of a large and able group, and I more or less did.

More subtly, he helped me to pass an important stage in maturing. Good students very often come to research with the idea that the basic essential principles reside in the minds of great physicists of the present and past, if not in the reviews and textbooks. What is left to be done, is then, merely technical. (Of course, many students come with the opposite attitude, that their elders know nothing of value: these are to be avoided.) The excessive respect for what is "known" or "in textbooks" gradually must be grown out of, and Kubo, first by recognizing the originality of some of my earlier work, and then by rewriting textbook ideas himself in related work, was my first teacher in this sense. Japanese readers will understand why I always addressed him as my "Sensei" in later life.

After a last swing south to Hiroshima and a stay to lecture in Kyoto we returned to the US. I saw little of Kubo after that; we kept in touch and met at meetings, but my main contact with Japan was hosting occasional visitors at Bell — notably Toru Moriya. My interest in detailed relaxation and spectral problems continued in my work with Feher, which led eventually to localization; and in magnetism to superexchange. In the '60s I drifted away to new areas, as did he, with the interest in far-from equilibrium problem which occupied his later years. We always looked forward to his beautifully handcrafted Christmas cards.

After many years of casual encounters, for example a couple of visits in to my various homes on his part, a dinner on my one return to Japan in 1970, an encounter at the Solvay meeting in '78, it was a great pleasure to be asked by him to give two public lectures on my visit to Japan in '89, one at his new "retirement" university and one at the Keidanren. Our rapport seemed unchanged. On that occasion he took me to be formally inducted into the Japan Academy. I had many years previously been successful with his nomination to the National Academy of Sciences in the US, a source of some pride to me. After that we had one more exchange of visits, he coming to our new home in Hopewell with his younger daughter (not the one with whom our Susan had long ago played) and son-in-law, on one of the continual trips of the heavy schedule he maintained nearly to the last.

It was a wonderful thing to have known this man.

II. History

Introduction

It is common for scientists to become interested in the history of the subject, if only because such a disproportionate share of credit (and tangible benefits such as prizes, honors, and tenure, not to mention patent rights) is supposed to go to the first on the scene. Everyone wants to protect his own intellectual heritage. I have to confess to a healthy dose of that kind of motivation, myself, some of which may show through in the following pieces. But the real interest always turns out to be not the individual contributions and the precise sequence of "who did what when?" but the remarkable intellectual and even sociological shifts that take place, the mini — or maxi — Kuhnian revolutions or paradigm shifts that in fact are usually so subtle and revolutionary that no journalist ever reports on them. (Is there any newspaper article of 1905 mentioning A. Einstein? — to toot my own horn, the APS, to its credit, used to give press coverage to invited papers, including one of mine which later won a Nobel; but no reporter bit, even the in-house ones.) Some of the papers in this section do try to identify such moments in the history I've experienced; moments which began, or ended, ("the fat lady sang") such intellectual shifts.

It is interesting that as with other branches of history broad generalizations are often possible — history is not a featureless landscape of independent random events but full of enormous turbulent eddies, rising plumes of enlightenment, in general unexpectedly massive or sudden events. All of a sudden, for instance, chaos (in the technical sense) was everywhere; where a few years before that classical mechanics of any sort was thought to be a dead subject. Within ten years of the founding of the Santa Fe Institute, there were institutes focusing on complexity and interdisciplinarity all over the world. The idea of emergence was not even in the physics vocabulary when I published "More Is Different". And so on.

Physics at Bell Labs, 1949–1984

Young Turks and Younger Turks

The American Physical Society met in Canada in 2004, I believe in Montreal. Bill Brinkman, my old and very dear friend from the Bell days, had come to Princeton as a post-retirement base, and he asked me to do a historical talk on the background of the glory years at Bell Labs, as part of a session he was to organize on industrial research. I think the title I chose was something like "Young Turks and Younger Turks". I will cover similar material elsewhere but not quite this way.

Physics at Bell Labs had its beginnings long before WWII. Not only had C. J. Davisson and Lester Germer demonstrated electron diffraction already in the '20s — and won the 1937 Nobel Prize with it — but there had been a number of other quite fundamental discoveries: Johnson Noise in resistors, for which Nyquist provided the theory, and radio astronomy, are examples. Perhaps even more remarkable were the fundamental advances in mathematics which came out of the prewar labs, in stochastic theory and, of course, information theory. Bell Labs had from its beginnings in the '20s made a point of hiring absolutely first-rate scientists, beginning with H. A. Arnold; and the generation that went through the war was no exception: it contained such future stars as Charlie Townes, Bill Shockley, Jim Fisk, Sid Millman, John Pierce, Walter Brattain, to mention only a few, and the mathematicians Claude Shannon and S. O. Rice.

But the discoveries I mentioned above all had in common that they had come across in the course of work motivated entirely by direct applications to communications systems, and while the Labs had the enlightened policy of allowing the scientist to carry such serendipitous findings to some kind of fruition, and to allow their publication, there was not a hint of encouraging purely curiosity-driven research.

In the late '30s, however, a very small nucleus with a somewhat broader agenda began to accumulate, on the initiative of M. J. Kelly, the research VP of the time; he seems to have had in mind from the first the possibility of a semiconductor amplifier, but he realized that his purposes were better served by starting a broadgage program in the quantum physics of solids. Bill Shockley was the core and was clearly hired for the purpose, but other names that might be mentioned are Jim Fisk, Stan Morgan, Walter Brattain, Gerald Pearson. This is the group, specifically Kelly, Shockley and Fisk, that I have read about as calling themselves the "young Turks", implying a revolutionary attitude, and from this time many date the origin of a research-style atmosphere in the Labs. I don't have a personal knowledge of this period, but two anecdotes might illustrate that they were looking at physics with a very broad perspective. One was that I remember studying, with Wannier, a long prewar paper by Shockley and a colleague named Foster C. Nix on order-disorder phenomena in alloys; second is the well-known fact that Shockley and Fisk took upon themselves the task of designing a hypothetical nuclear reactor in 1941 or so, with no knowledge of the Manhattan project, and succeeded so well in reproducing Fermi's graphite reactor that the government became quite disturbed when they tried to patent it after the war. Neither project seems to have had anything to do with the telephone industry. Physics on the applied level contributed enormously to winning the war — for instance, the Labs were the site of choice for the development of the English invention of the magnetron microwave generator, which was put in the hands of a small group under Jim Fisk. This device was instrumental in winning the Battle of the Atlantic and in the Allies' continuing superiority in radar. Scientifically more significant, the Si crystal detector, which was a key component of every microwave radar set, underwent steady development. But as the war wound down in '45, both the management and the working stiffs began to see fantastic prospects ahead of them, due to the incredible new technologies and materials they had available from wartime developments.

Charlie Townes, for instance, immediately began to apply his wartime microwave radar skills to a series of studies of the spectroscopy of molecular gases like NH_3 and OCN. To an obscure graduate student at Harvard these results were just what my thesis needed, and they seemed to me clearly much more accurate and professional than the competition at Duke and Oxford; I developed therefore a determination to go to Bell Labs if I possibly could. But it is an indication that the atmosphere at Bell for pure research had not yet really become totally friendly that

Charlie had, by the time I got there in 1949, departed for Columbia where he felt more comfortable following his own scientific imperatives — although I believe he had received from Bell all kinds of assurances of total freedom.

Mervin Kelly, Bill, Jim and their allies immediately began a program of aggressive hiring across a number of fields of physics. Quite contrary to the impression which is left by some books about that period, this hiring was not focused only on the dream of a semiconductor amplifier. A big department called "Physical Electronics" made a series of magnificent hirings such as Molnar, Mckay, Hornbeck and Conyers Herring, the three former seguing into high management positions when the physics of the electron tube became less urgent. The then new insulating magnetic materials, the ferrites, led to the hiring of John Galt and Charles Kittel, at least, and Bernd Matthias and I, and at least two others eventually, were aimed at ferroelectricity. It may have been an indicator of Bell management's interest in solid state physics generally that I had read, as a grad student, about just these topics in enthusiastic factual pieces by J. J. Coupling (aka John R. Pierce of the Bell Labs) in *Astounding Science Fiction* magazine, sandwiched between stories from the Golden Age of sci-fi, by writers such as Azimov, Heinlein, and Van Vogt.

I can by no means give you a complete picture of this period of expansion — many other scientists appeared in the two early physics research departments, either from within the Labs' war effort — like Homer Hagstrum, Herb Mcskimin, Alan Holden, H. J. Williams — or as new employees. Then, as soon as it became evident that not just the transistor, but a number of the other initiatives such as ferrite cores and steering magnets, had real economic potential, and the Labs' gamble on solid state physics was likely to pay off, a steady stream of new colleagues — and soon a number of spinoff departments in disparate fields — appeared. But I don't want to give the impression that the growth was an untrammelled exponential — much of it was internal cannibalization, and in that early period there was also a lot of exodus to the development side or up into management or both. I have learned on good authority that the physical research department hardly grew in actual size after the '50s.

But something else very interesting happened during that decade to a decade and a half, something rather fascinating from the point of view of the history and sociology of science. When I arrived in early '49 the Labs still had mostly the characteristic mores of an industrial laboratory — a very enlightened one, with no time clocks and no rigid requirements of shirts, jackets and ties as was the case at IBM at that time. But we all worked more or less from 8.45 to 5.15 — our equivalent of 9 to 5 — and what we did was expected to be more or less what our supervisors assigned us to do, which they in turn justified to the higher ups in terms of its relevance to the telephone business — again, with a very enlightened view of what might be relevant, as I explained above — and all of us underwent a compulsory tour of indoctrination where we learned how the telephone business actually worked. There was even a course meant to bring the engineers who came with mere bachelor's degrees up to snuff, and one of my less pleasant early assignments was to teach atomic physics in this "Kelly College", as it was called. Things were very hierarchical, in terms of salary structure, privileges, and responsibilities. And management was unquestionably the business of management, you never knew what decision had been taken until it happened, including your own fate. (I was all but fired after the first year, but learned that only years later.) Relative salary levels were utterly secret. Papers were prepared first in internal memorandum form, and underwent review by management for scientific content, and then by the patent lawyers. Some very sound work of mine never made it beyond the memo stage — but after all, everyone who mattered was on the internal distribution list.

But there were compensations: it was also paternalistic. A senior scientist was assigned to listen to my first effort to give a ten-minute paper at the APS meeting and give me helpful advice; and the management style was, and remained for many years, to use the lightest touch and absolutely never to compete with underlings. (This was the taboo that Shockley transgressed, and was never forgiven.) Corresponding to the hierarchical management structure, there was almost no requirement to justify one's own work — that was your supervisor's responsibility — no one ever wrote a proposal, if you can believe that.

Incidentally, seducing the female help was, and as far as I know still is, absolutely forbidden. That was a rule; but the culture was such that divorce was very rare, oddly enough, even among the newly arrived scientists.

So what happened? This is clearly not the Bell Labs that many people here are familiar with, the Bell Labs which could be thought of as the eighth member of the Ivy League, except that it didn't play football and the scientists had more freedom. Thinking about it after all this time, it almost seems inevitable. The key is that in that early burst of hiring after the war, and the next round brought on by the exhilaration of success, the Labs had done far too well: they had all but cornered the market.

At the same time, the rest of the world was more and more waking up to the fact that solid state physics was something worth pursuing. Finally, the people they had hired were of a type and quality that was certain to find the status of docile wage slave, no matter how paternally coddled, a bit uncomfortable.

There were two alternatives. One was to let us all go, once we had realized our value in the outside world — to replace us with more docile, if less creative, scientists who would do what they were told. This is the alternative that Bill Shockley forced on the Labs in the case of John Bardeen, and that seems to have been chosen, to their serious detriment, elsewhere in the semiconductor program. It is very much to the credit of the Labs management of the time — people like Stan Morgan, Addison White, Sid Millman and soon W. O. Baker — that they chose the other alternative, namely to switch to a very different management style in order to hold on to what they realized was an irreplaceable asset.

Except in one instance, I don't think this was a conscious choice — it was a response to a series of incidents, some of which I'll recount here. After the loss of Bardeen the Shockley–Morgan department was divided up — the implication being, in order to save us from Shockley's insatiable appetite for peons — and those of us not working on semiconductors remained with Morgan, the magnetic types having as a kind of straw boss Charlie Kittel. But one could hardly break into the exciting new world of NMR and EPR without a magnet, even though telephones needed no fields bigger than .1 tesla; so almost the first real instrument purchase was a Bitter magnet. I still have in my files the memo of the scheduling meeting in 1950 in which we all shared out the early experiments — no hierarchy here. But then, in a few years, we lost Charlie to UC Berkeley, a replacement for their losses to the notorious Loyalty Oath.

How did we get into low temperatures? All prewar work had been at room temperature, temperatures below 250 K seldom being encountered by telephone equipment. This too is a story of response to outside pressure. Bernd Matthias had found us five or six fascinating new ferroelectrics, but actually more importantly he had trained a high-school graduate TA named Remeika in the crystal-growing techniques that eventually made Remeika famous. But Bernd in 1951 submitted to the blandishments of the University of Chicago, and actually spent a year there on leave learning, with John Hulm, the rudiments of superconductivity. In order to entice him back, we had to let him continue in what then seemed like the purely academic field of superconductivity, to buy an A. D. Little Collins liquifier for him and provide another TA to keep it running — and of course, its output soon became the mother's milk of a dozen experiments. It wasn't a sabbatical leave, but he treated it as one. Bernd, incidentally, cast the first stone in breaking our dress code — when the VP's administrative assistant objected to his not wearing socks, he was told to get lost. Bernd was very much the kind of person I mean in my title by the "Younger Turks" — he was simply not about to live by any conventional code. Where I managed by simple naïvete and dumb luck, assuming that the Labs would not be so stupid as to fire me, Bernd had incredible networking skills, long before the term had been invented, and he came back from Chicago close friends with the whole Fermi circle — Urey, Goldberger, Zachariesen, Wentzel, and several others; and he was not above practicing these skills on whomever had financial control over him.

Travel, especially overseas, was in the traditional Bell Labs a reward for long service preferably in management, and one was expected to travel in a group, first class, to carefully specified sites, and to come home with a formal written report of all the technology one had witnessed. When I was invited to be a Fulbright lecturer in Japan and to attend a posh international conference on the way, I was oblivious to this tradition, and blithely assumed that acceptance was up to me — hence moved a little closer to the sabbatical idea — though by no means was it a leave with pay. Finally, it was Conyers Herring who first managed a formal sabbatical leave, when he went off to the IAS in 1954 or so — and while I know of no competitive threat, it was relevant that when the Labs first began to mend its uncompetitive salary structure, they began with a 40% raise for Conyers, who was obviously irreplaceable.

I remarked that there was one case where the change in culture was totally deliberate. The population most at risk seemed to be theorists — we had already lost Bardeen and Kittel, were on the point of losing Wannier, and also Harold Lewis, who was another odd story. Lewis came to us from the IAS, presumably via Conyers — he had been an Oppenheimer student and had had a job at Berkeley, which he gave up because of the Loyalty Oath. We had had our own loyalty "questionnaire," which a very few of us had not signed — Wannier, me, and perhaps a few others — but with no perceptible consequences. Perhaps that was when we began to feel even a little bit superior to the academic world. Anyhow, Lewis was hired with hardly a hiccup — after all, he kept his Q clearance through the whole oath nonsense, and the Labs was not naïve on matters of security and secrecy. But he was more of a danger to the Labs than they knew — he had been an academic and made us aware of what we were missing. When we requested in 1955 a separate "subdepartment" for theorists, to our surprise the powers that be — at that time Addison White and VP W. O. Baker — merely asked how we would like it to be. The resulting design was almost all Harold's — postdocs, a rotating boss on whose identity we were consulted, sabbaticals, a travel budget under our control, and a spectacular summer visitor program, which for a few years, until other summer programs opened up, attracted an elite bunch. One of the reasons for our success with management was the fact that for several years we had had Walter Kohn and Quin Luttinger as regular summer visitors, and they had become so useful that our bosses desperately wanted to attract them permanently. In the case of Walter, not only did they fail, but the shoe ended up on the other foot — he went soon after to the new university at La Jolla and used his knowledge of Bell to hire away three of our stars, Matthias, Feher and Suhl, starting off his new department with a bang.

One more story: how did the Labs ever get a biophysics department? That was done by perhaps the brashest Young Turk of them all, Bob Shulman. Soon after the theory department had gained its extraordinary privileges, the rest of Physical Research began to demand equal rights. Bob managed on that basis to wangle a sabbatical leave to take a visiting professorship in Paris, lecturing on magnetism of transition metal fluorides, his subject of the moment. The same year, 1961, I had finished my stint as department chair and was invited to lecture in Cambridge and be a visiting fellow of Churchill College. Bob had said vaguely he might see me in England, and then set off across the Pacific with family in a slow boat towards Paris. What followed resembles nothing so much as the story of the Boll Weevil song — after three weeks in Paris he offered himself at the door of Crick's operation in the Cavendish's courtyard and was given space and a lab partner; next he was renting Leslie Orgel's centrally heated house — and then I got him dining rights at Churchill, where he got along like a house afire. He even brought home a Jaguar. By the end of the year he was a full-fledged molecular biologist, and he soon had talked management into letting him attract a few friends and start doing resonance on biosystems. That department outlasted most of the rest of Bell research, for good reason, though Bob left it 25 years ago.

There are many more stories I could tell — how an eminent astrophysicist finally wore out the patience of our long-suffering management in the rebellious '60s — the Russian visitors, the yellow signs, and how

Bob and I got our phones tapped — and, more seriously, how we got into nuclear physics. But I hope I've given you a bit of the flavor of those days, and also of the social and economic system which made it possible.

For three decades it seemed the research end of Bell Labs couldn't turn around without inventing something extraordinary. The transistor of course. The first three-level maser, high field superconductivity, the laser of course in three or four manifestations, the semiconductor one the most useful, but also the LED, fiber optics, the communication satellite, liquid crystal displays, the cellular telephone system, *in vivo* nuclear resonance, MRI, MBE, the Josephson effect, you name it. As long as the company remained a unity whatever we invented was bound to come back and be useful in the end. This concealed the fact that the company was extrordinarily poor at economically exploiting its technology. Almost all of the above were either never exploited by Bell, or only after being put through stages of development by others; the story of semiconductor technology told in "Crystal Fire" is not at all atypical. There were plenty of cases, for instance, where we in the research department ended up manufacturing a device for the telephone system because our development engineers couldn't or wouldn't do it.

Why this was so is not part of my story. It was a management failure on very many levels, mostly the highest and best paid. One part of the blame has been put on the undoubted fact that after we in the pure research nucleus managed to change the mores, because our sensible managers felt we had earned it and would be gone if we were not indulged, the rest of the Labs insisted on doing the same, because creative "research" seemed to earn all the goodies. There was not enough acceptance of the necessity for peons as well as Young Turks. This is undoubtedly true, to an extent, but the decline of the Labs was far more precipitate than that single reason can explain, and I personally believe that if managed through the post-'84 crisis with the flexibility and intelligence exhibited in those early days, rather than with greed and executive suite hubris, it might still be with us.

It's Not Over Till the Fat Lady Sings[*]

For 46 years prior to the BCS paper in 1957, superconductivity had baffled the best minds in theoretical physics. For example, Bob Schrieffer has described Feynman's self-confessed bafflement in 1956 very graphically. After BCS, it took remarkably little time, as such things go, for most experimentalists, especially those not previously in the field, to accept the basic tests given in the first paper and in its immediate aftermath; it was much longer before the theorists' skepticism was quelled. A few very senior theorists remain skeptical to this day. I would like to describe the processes by which most of that skepticism eventually was dispelled, and the healthy additions and changes to the BCS theory which were added in the course of this struggle. In fact, the changes were so great as to inspire my title (borrowed from the Philadelphia Flyers): the game was not over in 1957 with the BCS paper; perhaps the fat lady sang "America" sometime in 1963; or perhaps — we learn this year — it is still not really over. By being over in 1963, I mean that at that point any *rational* objection could be answered to the satisfaction of the answerer at least, if not the objector.

Most theorists' objections to BCS focused initially on gauge invariance, quite properly, because many previous theories — for instance Bardeen's early one — had foundered on that. The question is very straightforward: London's equation as it comes forth from perturbation theory with an energy gap reads not

$$\nabla \times J = -1/\lambda^2 H$$

[*]Talk delivered at the APS Meeting on the History of Superconductivity, New York, March 1987.

but

$$J = -1/\lambda^2 A$$

which is not gauge invariant; and this form is easy to derive incorrectly from a variety of non-superconducting states. These points were made especially by Wentzel and by Blatt and Schafroth, following Buckingham's lead; but the question was raised again and again in public presentations — as Nambu recalls doing at Schrieffer's talk in Chicago.

The question is in fact a non-trivial one. The first response chronologically, and possibly the only fully correct one, was probably my own first paper, which I discussed with Bardeen before BCS actually appeared (it is acknowledged in a somewhat backhanded way in the BCS paper itself). A similar argument appeared independently in Bogoliubov's very rapid series of papers already appearing as a book in late 1958. Bogoliubov, as well as Nambu in a more complete (and technically very useful) paper in 1959, argued that the system would have a branch of longitudinal acoustic-like excitations (we would now call them zero sound) which would cancel out the longitudinal currents but not the transverse ones, and satisfy the relevant sum rules. From the start I accepted that zero sound waves would appear and be relevant in a neutral BCS gas but that what we now call the "Higgs mechanism" would operate to eliminate them in favor of plasma modes for the charged gas. Both explanations relied on what we now call the "broken symmetry" of the BCS state. This broken symmetry aspect was simply not available in the original BCS theory — as can easily be seen from the Ginzburg–Landau formula for the current,

$$J \propto \psi^* (i\nabla - (e^* A)/(hc))\psi$$

where the phase of the order parameter, which is absent in BCS, plays a key role. The derivation of this formula depends vitally both on correct treatments of collective excitations *and* on allowing fluctuations in the order parameter in these collective excitations.

In fact, with surprising rapidity this explanation of the gauge problem was accepted. The controversy had, however, an important byproduct: it brought Nambu into an awareness of the structure of BCS and as a result he introduced the concept of broken symmetry into particle physics: a concept responsible for many of the parts of the "standard model": at least for the "electroweak" theory and for many of the ideas of grand unification etc., now current. It is important to realize that particle physics' borrowing of broken symmetry was repaid by causing condensed matter people to refine and conceptualize their vague notions of broken symmetry, which had been floating around previously and which had been used several times before, as in my antiferromagnetic ground state paper of 1952. The failure of the Higgs mechanism to reappear in particle physics until much later is part of the history of particle physics.

Within condensed matter physics the main opposition to BCS, led basically by B. T. Matthias, with the help of Fröhlich, Bloch and others, was to the phonon mechanism and to the lack of quantitative energetics and of predictive power for the chemical occurrence of superconductivity. At first this focused on the isotope effect, which soon was measured well enough to bring out real or imagined deviations from the BCS −1/2. Swihart qualitatively, and Morel and I quantitatively, set out to make a roughly realistic calculation of T_c's from first principles, and soon realized that the BCS equation might be taken fairly literally as an integral equation in the time domain rather than in space, which, using fairly standard dielectric screening theory including phonons, predicted the general run of T_c's and isotope effects, albeit with rather broad limits of errors which by no means satisfied Matthias. As an afterthought, let me note that it was on this work that Cohen and I later based our ideas on the upper limits to T_c in phonon — or for that matter, exciton — based versions of BCS; and I have never seen a serious answer to our arguments.

The quantitative theory of energy gaps which arose from this work came as the result of an informal collaboration between Bob Schrieffer and myself, which, I remember well, had its origins in a long and very boring bus tour we took together of the Polders in Holland, on a rainy day, during the Utrecht many-body theory meeting in 1960. I told Bob of our physical ideas about integration in the time domain, while Bob informed us of his study of the Green's function formalism of the Russians, especially Eliashberg, which was the correct way to do this *and* to express the tunneling current. At first we went off in different directions, Morel and I with our systematics and Bob with his early work on online integration of the gap equation (one of the first on-line scientific calculations ever!) but by summer of 1962 John Rowell had invented differential tunneling spectroscopy and the lines reconverged. Some time that summer John and I went down to Penn with his data on lead, and a suggested rough model, and the result were the twin letters of Schrieffer, Scalapino and Wilkins, and of Rowell, myself and Thomas which founded the truly quantitative theory of T_c and the coupling parameters, which was later carried to such exquisite precision by McMillan and Rowell. Nowadays, I feel oppressed by that success: it is very hard to break the new orthodoxy based on that work in order to get honest consideration for new mechanisms.

The third objection, which seemed to be best expressed by Felix Bloch and H. B. G. Casimir, was very fundamental: had the basic phenomena of superconductivity really been explained? Felix kept arguing that BCS did not explain the most fundamental observation, that of persistent currents, which Felix himself had shown could not be an equilibrium state, And Casimir kept asking "how on earth can the voltage be exactly *zero* along a mile of dirty lead wire?"

The answer to these questions, in principle, had oddly enough been given essentially simultaneously with the answer to gauge invariance. When I saw Landau in December 1958, he remarked that Gor'kov had derived the Ginzburg–Landau theory from BCS re-expressed in Green's function language and that since Ginzburg–Landau was gauge invariant there was no problem. He was not correct: while Gor'kov's derivation solves *Bloch's* problem, it does *not* solve gauge invariance, which involves a correct microscopic, dynamical treatment of collective modes, not basically available via G–L in spite of all the more modern claims to the contrary.

What ensued is one of the most confusing stories in modern science, a story which has permanently cured me of trying to apply logic to history. What should have happened logically is something like the following: Gor'kov should have noticed that his derivation implied that the charge parameter e^* in the Ginzburg–Landau theory was $2e$, unequivocally. Then, Abrikosov should have noticed that *his* big paper of 1956 on type II superconductivity contained an actual calculation of a quantized flux line and said: 'aha!, the flux quantum is $(hc)/(2e)$!' Then everyone should have noticed that in order to kill a persistent current you have to pass flux quanta through the superconductor, which implies an energy barrier $\sim \bar{e}v/\text{Å}$ of flux line, a number easily derived from Ginzburg–Landau.

None of the above happened. Landau idly speculated that e^* in G–L had to be integer, but nobody understood him. He also noticed the analogy between Abrikosov's lines and Feynman's quantized vortices but Abrikosov's dimensionless numbers concealed the value of the flux quantum. And no one went back and read London's arguments on persistent current. Only Onsager had it all correct, and nobody understood him, because he spoke so cryptically in a deep Norwegian accent.

Meanwhile, a race to do the flux quantum experiment involving at least five laboratories was won in a dead heat by Doll and Nübauer, and Fairbank and Deaver, in early 1961. Again, at Utrecht prior to that experiment, while planning it: Fairbank asked every theorist he could find what value he would get, if any, and had no unequivocal answers. But oddly enough, of all the papers explaining that value, no Russian pointed out that it had already been derived and the flux line described.

Full understanding came eventually from two new and rather unlikely sources: Brian Josephson and B. B. Goodman, and one could say that by summer 1963 the whole story had become clear and was told at the Colgate meeting at which Kim and I spoke about flux creep and flux flow.

Josephson: as a student in my lectures at Cambridge in 1961–62, he became fascinated by the concept of the phase of the BCS–Ginzburg–Landau order parameter as a manifestation of the quantum theory on a macroscopic scale. He seems always to have hoped that normal causality will somehow break down via quantum mechanics, leaving room for paraphysical phenomena. Playing with the theory of Giaevar tunneling, he found a phase-dependent term in the current which none of us could make go away. Pippard says I first wrote down $J = J_o \sin \Phi$ but I'm not sure; what I do think is that I may have been the first to say "why not!" Josephson, however, worked out all the consequences in a gorgeous series of papers, private letters, and a privately circulated fellowship thesis.

When I returned from England in June 1962, John Rowell told me that some junctions showed what might be the Josephson effect, and by December we had firm evidence for it. The observational problem was probably the noisy environment of most laboratories at the time. Josephson's equation

$$h(d\varphi)/(dt) = 2e\Delta\mu$$

which, as he pointed out, is implicit in Gor'kov's theory, then gave us the real and final answer to Casimir's question about the "mile of dirty lead wire". So long as φ is time-independent, there can be no voltage drop $\Delta\mu$.

Persistent currents had to wait until they were actually shown not to be persistent. Goodman finally reawakened us in the West to Abrikosov's beautiful paper, and we began to understand why such large critical fields were possible in the type II flux quantum array state — but not why they could carry such large currents!

Again in that multiply eventful fall of 1962, Young Kim demonstrated that the critical currents *did* decay, and I showed that this was a consequence of the pinning of flux lines, their slow release, and Josephson's equation, so we knew how and under what conditions currents persist. Shortly thereafter Kim demonstrated the phenomenon of flux flow: large diamagnetism, a Meissner effect, but plenty of voltage and no persistent current, showing in essence that Casimir and Bloch were correct in separating these phenomena out as independent, in some real sense, from the original BCS theory. They involve topological considerations about the order parameter and the broken symmetry whose most *general* structure was not understood until much later still, about 1975.

Associated with this story is a bitter little one for Kim. We wanted, when given the opportunity by the AIP's newsroom, to make a little public relations fuss about the decay of persistent currents, especially because it was to be done at the Seattle APS meeting, and Kim was from Seattle. It may not have been the most important discovery in superconductivity that fall, or even at BTL: the Josephson effect had been found and about that Bell Labs never suggested making any fuss. But it was the first clear solution to the oldest problem in superconductivity and possibly in condensed matter physics. Matthias, however would have none of it and our superiors permitted no release. Bernd felt this result sullied in some way his beautiful materials.

It is in this sense, that the real explanation of the fundamental properties of superconductivity was not explicit at all in the original BCS theory but required all these developments, that it was not until early 1963 that an honest answer could have been given to Bloch and Casimir. Thus we have to say the game was not over in 1957, and that the fat lady really sang "America"; roughly in early 1963. Oddly enough, this was also the year that Bloch's presidential address to the APS, published in *Physics Today*, attacked the theory of superconductivity as incomplete: a masterpiece of bad timing to which I wrote — and then I believe destroyed — a strong response.

Of course, much advancement followed that point, especially the great work on SLUGS and SQUIDS and other Josephson devices. But until this year, it has seemed that the really vital part of the game was over.

Reflections on Twentieth Century Physics
Historical Overview of the 20th Century in Physics

Introduction

To write a philosophical overview of this century of physics is a more than daunting task.

It may be that with this century the history of science and technology will be seen to so overshadow and determine the conventional history of the world as to be inextricable from it. The ramifications of physics alone determined the outcome of the century's major war and dominated the politics in the half century since that war, through the physics-based revolution in communications as much as through the revolution in weaponry. With luck the politics of the next century will focus on science-dominated problems: population, energy and global ecology. Technologies based on new science — the Green Revolution, the Pill, increasing control of many diseases, the electronics industry, aerospace, and the many uses of the computer — have dominated world economics and sociology (a wonderful reference on this point is Pico Ayer, *Video Night in Kathmandu*). I also sense seeds of a coming revolution in modes of thinking which certain scientific discoveries — fractals, chaos, complex adaptive systems such as neural nets — are preparing for us. Leaving aside this wider context of physics I turn my gaze inwards, to a great extent, to look at how physics grew and changed, seeing how the world context affects physics and physicists but ignoring the very important feedback loop of how we affect the world.

Even so, I am left with a great variety of choices as to how to structure what I have to say, whether to focus on the great theoretical discoveries such as relativity, the structure of the atom and the nucleus, quantum mechanics, the Standard Model, broken symmetry, chaos, the Big Bang, or on the great technologies such as radar and vacuum tubes, the Bomb(s) and fission, macroscopic coherence and the laser, band theory and semiconductor electronics, diffraction (x-ray and neutrons), radio astronomy, SQUID interferometry, NMR and MRI, atomic microscopy, computer simulation and computer-aided experimentation. Another focus could be socio-historical: how, when, by whom, and out of what context these developments evolved? I will come back and touch on each of these structurings later, but first I want to talk about two related themes which run through all of physics in this century.

The Flight from "Common Sense"

The first theme is a complete reversal in our view of nature, much more pervasive than we realize. Maxwell, the ideal nineteenth century physicist, discussed his equations for electromagnetic radiation, not quite with tongue in cheek, in terms of physical entities actually present in the "ether" to carry his waves about. In his "*anna mirabile*", 1905, Einstein discovered the physical unreality of the ether, yet expressed his theory of special relativity in terms of very solid and classical "metre sticks" and "clocks" and "observers", and to the end of his life he believed, in some sense, in the reality and primacy of the directly perceived world, and in space and time as he, Albert Einstein, saw them. I would argue that the result of a "Century of Progress" (the motto of the 1932 Chicago World's Fair) in physics is that no theoretical physicist can take himself seriously if he thinks that way in the 1990s.

Again and again, in many and in very completely structured ways, we are led to conclude that "what we see" is *not* what is really there. Historically, this revolution began, I think, with Rutherford's discovery of the nuclear atom. Solid bodies were seen not to contain their mass uniformly, but to be mostly empty space. This was soon confirmed by the analysis of x-ray diffraction. But the great wrench in our perceptions of the world was caused by the quantum theory — a dislocation which is yet to be mentally healed even for many physicists. The first problem was the well-known and much-discussed question of measurement theory and the uncertainty principle. However, although less evident, equally important is the realization that the vacuum has properties: it is chockfull of fluctuations which can be felt by particles passing through it, as was proven experimentally in the late 1940s, in a quiet second revolution in our thinking about nature which still engages us. This, perhaps as much as the problem of measurement, left Einstein philosophically at sea: how could a vacuum with properties remain relativistically invariant?

Condensed matter physicists, then known as "solid staters", had taken the idea of a vacuum with properties and turned it around; using Landau's concept of "elementary excitations", and Heisenberg and Peierls' idea of "holes", solid matter had taken on the aspect of a vacuum through which excitations traveled like elementary particles, without interacting except with each other. We treated the quantum theory of crystal lattices as though the lattice were itself a vacuum. Soon, Nambu, and then Goldstone, Ward, Salam and Weinberg inverted this idea to invent the field-theoretical version of our concept of "broken symmetry"; that is, there could be a vacuum containing not only fluctuations but actual finite values of the averages of physical quantities, in this case of a quantum field. This left the field theory of the *physical* vacuum having even different symmetries from that of the *real*, underlying theory with its *real* vacuum. Yet another counterintuitive step was necessary before we were able to construct the successful "standard model" of the elementary particles and forces. This was the introduction of the non-Abelian gauge interactions of quantum chromodynamics, again making the underlying physics of the strong nuclear interactions utterly different from that which is directly measured. Of course, great simplicity and generality derive from these manipulations since there is only *one* type of equation, in which interactions and symmetries are equivalent. The basic postulate of the Standard Model is that all interaction obeys the gauge principle. It is not clear whether we will find still greater dislocations of the conceptual structure of physics as we proceed toward higher energies in the twenty-first century. It is already a serious proposition that the real theory is of strings, not of particles, and has no clear statistics.

Emergence as the God Principle

What this brief jaunt into the philosophical structure tells us is that the structure of physical law can no longer be assumed to correspond in any way with our direct perception of the world. The philosophical phrase is "emergence at every level"; the properties of space, time and matter which derive from common sense are not the "true" properties of the underlying theoretical structures as we have come to understand them. This has left physics increasingly estranged from the common wisdom, an estrangement which may have disastrous consequences for both scientists and the lay public. The relatively simple and trivial wrench for the imagination which the beginning student of physics encounters, on being taught to substitute Newtonian intuition for Aristotelian common sense, is as nothing compared to this complete break between direct perception and the underlying theoretical concepts of the physicist. Modern physics has become very remote from the common man or woman.

The second theme is that the process of "emergence" is in fact the key to the structure of twentieth century science, on all scales. This fact has been emphasized in a very insightful article by Sylvan Schweber in the November 1993 issue of *Physics Today*. Broken symmetry, an emergent property, is also the central concept of solid-state (now "condensed matter") physics, including the symmetry-changing phase transitions sorted out in the 1970s by Kadanoff, Widom, Fisher and Wilson, as well as the broken symmetries of ferro- and antiferromagnetism and, crucially, the broken gauge symmetries of superfluidity (Penrose, Onsager and Feynman) and superconductivity (BCS, PWA and Gor'kov), manifesting the quantum dilemma at the macroscopic scale. Emergence is also coming to be understood as the process by which our biological and social world has developed from its physical substrate. As sources of an evolving complexity in living matter, the nineteenth century's "heat death" and its "élan vital" are inappropriate descriptions of the apparently inevitable (at least on Earth) emergence of life, first primitive and then, through morphogenesis, with increasing levels of complexity leading on to consciousness, communication, and emerging social complexity. This, then, is the fundamental philosophical insight of twentieth century science: everything we observe emerges from a more primitive substrate, in the precise meaning of the term "emergent", which is to say obedient to the laws of the more primitive level, but not conceptually consequent from that level. Molecular biology does not violate the laws of chemistry, yet it contains ideas which were not (and probably could not have been) deduced directly from those laws; nuclear physics is assumed to be not inconsistent with QCD, yet it has not even yet been reduced to QCD, etc.

This hierarchical structure is described well in the above-mentioned article by Schweber, which in turn quotes extensively from an earlier (1967–71) piece by the author. His conclusion, as mine, is that philosophically such a structure as the Standard Model, or the laws of chemical bonding, breaks the chain of reductionism and makes further delving into the underlying laws somewhat irrelevant to higher levels of organization. Schweber's article may be seen as a contribution to the great philosophical debate over the SSC and some other large science projects which have dominated the final decade of this century, in rebuttal to such books as *Dreams of a Final Theory* by Weinberg, or *The God Particle* by Lederman, which express a strong belief in the importance and relevance of reductionism. To me it seems to argue effectively that the "God Principle" of emergence at every level is far more pervasive in our understanding of the universe than any possible "God Particle" representing a hypothetical milestone in the reduction to ever simpler and more abstract laws of the dynamics of the interiors of subatomic particles.

This mention of the SSC debate is a good point at which to change subjects and have a look at the practical and sociological state of the world of physics, since that debate and its outcome have seemed to symbolize a state of affairs which has been journalistically expressed as "the end of the age of physics."

The First Half-Century and World War II: The Triumph of Physics

The twentieth century opened with the Western world in the throes of a massive and accelerating technological transformation. Science was immensely popular, although very few understood that the fundamental discoveries in physics and chemistry of the nineteenth century were behind the rapid development of the internal combustion engine, of electric power and its use, of wireless and telephonic communication, and of flight in the first decades of this century. The glamour figures of science were the practical men, engineers like Edison, Steinmetz, the Wrights, Marconi or chemists like Langmuir. World War I left both of these professions highly regarded as practically useful even in that lethal business, and had little effect on physics except in killing off some of the brighter youth. (It is interesting to note how many of the significant figures of later twentieth century physics, e.g. Wigner, Teller, and Bardeen, started their training in this period as engineers, chemical engineers or chemists.) Chemists and engineers were also the first scientific professionals to be heavily supported by industry. Nonetheless physics shared in the general popular adulation, as Einstein's status as a folk hero attests. The popular appreciation of physics and other sciences was sufficient to attract increasing support for research from private philanthropy: the Nobel Prizes, with their great prestige; the Rockefeller foundation and later others; and private donations which Bohr, Rutherford, and later, and on a larger scale, Millikan and Lawrence were able to attract for research support. Until World War II there was no question of massive governmental support of physics as agricultural science was supported at the Land Grant universities in the USA, or geological science through the Geological Survey (in the USA; similar mechanisms operated elsewhere).

Physics was just starting to need the kind of funding only governments could provide. Incidentally, in the decade just before that war another significant innovation took place: some industrial organizations in the USA at least (GE, AT&T, RCA) and some military establishments, seeing the open-ended possibilities of what we would now call "High Technology", began to fund relatively undirected investigations in physics and other sciences. The seeds of the post-1940 explosion of radar and other technologies were planted primarily in these beginnings.

The events of World War II caused an enormous change throughout the Western world in how physics was perceived and how it was funded. We should keep in mind that a number of physics-related developments were incubated in that war: (i) radar, with a concomitant radical acceleration in all the related electronics and communications fields, such as microwave techniques, solid-state diodes, signal processing (radar was the practical success of the war, contributing perhaps as much or more than the Allies' phenomenal code-breaking successes to Allied victories); (ii) missile technology, which was developed by the Germans almost exclusively but had little effect on the war's outcome; (iii) jet aircraft and other aerodynamic developments, both sides contributing, with the Germans' superior but desperate effort coming too late; (iv) primitive electronic calculators which had zero operational value; (v) nuclear fission and fusion, which, despite its very high profile, probably had only marginal effect on the war's outcome, as we now know, but has dominated military strategic maneuvering in the postwar half-century. In many countries nuclear fission has contributed significantly to electrical power generation.

Governments emerged from World War II convinced that investments in scientific research were vital to their military strength and were economically valuable, and the major participants on both sides developed, as they recovered economically, systems of national laboratories as well as schemes for supporting science in research universities. Industrial laboratories also enormously expanded and proliferated, although much of that expansion was also funded in the USA by government (mostly military) contracts. Physics was a major beneficiary of all this activity. Where, before the war, the major employment in the profession had been a relatively small number of academic jobs, there now came to be literally tens of thousands of individuals who considered their profession to be research in physics, within academia, at a national laboratory, in a private foundation, or in industry.

For some decades physics more than lived up to these expectations on the part of government. Within less than three years, the first transistor was operating, invented by a group at AT&T Bell Laboratories formed before the war, but expanded specifically to discover semiconductor devices. Equally quickly, the new science of radio astronomy, using radar technology, began to give us the first of several new windows for looking at the universe: a window through which we now see the radiation from the birth of the universe, for one. The fusion bomb made its dreaded appearance, and in combination with emergent ballistic missile and jet technology has held the world in a state of deadly apprehension for half a century. These were only the few of the first fruits of a cornucopia of technological and scientific breakthroughs which transformed our lives and, perhaps as significantly, transformed our understanding of our world, out of all recognition. New semiconductor devices transformed the electronic computer, which, plus an unlikely mix including the sophistication of quantum electronics, leading to the laser, empirical inventions such as xerography, high space technology, and materials technology such as glass fibres, has given us a completely new and rapidly evolving worldwide information network. Physics has transformed other sciences: molecular biology began in departments of physics, physical techniques sparked the plate tectonics revolution. Physics has transformed medical diagnosis with the CAT scan and MRI, and will do more in this area.

The Age of Big Science

I do not need to further expand the list of the achievements of the second half of our century. What I do want to do is to discuss some generalities about their history and the sociology which spawned them. My first point is that this period of rapid expansion was also a period of centrifugal fragmentation of physics. A somewhat self-appointed elite had already broken off in the 1930s from the study of atoms and molecules to focus on the nucleus; this was the group who formed the Manhattan Project and its offshoots during the war. After the war they had the choice of two paths. Many remained primarily concerned with weapon design and military technology, and with advising the government, primarily in military matters. Possibly the most visible and controversial example has been Edward Teller. Many others returned to their universities or to national laboratories and followed the natural evolution of their subject from nuclear physics to mesons and other subnuclear particles, or into modern astrophysics, in either case learning to spend money in gigantic amounts and to build giant accelerators. A large coterie followed both paths in different mixes, and formed a most influential group of scientist advisors, including, for instance, a considerable number of members of such groups as JASON and the Institute for Defense Analysis.

It was less than a decade after the war that a definite culture which could be called "Big Science" grew up, largely out of this nucleus which had been involved in the Manhattan Project. There were other contributors: for instance, the hope of fusion energy by magnetic or inertial confinement came to claim large subventions from the US, British and Soviet Governments. And, at least at first, the large research fission reactors at Brookhaven, Oak Ridge, Harwell, Chalk River, and eventually Grenoble were seen as "big science" because they cost on a comparable scale, but they were soon taken over as "user" facilities by large coteries of "small" scientists. The military fission program, now known to have been callously misrun, fell out of the hands of nuclear physicists, although it still taints their reputation. Finally, so-called "space science" developed rather independently as a creation of the "military–industrial complex" on the coat-tails of which many physicists came eventually to ride, with microgravity experiments of uniformly dubious value as well as some quite meaningful astrophysical probes and investigations into solar system physics.

That Big Science culture in the USA, and similar groups elsewhere, tended to have separate, direct access to government and hence to funding sources. It was independent to a great extent of the rest of science, of which it was never a majority component except in funding. In the USA NASA, the Department of Energy (earlier the AEC), and military support operated outside of the standard peer review mechanisms. (The DOE funded other science, but from a separate budget.) The sums involved precluded private support, either by industry, universities, or private foundations; the former came to view Big Science not as an investment but as a "cash cow", a source of very helpful overhead charges and of enlargement of the bureaucracy, if not of actual profits.

Eventually one could identify an entity (which perhaps was never more than a certain state of mind and set of common interests) which could be thought of as the Big Science subdivision of the military–industrial complex. Whatever one may call it, towards the end of the century a number of controversies or problems arose which indicated that Big Science no longer had unlimited approval from the public nor unlimited access to the public purse. In fact, it is this series of events which have been referred to as "the end of the age of physics".

The most conspicuous event has been the collapse of funding for the SSC, but there are a number of related, still unresolved situations:

(i) The US Space Station. This continues, with diminished funding and downsized mission. General scientific opinion is overwhelmingly negative and seems likely to carry the day in the end.

(ii) Fusion. Inertial fusion, the darling of Livermore Laboratories, is being cut back. Magnetic confinement is at last talking about realistic time scales (2040 is the last I heard) and serious engineering questions; clearly it too no longer has a free ride.

(iii) The Laser Interferometric Gravitational "Observatory" valiantly maintained by the support of one US Senator, seems likely to be another failure hastening the demise of Big Science, though it is not so big. Other Big Science projects (ANS; B-factory) are also in trouble in the USA.

Apparently unrelated, but not necessarily so in the public mind, are two essentially political events: SDI (the Star Wars enterprise) and the end of the Cold War. Physicists of most stripes in the USA opposed — to their great credit — the self-serving, unrealistic, sales pitch for SDI. Nonetheless the public saw it as a promise of safety through high technology, and it was certainly a coterie of the Big Science complex who supported it. The same political mechanism — direct intervention at the White House level — was used by its advocates as was used to support the SSC.

The end of the Cold War is, not incorrectly, seen as an enormous setback for the military–industrial complex. Whatever the personal feelings of the individual scientists involved, Big Science was maintained in its unique position by its ties to that complex, and it will in the end lose power as that complex loses power.

In Europe Big Science still maintains most of its influence. CERN shows no weakening of its stability, and other European large projects continue to thrive. Nonetheless it is possible to detect a public disenchantment with large investments and an increasing concern with economic and political problems which must eventually lead in the same direction as in the USA. The "age of Big Science" seems likely to end with the twentieth century.

The Flowering of "Small Science"

Big Science is, however, only a fraction of all science, even — or especially — of all fundamental or "basic" physics. A second type of physicists doing a second kind of physics was given a jump start from wartime advances and postwar funding. The specific nuclei of the resulting growth were distinct from the Manhattan Project core of nuclear, particle, and fusion physics.

(i)   The availability of new ranges of coherent sources of electromagnetic radiation, of new types of circuitry, and of ultrasensitive detection, began a continuing trend to cover the entire range of spectroscopy by coherent methods: at first EPR, NMR, and microwave gas spectroscopy. The range took an enormous leap with the invention of the laser in the early 1960s. ˙

(ii) Neutron diffraction and scattering, using the fission reactors left over from the Manhattan Project, and those specially built for the power reactor program, had a large influence on solid-state physics research and formed nuclei of specialists at each of the national centres. These soon became quite intellectually isolated from the Big Science projects, often at the same centres. This isolation was signalled by one of the two or three most conspicuous omissions in the list of Nobel Prizes: neutron diffraction and scattering, allowing the experimental proof of antiferromagnetism and the study of phonon spectra. (In 1994 this omission was finally corrected.)

(iii) Wartime research on semiconductors and other materials, primarily carried out at industrial centres, together with the new sophistication in electronics, led to a great increase in the sophistication and depth of research on semiconductors, which of course spawned the invention of the transistor. This was only the first of a stream of developments of useful solid-state materials and devices, for example, the discovery and development of insulating magnetic materials.

(iv) A very important stimulant to fundamental physics was the development of a liquid helium cryostat affordable by almost any laboratory. This opened up the temperature regime in which quantum effects on almost all properties of condensed matter were of importance. Superconductivity, superfluidity, and fermi surface effects in metals were now open to investigation at many laboratories throughout the world.

(v)   As the world's technical sophistication advanced, advances in relatively old kinds of measurement capacities developed: the electron microscope and x-ray diffraction using new, powerful sources being examples which were very important in molecular biophysics and in metallurgy on the atomic scale, as first applications.

The wartime roots of these developments were, on the whole, separate organizationally and geographically from those of Big Science. The twin institutions in the USA whose contributions in wartime had the biggest impact in this area were Bell Laboratories and MIT's Radiation Laboratory with contacts to Harvard's Radio Research Laboratory. It is significant that the Collins liquefier and the Bitter magnet were MIT developments, that the transistor arose at Bell, and that NMR as a useful, measurement tool grew out of Harvard. (I emphasize that I am talking not about organizational but about intellectual history. The roles played as government advisers by administrators like Conant of Harvard or Buckley, Baker and Buchsbaum of Bell Laboratories in the arms race and in the government support of both kinds of science are irrelevant to the point.)

From Los Alamos Big Science diffused first to American universities such as Berkeley, Chicago, Princeton, Columbia, Stanford and Caltech; similarly small physics tended to spread from Bell Laboratories, Harvard, and MIT to a distinct set of universities, such as Cornell and Illinois, and became strong overseas in the British university system and at Harwell, and eventually Leyden (and Philips Eindhoven and Paris on the continent) as well as Japan, the latter for obvious reasons. Many physics departments accommodated both: Chicago and Harvard, and, at first, Columbia. But it was clear in some physics departments that one group or the other dominated hiring and promotion, often for decades, as it did (to emphasize the Big Science examples) at Caltech, UCLA, Columbia (with a separate organizational set-up for some small scientists), Princeton, and Stanford, and on the continent of Europe generally, especially in Italy and Germany, and, when it joined the scientific world, China.

In the USA through the 1950s the American Physical Society tended to have relatively large general meetings with physics of all kinds represented, but by the mid-1960s the March meeting, originally set up to give Small Science a forum, had grown into the equal of any APS meeting and by the 1980s it dwarfed all of the others. Almost in self-defense, the Big Scientists took over the April meeting in Washington, and then began to have separate "nuclear" and "particle" divisional meetings. Meanwhile, on the international scene, the International Low Temperature Physics Congress had become a giant convocation of many kinds of Small Science, and other specialities (magnetism, semiconductors) developed their own giant congresses; an entirely separate, very internationalized system of congresses grew up, starting from the Geneva congress where the

Cold War opponents first made open contact in what seemed, then, the sensitive fields of nuclear and particle physics.

The growth in these two areas took place in very contrasting ways. The Big Science experimental project has become larger and more collaborative, finally involving, in some cases, literally thousands of scientists on a single experiment. The great majority of these people have no truly independent role to play. On the other hand, theoretical work in this field became more and more speculative, esoteric, and abstract, with only a small fraction of theorists actually involved in detailed interaction with experiment. Some of the best known of present-day theorists, Witten, Penrose, Hawking, Schwarz for a few, have eschewed the prediction of and explanation for experimental fact as a primary goal of theoretical physics.

The evident relevance of much Big Science to the deep, quasireligious human impulse to know the ultimate origin of life and of the universe and the ultimate stuff of which we are made, aside from affording at least some access to public money, also assures it an indefinite supply of eager young recruits, especially to the theoretical effort. These recruits, so numerous as to overwhelm all available research posts even in "good" times, in turn dominate physics departments in smaller colleges and in the Third World, which helps to perpetuate the supply.

The Saturation of Small Science: The End of Entitlement

The gigantic growth in Small Science (particularly physics), on the other hand, took place to a large extent because of a perception both in industry and in government that the products created were economically useful. The result was funding on a scale which absorbed essentially all new recruits. Through the first three postwar decades this was unquestionably the case; we hardly need to repeat the litany of practically useful materials and devices which came out during these years. But in this half of physics, too, there were ominous indications of trouble ahead.

It is a great boon that this kind of science can produce both practical, useful devices and methods as well as intellectually exciting scientific knowledge. But this fact can also lead to confusion. The scientists' natural preference is for the more intellectual work which is reinforced by the fact that the pure researcher is more visible, more honoured, more mobile (he publishes in the open literature) and often better paid than the worker on applications, and certainly better paid than the production engineer who supervises actual manufacture. In academia and national laboratories, and even in some of the great industries, it has seemed as though most departments have been trying to compete with the pure research divisions within the organizations, and with university departments of physics and engineering science, as well, even when the nominal function was to do device development, manufacturing design, or even marketing. The fresh PhD, coming from the relatively "pure" atmosphere of university research, has found it natural, easy, and often at least financially secure to continue along the lines of his thesis and in the "research" mode. The sociology, again in the USA, encouraged him to feel this as an entitlement (encouraged, as well, by some unfortunate propaganda from scientific administrators). It is interesting and important to realize that this sociology was also rampant in the Soviet Union and the Eastern Bloc as well as in certain parts of Western Europe, with variations. In the Soviet Union and East Germany, large institutes supported by governments grew up, combining (like our DOE laboratories) military and big space responsibilities with Small Science. Employment in these institutes held great personal advantages relative to the general economy. Reunification, in Germany at least, has found these institutes overpopulated relative to any reasonable mission by factors (conservatively) of 3–10. A similar situation seems to hold in the FSU. In Germany, at least, the cause of the collapse is *not* by any means entirely the abandonment of military and space missions, or of the Cold War mentality, but primarily a serious neglect of the economic, productive aspect of technology; Europe, the USA and the Eastern Bloc differed in degree in this, rather than qualitatively.

Throughout the period the same rapid growth and apparent boundless opportunities meant that Small Science, particularly, seemed an ideal entry point to the advanced societies for immigrants from India, from the Soviet system, and from the Orient. This group of people had an additional motivation to stay in research: visa regulations which allowed for postdoctoral stays and were unfriendly to changes in jobs, and even more so to changes in field or employer. (And often language, or other cultural factors, could make such changes difficult in any case.)

The result was an increasing overcrowding of research as a profession. One may question — and leaders of various kinds did — how much research is too much, or even whether there *could* be too much research, research being viewed as an absolute good. But conditions within the profession, viewed objectively, made it quite clear that aside from the inevitable funding crunch there were dysfunctions in the system.

There was a very sharp change in the nature of a research career. The "promising" young scientist's publication rate grew by factors of five to ten; the number of applications a young researcher might make for post-doctoral work or an entry-level position from two or three to 50. Senior scientists were overwhelmed with receiving and sending reams of letters of recommendation, which thereupon became meaningless. The numbers of meetings in a given specialized subject, and the number of subjects with a formal meeting list, both grew by factors of ten or more. In many subjects one could "meet" nearly 52 weeks in the year, somewhere in the world, and leaders in the field were invited to all. Meetings almost inevitably led to publications. Most publications became tactical in this game of jockeying one's way to the top; publications in certain prestige journals were seen as essential entry tickets or score counters rather than as serious means of communication. Great numbers of these publications were about simulations of dubious realism or relevance. Essentially, in the early part of the postwar period the career was science-driven, motivated mostly by absorption with the great enterprise of discovery, and by genuine curiosity as to how nature operates. By the last decade of the century far too many, especially of the young people, were seeing science as a competitive interpersonal game, in which the winner was not the one who was objectively right as to the nature of scientific reality but the one who was successful at getting grants, publishing in PRL, and being noticed in the news pages of *Nature, Science,* or *Physics Today.*

In many subjects the great volume of publications, the fragmentation into self-referential so-called schools who met separately, and a general deterioration in quality which came primarily from excessive specialization and from careerist sociology, meant that quite literally more was worse. This was very obvious in the Soviet Union, in many fields. In the field of high T_c superconductivity with which I am familiar, enhanced national and regional funding has simply proliferated non-communicating schools and subcultures rather than focusing meaningful effort: it seems certain to delay solution of the scientific problem. One wonders whether AIDS could not suffer the same fate. Fortunately in both cases practical, as opposed to scientific, progress is more easily judged and managed.

Observation of the actual working of modern science in this period has led some sociologists to attempt to apply deconstructionist ideas to it: to maintain that the "truth" is purely sociological and determined by power relationships, not by nature. Even the president of Radcliffe University seemed to be lending her support to this nonsensical point of view. These sociologists are seeing a temporary and, I believe, aberrant *sociological,* rather than scientific, phenomenon; in fact one does not know of any case in which scientific error caused by political, economic or sociological pressure has not eventually been corrected. If this were not so, how could Darwin have prevailed? Or Copernicus? The truth has great power to prevail; it is fixed and solid, while error is variable; reproducibility wins over visions and poltergeists.

To complete the story of the last decades of the twentieth century, eventually the inevitable collapse occurred. It was not through failure of the research engine in the West, but that this engine produced device after device of which Japan — which has not had a great record of original innovation — cornered the great part of the market share and then often lost it to even younger societies. The liquid crystal computer display may have been the straw that broke the camel's back; the entire technology is American, the American market share is approximately zero. Quite sensibly, the American corporations decided, one by one, that it was essential to shift the motivation system away from the entitled "pure" research career. This coincides with a great wave of immigration from the FSU and Eastern Europe, and with a (however glacial) downsizing of the national laboratories. The US Congress is making threatening noises at the National Science Foundation which supports much university research, but these have not yet resulted in seriously damaging that research. Many students and a few of the sharper postdoctorates are moving out of the research pipeline. Perhaps by the end of the century we will see the 1980s and 1990s as the bad old days and will be back with a reasonably efficient physics establishment. What seems to be the worst danger is allowing the pendulum to swing too far the other way.

Concluding Comments

Let me finish this essay with some comments about recent developments and trends in physics, most of which are more optimistic than otherwise.

One healthy trend is growth in the level of complexity of systems which are being studied with the quantitative precision, instrumental ingenuity, and mathematical sophistication which characterize the best of physics. A wonderful indication of this trend was the Nobel Prize award to de Gennes, who has spent decades on "soft physics", the classical physics of polymers, colloids, liquid crystals, glues, and so on. The field is growing and many promising results are appearing. Biophysics is a mine of unsolved, or only qualitatively, solved problems like biocatalysis, cell microstructure, neural function, and a relatively small coterie are moving in these directions, using very sophisticated technology in many cases.

In the past physics has tended to disengage itself from such fields as molecular biology once they are well started. This has often been a loss both to physics and to the new field, since the attitudes of intellectual rigor and quantitative precision which physics brings can remain relevant even when the matter under study is biological. I hope that physics will more and more retain an interest in these new fields of biophysics.

An explosive advance in non-linear science is leading into new problems and results in driven, complicated non-linear systems. A danger in this area is the temptation to substitute simulation for its own sake for the appropriate subject matter of physics, which is matter and energy in the natural world. Where a simulation or a computation is relevant to the real world, it is physics, but otherwise one tends to lose the guiding criteria of reproducibility and experimental verification which are at the intellectual centre of physics. These fields, both biology and non-linear dynamics, are as yet not promising ones for the "careerist" sociology, though one finds a certain level of hype surrounding some developments.

The beautiful, flexible kit of mathematical tools which the theoretical physicists have developed, first to build the Standard Model and to solve similar quantum many-body problems in condensed matter, and then in attempts to go farther, should not be allowed to die in the general downsizing of both big and small physics which must occur. (I do not mean here *mathematical* physics, which has tended to become separate from experimental physics and sterile.) The substrate considered may be totally different, but I must believe that it is both beautiful and important that the same mathematical constructs illuminate liquid crystals and quantum chromodynamics; that phase transitions in the early universe are echoed, the one in superconductivity, another in droplet metallization in semiconductors. In theory, more is probably worse and less better, but zero is totally unacceptable. There remain many exciting problems in fundamental physics which will not die with the SSC and its compeers. Big astrophysical science seems set for decades, for instance; and fundamental condensed matter physics is full of problems, if not of solutions.

By the rather mordant picture which I have given of the sociology of physics I do not mean to imply that physics is not intellectually very lively, even in some of the areas one might have expected to have been thoroughly exploited. For instance, it has become increasingly clear during the last decade that the quantum many-body problem encountered in the bizarre new superconductors, in a broad variety of intermetallic compounds with rare-earth atoms, and in layer, surface, and chain structures remains full of surprises, and essentially unsolved. In the opposite case is astrophysics, where each new type of probe leaves us with a new and exciting puzzle to solve, and even the overall structure of the universe remains controversial, much less whether or not there are recognizable traces of phase transitions in the early universe. What is depressing is how few of those who are supposed to be investigating these mysteries are doing so meaningfully, because of the unsatisfactory career-dominated sociology which has arisen.

A third important point is that much of the present malaise in Small Science comes about because in great part it had historically been driven by Cold War and industrial needs for electronics technology. The electronic and communication industry were its chief customers. These industries have correctly and healthily realized that this pipeline of "hardware" technology is full, and that software and economics will be the limiting factors in future. The "hardware" problems of the future can be expected to lie in at least two new directions: biomedical technology; and energy and the environment. Great contributions in the past decade have been made by physics to medical diagnostics, and, in a quiet way, to biomaterials. The impact of advanced physics on energy and the environment is only beginning. Satellite studies of all kinds, and fission reactors, are so far all that is of major importance. But solar photovoltaics, batteries, energy storage, transmission, (suitable superconducting cables are not very far away), and of course fusion, are all in the future, as are minerals exploration and trace analysis. These are not fields which have a Bell Laboratories or an IBM to lead the way; new mechanisms to support — and manage — research in these fields will probably evolve.

We conclude, then, with the thought that there is no dearth of new directions for physics and physicists to take. However, there are real problems in the field. One is the increasing estrangement of even the best-educated public from physics which is inherent in the phenomenon of "emergence". When we are producing economic goodies for the public, we are tolerated; but we cannot expect them to support our pure curiosity with extraordinary generosity even if we make every effort to explain ourselves, and popularizing and explaining physics will be an increasingly difficult task.

A second problem we have is to survive the present era of over-population of research without a complete breakdown in our system of ensuring quality and integrity in the research process, and without having some inappropriate system of standards imposed from outside. Also, scientists must take more responsibility for helping to decide what projects should be funded (even outside their own specialities) and which fields are intellectually sound, as well as relevant to the rest of science. If we do not make more of an effort to regulate ourselves and to curb our tendency to assume an entitlement to go on doing things as we always have, the public tolerance may properly become overstrained.

21st Century Physics

A century is a long time. The past one has produced complete revolutions in almost every subject of natural science, and two or three or more in the most active ones such as physics.

Perhaps the greatest of these is the total change in philosophical attitude to the underlying structure of matter which came with the quantum theory. Here I mean not only the relatively minor (though essential) changes in point of view associated with the "uncertainty principle", but the complete abandonment, at the microscopic scale, of identifiable individual objects — of the particle concept and the idea of separate material bodies, instead replacing particles with quanta of fields. There is also the wonderful way in which, at the micro level, symmetry can be made the fundamental principle: the tripartite relationship of conservation laws, symmetries, and interactions which follows from the gauge principle as the universal source of interactions.

Yet the basic revolution in the life sciences is hardly less fundamental: the reduction of the whole of the actual machinery of life to knowable — and in many cases known — molecular mechanisms. We know not only the molecular mechanics of heredity and evolution, but those of development and morphology, and even of the immensely complex nervous and immune systems.

In each of these cases, as well as in others only less fundamental such as the plate tectonics revolution in geology, I would venture to say that these revolutions are complete and irreversible. A few cranks believe, perhaps, that somewhere in the higher nervous system there is new physics to be found; a vocal minority probes, without attracting much meaningful attention, a Bohmian or a discretized revision of the basic quantum scheme. I do not think that either represents what will happen in the 21st century, basically because the structures built upon the new principles are so satisfactory.

The 19th century was almost equally revolutionary: Darwin, electromagnetism, paleontology. What the magnitude of these revolutions tells us is that physics in 2100 is extremely unlikely to be recognizably the same subject as we now understand it. The author who attempts to look beyond the first few decades, beyond essentially the problems and concerns we already have on our plates, is making an impossible leap in the dark beyond unknowable revolutionary changes, and must admit to full knowledge of that fact.

In physics, as in biology, this has been the century of reductionism. Starting from an epoch when the positivist philosophers could seriously question the existence of atoms, chemistry (though not prone to the same antiatomistic stupidity) was a jumble of empirical observations, and biologists had no inkling of a molecular basis for their subject, we have produced a working model for matter at any scale we can reach: the ubiquitous and frustratingly successful "Standard Model". We have also produced a workable cosmology, starting a century ago when, again, we hardly knew that the cosmos as a whole existed, and even the local galaxy was yet to be comprehended. What I am going to say in the end is that my guess for the next great revolution that will take place is a shift of emphasis to the antithesis of reductionism: to the understanding of complexity. But in the first decades of the century, there are still uncompleted items in the reductionist agenda to be disposed of: so let me discuss these first.

1. Particle Theory

I may disappoint my condensed matter colleagues in that I see signs that the string theory approach to the ultimate "Theory of Everything" may be bearing fruit. One is that the ugly high-dimensional supersymmetric theories are turning out to be only one formal approach to something more general which, from some aspects, looks much more like the world we live in. A second related point revolves around the fact that strings make it possible to understand the Bekenstein limit on the entropy of a black hole, which in turn suggests that quantum gravity is, in some sense, able to create its own "convergence factor": to prevent the infinite subdivision of space and time. But between here and the Planck limit there is still the probability (rather than a mere "possibility") of enormous surprises. The great variety of unexpectedly large or small parameters in the Standard Model (the neutrino mass the latest to appear) makes it almost certain that there are several further stages in the hierarchy. The first of these will appear in the next decade, certainly, when we will probably discover unexpected complexity in the Higgs phenomenon. (A personal guess: but no supersymmetry.) The most serious disconnect between the Standard Model and quantum gravity is in the fact that the former fills the vacuum with properties, fluctuations, and even finite fields, which seem to have no gravitational consequence. This has been a major anomaly from the start, the basic dilemma in the no-man's land between cosmology, phenomenological particle physics, and quantum gravity. It is the biggest anomaly, telling us, if nothing else does, that the Standard Model is logically incomplete. Nonetheless, we must assume that by 2050 a TOE will probably be in place, in some sense.

2. Astronomy and Cosmology

This is the area of "fundamental" (translation: reductionist) science which is enormously active and exciting right at the moment, and shows no signs of running out for a few decades to come. I am far from the appropriate person to make serious predictions in this field, but it is easy even for an outsider to appreciate that this is the great age of experimental cosmology, with an unbelievable rate of discovery. Let me make only one provocative suggestion: that when we are all done, it will turn out that there is no exotic form of "dark matter", merely a comedy of errors in a field where it is practically de rigueur to underestimate one's limits of error.

As for pre-observational cosmology: inflation, cosmic foam, and the like — here progress will be extremely slow and dependent upon developments in particle theory. Inflationary cosmology has consequences in observation, for instance in understanding the unbalance of baryons vs. antibaryons, and the possible fractal distribution of baryonic matter, and in that sense is an experimental subject; but the wilder shores of many-universe physics remain the only free space in all of science for unbridled speculation, at least until the 22nd century.

But the positivist error of "I must not speculate on that which I cannot measure or touch" has been refuted in the past century again and again: by the discovery of the cosmic background radiation, most spectacularly; by general relativity; by Dirac's prediction of the positron, and by many other examples. I happen to be conservative enough that I do think that a restricted version of positivism is a valid guide to research strategy, if only in the form of "Occam's Razor": "one should not multiply hypotheses unnecessarily". Where a speculation simply adds irrelevant degrees of freedom without contributing to the net body of things explained, or, even more broadly, without the motivation of doing so, I believe it is pointless. Much of the many-universe world of cosmic foam, worm-holes etc., seems to me to be in this category: there seems no explanatory motivation. (Whereas, for instance, string theory has the hope at least of motivating the many unmotivated symmetries and near-symmetries of the Standard Model.)

3. The Opening to Complexity: Emergence and Antireductionism

A. *The Background as of the Millennium*

As I remarked above, the 20th century is the century of the triumph of reductionism, in that we have adequate theories at the microscopic level both for the behavior of matter and energy at ordinary scales, and for the molecular basis of biology. But interestingly, in both cases one may look at the intellectual structure in two ways: as a reductionist search for the fundamental theory, given the observed behavior at the more acces-sible level; or, conversely, a lesson in how the observed behavior emerges from simpler laws involving simpler and smaller entities. In physics, the two great reductionist discoveries (aside from the quantum theory itself and the Rutherford-Bohr atom) were the renormalization program and its ultimate embodiment in the phenomena of confinement and asymptotic freedom; and the concept of broken symmetry as introduced into particle theory by Nambu and Weinberg et al. Each of these two developments can be seen conversely, as starting from simpler and more symmetric underlying theories of the substrate, and making different and more complicated emergent entities from them. At the level where the quantum theory intersects the theory of ordinary matter — i.e., the physics which has come to be called condensed matter physics — the concept of broken symmetry is all important: it manifests itself again and again, in the emergence of the phenomena of crystallinity, superfluidity and superconductivity, magnetism, and more complex condensed phases such as those of liquid crystals. The very creation of our classical world from the microscopic quantum world is an example of emergence: The fields of quantum field theory only became distinguishable objects, or classical fluids, and we only became capable of the measurement of such conventional quantities as space, time, and orientation, as a consequence of the fact that macroscopic assemblages of atoms break quantum sym-metries and become classical objects. Thus the ideas of emergence and of reductionism are simply two sides of the same coin, philosophically. But it is plain to see that the next century will be dominated by the former, since the reductionist program is all but completed, and since most of the interesting problems are on the complex side of the hierarchies.

B. *Unfinished Business: Open Problems in Quantum Physics*

As we saw to be true of particle physics, and even more of astrophysics, we cannot boast that in fact we have come to the end of reductionism — yet. With the 1998 Nobel Prize in Chemistry, it may have seemed, for instance, that the labors and controversies of Heitler–London, Hund–Mulliken, Pauling, Wigner and his students Seitz and Herring, in understanding the chemical binding of materials, had come finally to a resolution in the "LDA" and the modern computer. Surely this is true for sufficiently symmetrical simple materials like the all-important semi-conductors Si and GaAs, or for covalently bonded molecules. There is not much mystery left in the dynamics of electrons in simple metals and semiconductors. But complex materials on the borderline between metallic and insulating, like the infamous cuprate superconductors and the reduced dimensional metals like the Bechgaard salts, or metals containing open inner shells, like the "heavy electron" metals UBe_{13}, $CeCu_2Si_2$, etc., remain very deep problems for the quantum theory of interacting electrons. These problems are being discussed elsewhere in this book, by Rice, Haldane, and Varma. Suffice it to say that, fascinatingly, for almost every difficulty and complication which has arisen in the quantum field theory of the elementary particles, there has appeared something as interestingly complex and counterintuitive in the quantum theory of "ordinary" matter; and cross-fertilization between the two domains of applicability of quantum field theory has been immensely valuable in the past and will continue to be for the next century. Broken symmetry and the Higgs–Anderson phenomenon; topological solitons and defects; the many aspects of renormalization group ideas; confinement and asymptotic freedom; chiral anomalies; all have had their day in both fields, and we can anticipate that applications of the new ideas of string theory and quantum cosmology will, in the end, cross-fertilize with the many-body theory.

The latter has two enormous advantages for the investigator: first, one knows that whatever anomalies appear are not to be resolved by changing the fundamental laws — the rules of the game — but are a consequence of emergence from these laws; and second, one has many experimental ways of probing the phenomena.

To be provocative, having made rash predictions about matters on which I have no expertise, I will stick my neck out on my specialty. The room temperature superconductor may appear — in a system and using a mechanism (not "excitonic") of which we had no idea but probably will be of little interest because of unexpected developments in thermoelectric cooling using, for an example, bizarre properties of low-dimensional systems. The cheap, 20% or more efficient solar cell will appear as a consequence of incremental developments and produce most of our power; this supplemented by the storage battery which in some decade will appear as an unexpected byproduct of some exotic material, will allow the gradual removal of the excess CO_2 from the atmosphere which, in the first decades, caused global warming, violent weather, and flooding in London, Calcutta and many other cities, and serious climate modification.

To return to less speculative ideas, an aspect of quantum condensed matter physics which does not have any obvious parallel with quantum field theory is the role of quenched disorder and of finite sample size or dimension (which, interestingly, has similar effects.) What all of these do is to spotlight the coherence properties of the quantum theory, the aspects in which it is most perverse and counterintuitive. Localization, Bohm–Aharonov, the Josephson effect, Quantum Hall and quantized conductivities, are past achievements; but as we reach the nanoelectronic era, we may find this field interacting with the new concepts of quantum computation to give us an emergent technology with intriguing possibilities as well as challenges. Whether embodied in Josephson arrays, in quantum dots, quantum wires, or others quantized conductance channels, in tunneling between edge states, in nuclear magnetic resonance schemes, or whatever, there seem to be an infinite range of possibilities. This is the kind of area where one has a sense of a revolution waiting to happen. It seems likely that the key ideas have yet to appear, not simply extrapolations to faster rates and smaller sizes of today's technology, nor yet faster solutions of today's problems (such as factorization) but some new mating of technology and algorithms which is unimaginable today.

C. *The Physics of Complexity*

With this last area we begin the breakout from reduction to emergence, from analysis to synthesis, from simplicity toward complexity, which I feel will be characteristic of the 21st century. There seems to be much more complexity to be studied at scales larger than the atomic than at smaller scales, and physicists will more and more be involved with problems of complex systems.

The first of these breakouts already got under way during the last decade or two of the 20th century: a renewed interest in the physics of "soft matter". Under the leadership of de Gennes, Edwards, Kadanoff, Langer and their disciples, substances like foam, gels, glue, glass, inks, and phenomena like friction and adhesion have been rescued from the chemists and given serious microscopic study (see chapter by Cates). Engineering topics like fracture, pattern formation, flow and failure of materials, and the like, are also the topic of a growing subculture of physicists. Further growth in these directions will be essential to the health of physics.

A related broad field is the general area of nonlinear dynamics, chaos, and turbulence. The contrast between the time-scale of the revolutions in microphysics and the slow, stately progress of nonlinear dynamics and hydrodynamics could hardly be greater. The greatest revolution in this field, in a way, consisted in understanding Poincaré's insights of 100 years ago into chaos, and properly using Lyapunov's and Kolmogorov's ideas from mid-century. The work of such classical figures as Rayleigh is still not irrelevant. This is one field in which the use of computers has begun to bring qualitative improvement, although at first, as I said, it merely brought out old insights and destroyed false assumptions: this was the great contribution of the revolution in thinking of the '60s through the '80s sparked by Lorenz, Mandelbrot, Ruelle, et al.

With all our sophistication in this field, there is still a "problem" of turbulence, on which top-flight minds continue to labor. It may be that they chase a will-o-the-wisp, a generalization which is not really there; but I have a feeling it is. The puzzle seems to be the confusing mix of scale-free and scale-specific phenomena, of boundary layers, vortices and plumes with definite scales, nonetheless leading to apparently smooth power laws. The idea of self-organized criticality catches only a part of the elusive generalizations which I suspect will appear in the course of the next decades.

One of the great challenges for the physics of the future is to find generalizations which account properly for the multiplicity and variety of scales which we see in the universe around us, in many of its aspects. The universe itself seems to exhibit scale-free behavior at least in some aspects such as the distribution of luminous matter, an observation with which astrophysicists seem not yet to have come to terms. So far, the large-scale structure problem has been dealt with in what I call the "unary counting" approach: like the Red Queen, "one and one and one and ...": galaxy formation, star formation, solar system formation, supercluster formation, the anisotropy of the background, all as separate problems. The community is remarkably lucky in that all of these stages can be observed, but perhaps this is also bad news in that it tempts one not to put the whole picture together. On can hope that a synthesis will appear in the course of the next decades.

The idea of self-organized criticality seems to me to be, not the right and unique solution to these and other similar problems, but to have paradigmatic value, as the kind of generalization which will characterize the next stage of physics. Physics in the 20th century solved the problems of constructing hierarchical levels which obeyed clear-cut generalizations within themselves: atomic and molecular theory, nuclear physics, quantum chromodynamics, electroweak theory, quantum many-body theory, classical hydrodynamics, molecular biology, ... In the 21st century one revolution which can take place is the construction of generalizations which jump and jumble the hierarchies, or generalizations which allow scale-free or scale-transcending phenomena. The paradigms for the first is broken symmetry, for the second self-organized criticality. Another paradigm is the theory of distributions of Levy flights, applied to economics by Solomon et al.

D. *Physics Outside Physics*

By 2050 physics as we know it will have disappeared as a unitary discipline. There will still be a heavy employment in physics based technology, and many exciting technological developments, but the research supporting this will probably be in engineering and materials departments. The NSF and similar agencies having abandoned funding the fundamental exploration of matter's properties, these areas will have to re-invent condensed matter physics for themselves, as they have already begun to do. Astrophysics will remain viable much longer; whether a particle specialty will be taught is not clear. Nuclear physics will have a short glorious period in the early decades with the breaking of the nuclear interaction puzzle stimulated by new experimental results on the quark-gluon plasma transition, but will then decay. Physics proper will, like mathematics now, become almost entirely a teaching profession.

The real physicists of the later 21st century will only be called "physicists" if the profession has re-invented itself: if it has taken the opportunity to carry into new fields its uniquely fruitful methods, modes of operation and, most of all, its mindset. I have seen the impact which physics-trained people can have on fields as different as computer science, management studies, finance, and evolution, and it can be remarkable. What has been unfortunate for physics is that many have felt that the way to have an impact on the community of scientists in a field has been to join them, to try as far as possible to conceal the fact that one is a physicist, thinking as a physicist does. Thus again and again the physics contribution is subsumed into material which is not thought of as physics (a typical example is the way in which the chemistry community as well as medical science and neurophysiology have internalized the whole field of nuclear magnetic resonance). Biology and medicine are the obvious targets for physics to aim at. Physics is providing an increasing array of diagnostic probes, for instance, a very recent one being SQUID magneto-cardiography; it is also providing remarkable techniques for manipulating even molecular structures within the cell. There is an increasing level of matings of physics, technology and biology, such as the remarkable DNA wafers developed so highly for the genome project. Physics methods for nanofabrication are entering biology in a big way for manipulating cells and even intracellular structures. In short there is a strong trend in the direction of using physically, electronically and computationally sophisticated methodologies in biology. It is clear that in the first decades of the 21st century this effort will expand until the distinctions between physics, technology and biology — especially medicine — will become very blurry. It very much behooves physicists to try to have this area remain labeled biophysics, or at least be done in interdisciplinary centers where physics retains its separate identity.

One area in which physics can play a vital role is in the great problem of the origins of life as we know it. I use the plural intentionally because I am sure that two or more stages were necessary: the chemistry of life is likely to have originated before self-replication of information, and the construction of the organism is a possible third stage with its separate, individual identity. This is, like consciousness, one of the major deep problems, already apparent in the 20th century, which may take most of the 21st century to solve.

An entirely different way in which physics methodology and personnel can expand into the sciences of complexity is via the theoretical route. Physics is uniquely the subject which has developed the specialty of theorist to a high level. There are theoretical chemists and in the past such greats as Gibbs, Onsager, Mayer and van Vleck operated as theoretical physicists do now; but more recently "theoretical chemistry" has become more of a service skill. Theoretical biology (where it is not "mathematical biology", as barren a specialty as mathematical physics) is an unfairly denigrated reality which has been growing rapidly in effectiveness, and we physicists should support and if possible join their effort. But if E. O. Wilson's dream of "consilience" among all the sciences is a reality, and I believe it is, then "theorist" will be a specialty no longer confined by a modifier specifying "biology" or "astrophysics" or even "economics" or "ecology".

I am aware that premature stabs at this kind of generality have been disasters. One thinks of "general systems theory", "cybernetics", "homeokinetics", "synergetics", and similar oversimplifications. On the other hand, we should note the wartime successes of theorists as operations researchers, and the present popularity of management consulting, on a scientific basis, which is draining away so many young theoretical physicists. But perhaps they will form the nucleus of just the corps of generalists I envisage. It is healthy to note that as a result of the foolishly restrictive refereeing practices in professional economics, physicists are publishing economics in *Phys. Rev. A* and forming an "Econophysics" society. In the spirit of the Santa Fe Institute and of a surprisingly numerous coterie of similar institutions around the world (I mention SFI without claiming either priority or uniqueness of it in particular, I hasten to add, but as a visible example) I see the most expansive area of theoretical physics or at least of theoretical science, in the 21st century.

Y. Nambu and Broken Symmetry

This was prepared for a small symposium organized around the awarding of the Franklin Medal to Nambu. I'm afraid I used the occasion to also toot my own horn a little bit but Nambu didn't seem to mind. I felt a little justified in this when the APS awarded the Dannie Heinemann prize to no less than seven people who had contributed to the Higgs theory, but not to me. ("Omitting the eighth" — see elsewhere in this book!)

I have been asked today to talk to you about the early history of the concept of broken symmetry as it applies to the theory of elementary particles, in particular about Nambu's role in the development of this idea.

Landau introduced broken symmetry in condensed matter physics (simultaneously with Tisza, who in this as other things lost out to Landau's PR skills), making the simple observation that a symmetry change, being discrete, implies a thermodynamic phase transition. The higher-temperature and more entropic phase, he assumed, should always be averaging over a larger group and hence be more symmetric, though there are a number of cases where this tendency doesn't work.

Although Debye in 1912, and Landau and Bogoliubov in the '40s, quantized phonons and thus dealt with Goldstone bosons, I believe it was I who, in 1952, first realized that a massless boson (Goldstone) excitation necessarily accompanied a broken *continuous* symmetry, in connection with studying the ground state of an isotropic antiferromagnet. The reason is that the true symmetry must be restored by the zero-point motions, which must therefore have divergent amplitude, hence zero energy. In real antiferromagnets, which normally do not have perfect spin-rotation symmetry, the boson is only approximately massless.

It was also Landau (and here his contribution is clear) who introduced the concept of elementary excitations into condensed matter physics. He pointed out that one could consider the ground state of an ordered solid (or quantum fluid) as analogous to the vacuum of a field theory, and the lowest-energy quantum excitations then would be like quanta of a field and could, at low enough temperatures, behave like a rarified gas of elementary particles. Thus he was generalizing Debye and Einstein's phonon theory of the specific heat of a solid. What Nambu did was to be the first to stand on its head Einstein's progression from light as a classical wave in a medium to light as a gas of photons in vacuum to quantized phonon vibrations in a solid; he proposed that the vacuum was not empty space but a condensed phase which broke some symmetry of space-time, and that some of the observed spectrum of the elementary particles were merely excitations moving through this pseudovacuum — thus borrowing back from condensed matter physics Einstein's original loan of the non-empty vacuum.

Of course I am giving the history the benefit of perfect hindsight. None of us quite knew what we were doing at the time except in rare flashes of insight, even perhaps Einstein. In particular, the real value of broken symmetry concepts only became apparent when something new and deeply puzzling arrived on the scene. This was the new theory of superconductivity of Bardeen, Cooper and Schrieffer. As soon as this theory was reasonably complete, in mid-1957, its apostles began moving out across the world to tell us all the new gospel; in particular Bob Schrieffer, to Chicago where he was to be a junior faculty member, and David Pines, to his old alma mater Princeton, undoubtedly to show them what they had missed by not keeping him on. At both places a lot of the discussion revolved around the peculiar fact that although the theory was manifestly neither Galilean nor gauge-invariant, it gave experimental answers which were right on the button.

Many established theoretical physicists concentrated on trying to disprove BCS — hence the complaints that the world was slow to accept this theory, complaints which I think are exaggerated. But at both Chicago and Princeton, there were younger theorist auditors (Nambu and myself respectively) who had no baggage to dispose of and therefore concentrated on trying to fix the theory and restore its invariances. Oddly enough, both of us happened on the same spinor representation of the theory in which its nature was particularly perspicuous, and which has been useful ever since. Nambu's version was particularly beautiful.

I have referred to Nambu as "having no baggage to dispose of" but I don't mean to ignore the fact that he already had a towering reputation among elementary particle theorists, starting as one of the Tomonaga group who independently solved QED, and following with many important results in the early meson theories. But neither of us had worked previously on superconductivity; and to quote him, we both became "captured by the BCS theory".

Both of us realized that the solution to the problem of invariance was the same as it was in other broken symmetry situations: that one has to take into account the zero-point motions of the collective excitations. In Nambu's paper, appearing shortly after mine, the Galilean invariance was restored by divergent zero-point motion of a Goldstone boson, just as I had done for the antiferromagnet's rotational symmetry. This then has to be corrected by taking into account the Coulomb interaction, raising the energy of the Goldstone boson to the plasmon energy. It is interesting that he was, therefore, perfectly aware of the Anderson–Higgs mechanism that I had used, but never remarked on it in the elementary particle context!

I have emphasized his influential paper on the gauge invariance because it was how we met — he visited us at Bell Labs in '59 and his understanding of the deep meaning of what we were doing had an enormous effect on my own thinking abut broken symmetry. I think I even learned the words from him, if not the idea.

But soon he followed it up with something even more radical — the true origin of the idea of a dynamical broken symmetry of the vacuum, namely the series of papers with Jona-Lasinio, of which the first was given in April 1960. Quite explicitly, he pointed out that the idea had resulted from his study of the BCS theory. What they did was to propose that the large mass of the familiar nucleon was not fundamental at all but was caused by a pairing interaction between more primitive massless particles which were half nucleons and half antinucleons. The mass was analogous to the energy gap of the superconductor, and the pi meson played the role of low-mass Goldstone boson which restores the original (and approximate) chiral symmetry of the nucleon system. As a confirming result he derived the Goldberger–Treiman relation and a number of other experimental results from this remarkable hypothesis.

To me — and I suppose perhaps even more to his fellow particle theorists — this seemed like a fantastic stretch of the imagination. The vacuum, to us, was and always had been a vacuum, so to speak — it had, at least since Einstein got rid of the aether, been the epitome of empti-ness, the space within which the quanta flew about. Momentarily, Dirac had disturbed us with his view of antiparticles as "holes" in a sea, but by redefining the symmetries of the system to include a new symmetry called charge conjugation, we made that seem to go away. I, at least, had my mind encumbered with the idea that if there was a condensate there was something *there*, with properties like superconductivity that you could measure. I eventually crudely reconciled myself by realizing that if you were observing from inside the superconductor and couldn't see out you wouldn't have anything observable except the spectrum.

(Of course, in a sense the energy of the vacuum remains an unresolved problem — but it's not clear it is worsened by being a condensate.)

The other mind-blowing extrapolation for a naïve CM theorist was to dispense with any details of the interaction mechanism except symmetry and a cutoff — I didn't realize how incredibly fortunate we solid staters were that we never had to deal with ultraviolet divergences, so we hadn't learned to ignore them.

So this is why it took a Nambu to break the first symmetry — he had become familiar with how superconductivity worked, but he didn't have the restricted horizons of the rest of us solid staters — his mind was open to both the particle and the solid state worlds.

We were privileged to have him visit Bell Labs several times in 1958–1960, and I talked with him at some length. As a result I borrowed liberally from his insights about broken symmetry, and they informed my two books and the article "More is Different" which has had some influence.

But it was not through him that I learned about the dilemma about broken symmetry and mass which confronted the particle theorists in the early sixties, but through various rather indirect channels — some contacts in Cambridge during the year 1961–2 that I spent there, mostly through the students as well as Richard Eden; visits to the Bell summer program particularly by John Taylor; and some of my Bell colleagues, John Klauder and others. The particle world had become fixated on the Goldstone theorem, which required a massless boson for any true broken symmetry. But in fact I, and following me Nambu, had pointed out that specifically in the case of superconductors there was no mass-less boson, because the gauge-symmetric photon combined with it to make massive vector bosons, namely plasma oscillations (plasmons). In my original paper I had even pointed out that the symmetry-broken field would reappear as a scalar excitation, which in the simplest case would be at the double energy gap — so I derived a Higgs Boson. (In a super-conductor, such an excitation was found, 35 years later, by Littlewood and Varma.)

In 1962 I set out to make my gauge symmetry ideas into a relativistic field theory, and wrote a brief article about "Plasmons, Gauge Invariance and Mass", which caught the eye of Peter Higgs, who translated it into more acceptable "particlese" and thereby became famous. (To give him his due, he has absolutely never failed to give me credit for it.) Later on Brout, who had been a close associate in '58 and '59, also published on the idea; that paper caught the eyes of t'Hooft and Veltman. The recent Wolf prize celebrates their alertness.

My reasons for telling these stories here is that I give Nambu Sensei full credit for opening all of our minds to the consequences of allowing Nature to tell us what field theory can do, rather than confining Nature by our own preconceptions. The intellectual thread runs unbroken from his work to the final triumphs of the Standard Model.

Nevill Mott, John Slater, and the "Magnetic State"

Winning the Prize and Losing the PR Battle

I am going to talk tonight about how solid-state physicists learned to understand the way electrons hold solids together. I will say that what the majority of students learn about this is not wrong but it's subject to a very large, common, and well-documented exception, which I call the Mott magnetic state. I'm going to emphasize that what is believed about this by many solid-state physicists, and widely taught, is very much a consequence rather of the personalities of the actors and the dynamics of their interaction, than of straightforward scientific logic. The exception is important because the physics behind it underlies a large part of modern condensed matter physics. Here is a brief outline of the talk:

Mott, Slater and the Magnetic State

1. Bands (Molecular Orbitals) vs. Atoms
2. Reconciliation via Wannier
3. The Mott Insulator (Magnetic State)
4. Slater — the "Solid State Dream Machine"
5. Superexchange: the Culmination of Mott's Idea
6. An What Happened After: Dream Machine ↑ Mott ↓
7. Respectability at last?

From the very beginning of the application of quantum mechanics to collections of atoms (molecules, solids and the like) people approached the problem from two points of view: one assigning the electrons to live on the separate atoms, and letting the atoms interact; and one putting the electrons into wave functions running over all of the atoms, correcting as best as one can for the excessive charge fluctuations that causes when one or several extra electrons happen to accumulate on one atom. The first method to be worked out was the former, by Fritz London and

Walter Heitler, and it was soon generalized by John Slater and Linus Pauling; both, but especially the latter, making it into an enormously successful heuristic method by which it codified an enormous range of chemical facts using ideas like "resonance" and "hybrid orbitals." Fritz London spent a few years trying to find ways to make the methods into something more rigorous, suitable to his philosopher's tastes, but abandoned an almost completed book when it became clear that Pauling's sloppy but effective methods were irrevocably ensconced in the chemical curriculum. Incidentally, his program may have been more or less carried out by our group in Cambridge many decades later, in a volume of the Seitz-Turnbull series called *Band Theory from a Local Point of View* (Vol. 35).

Very soon after Heitler and London came Hund and Mulliken, with the method of "molecular orbitals", where the electrons are placed in wave functions extending over the whole molecule — simply the H_2 molecule, in the first instance; but Mulliken, in particular, and his school in Chicago generalized this, again, into a successful method of understanding most of the observed facts of quantum chemistry. (There was an important side-branch in Oxford as well.)

In the field of quantum chemistry, over the years, the difference between the two methods generated much heat, but looking back from a modern perspective this heat looks pointless. Slater himself showed how to reconcile them, in a beautifully written passage I remember well from "progress reports" of his group which he wrote in his discursive, very pedagogical style semiannually (at least) in the early '50s. He follows the progress of a pair of H atoms as he brings them together and shows how the atomic description passes smoothly into the m.o. one, the atomic description being more accurate at large distance, the m.o. one at small. But they flow smoothly into each other, which means that either can be used as a starting point and corrected — renormalized — to give the right answer.

My own professor, John Van Vleck, was a natural conciliator, and showed how the two led to similar conclusions and could be considered as aspects of the same intellectual structure, in two chemical problems; methane, and the "ligand fields" of transition metal ions.

I know my belittling of these deeply felt differences goes against the grain of historians of science, who take the attitude that it distorts history not to look at it from the contemporary point of view; but my take is that there are enough *real* controversies in science that we can afford to minimize those which we now realize to be without content.

Solids, too, at first were approached in the two different ways. The ionic crystals, for instance, like rock salt, were supposed to be made up from filled-shell Na^+ and Cl^- ions, attracted by their opposite charges and repelled by overlap (van der Waals') repulsion. Metals, from the start, required the molecular orbital theory, which for the case of solids is energy-band theory in which all of the electrons occupy Bloch wave states each of which extends throughout the crystal.

Eugene Wigner and his Princeton students (Seitz, Bardeen and Herring) had made the band theory into a quantitative approach to the binding of metals. The very early and influential work of Per-Olov Löwdin, starting with his thesis about 1951, showed how one could formally fit the "atomic" treatment of insulators like NaCl into the band theory by using the Wannier-function transformation from Bloch waves to localized wave functions — thus the solid state theorists were the first to free themselves from the atomic orbital vs. molecular orbital controversy for ordinary systems. The rule is, from the point of view of hindsight, that for metals one should (almost) always take the band (or m.o.) point of view, while for conventional insulators and semiconductors it is a matter of taste and convenience, with most modern calculators pre-ferring band theory, (but local methods are available, as the Cambridge group showed in 1980). Whichever method is used, it must be thought of as the starting point of a perturbation theory which will give us the essen-tially finite "renormalization" corrections due to more detailed treatment of the interactions left out in arriving at the one-electron solution for the set or orbitals.

In solids, however, as opposed to finite-size molecules, the idea that one can *always* arrive at a satisfactory band theory description is not true.

At this point it may be useful to inject a note of the personal into this narrative. I had the good fortune to spend the vital early years of my career at the Bell Labs, learning solid state physics at the feet of several of its originators: J. Bardeen, C. Kittel, G. H. Wannier, C. Herring, as well as the many visitors who wandered in and out such as H. A. Kramers, Peter Debye, and Herbert Fröhlich. Herring and Wannier — who was my first office-mate — were very thoughtfully deep, and in this group one was far more aware of the basic conceptual structure of our subject than one could have been anywhere else except, possibly, in Landau's group in Russia. But also Herring was one of the few Americans who made the considerable effort to keep up with the Russian literature, which was of great value to us.

It was from him and Wannier that I learned that in solids, as opposed to molecules, there could be true discontinuities; this is the basic reality behind the existence of polymorphism — different thermodynamic phases of the same substance. In the thermodynamic "$N \to \infty$" limit, state functions can have true mathematical singularities; discontinuities which can separate different phases. That is, a molecule never undergoes a true discontinuous phase transition: phase transitions are to be thought of mathematically as a property only of the limiting case of a large system.

Already in 1949 Nevill Mott had discussed the existence of such a possible "transition" to the atomic limit for solids, motivated by the energy cost necessary to create free charged ions from well separated neutral atoms.

$$U_M = E(M + M \to M^+ + M^-)$$

In this he followed earlier hints from Peierls and from Vonsovky. At large distances almost all the atoms will remain uncharged because of this energy cost. As they are brought closer together, a free electron or hole state spreads out into a wide band by virtue of the quantum mechanical hopping matrix element T, and eventually T gets big enough to overcome U (he estimated $T = 1/8 \ U$) and metallic energy bands will form (Figure 1). This argument was repeated and refined in a more complete paper in 1956; and during the mid 1950s he dropped by the Bell Labs and talked to our group several times (mostly while visiting people concerned with his other interests such as W. T. Read). We became familiar with the essence of his arguments.

The reason we were interested in them had to do with magnetism. At first under the mentorship of Charlie Kittel, and then on my own after he left, I had become a kind of "house theorist" for the various magnetism groups at Bell. We were very interested in the newly demonstrated phenomenon of antiferromagnetism and its practical relative "ferrimagnetism" as in magnetic ferrites and garnet structure oxides; as well as in electron paramagnetic resonance as applied for instance to the all-important doping impurities such as (P, As, Ga) in semiconductors. The reason magnetism and the Mott phenomenon go together is that almost any magnetic material has to have, by virtue of being magnetic, an open shell of atomic wave functions at each site, and these should normally mix quantum-mechanically in a solid and form a band: hence in very general terms, the paramagnetic insulator contradicts the basic idea of band (molecular orbital) theory: that any substance with a

THE MOTT MECHANISM

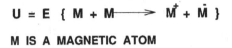

$$U = E \{ M + M \longrightarrow \overset{+}{M} + \overset{\cdot}{M} \}$$

M IS A MAGNETIC ATOM

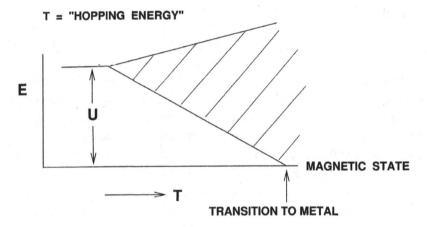

Figure 1

partially filled band is a metal. That the wave functions <u>do</u> mix and have a finite overlap was an experimental fact, which I was just learning: My first real success at Bell had come in 1950–51 when I applied an old paper of H. A. Kramers, which Charlie pointed out to me, about a phenomenon he called "superexchange" which allowed the coupling of the spins of relatively distant magnetic ions, through overlap of the wave functions with those of intervening atoms, to explain the antiferromagnetic arrangements seen in the early neutron scattering work of Shull's group (Figure 2).

This work among other things earned me the cachet to be invited to a couple of specialized meetings which were very important to my further thinking. One of these was a small meeting at Brookhaven on neutron diffraction and scattering, dominated by Slater, in 1952. Slater

"SUPEREXCHANGE"

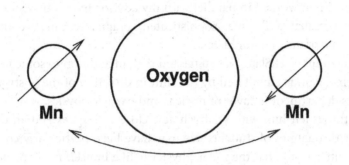

THESE SPINS COUPLE ANTIFERROMAGNETICALLY

HOW?   (KRAMERS)

Mn wave functions mix with O - - - -≫ other Mn

Figure 2

had a theory of antiferromagnetism in "my" oxides and fluorides, (presented at this meeting) in which, he opined, the insulating nature of these compounds was not caused by Mott's argument, but caused by the splitting of the bands by the doubled periodicity of the Hartree-Fock fields caused by antiferromagnetism. Hence at absolute zero, he claimed, there was no contradiction of band theory; bands calculated in this "unrestricted" way should lead to insulating behavior. His idea, then, was antiferromagnetism came first, insulating behavior second.

I remember riding back to New York on the train with Cliff Shull and discussing this situation with him. Of course, it was not all as clear as it is in hindsight, but Cliff's neutron scattering measurements had begun at room temperature with the disordered, paramagnetic spins in MnF_2 and MnO, which he could clearly see; yet these crystals are not metals at room temperature *either*, where there is no periodic exchange field to split the bands. So Slater's idea was clearly much too oversimplified; but that date was so early (probably ~ summer '52) that I had not yet comprehended Mott's idea.

It is perhaps worthwhile in a historical talk to say a bit about Slater's personality at that time. He seemed so dogmatic, arbitrary, and simple-minded that it was hard to reconcile him with the same Slater with whom I became familiar through his papers of the early days of quantum mechanics, and throughout the 1930s–a period, incidentally, when he often worked in parallel with my advisor John Van Vleck. These papers revealed a flexible, sophisticated, pragmatic mind, completely unlike that of the postwar Slater.

A biographer of Slater has suggested that, in addition to some personal problems, he was embittered by the shift in direction of the postwar MIT physics department in favor of nuclear and particle physics, which among other things left him with relatively few close colleagues and an unspectacular throughput of students. He may have been particularly conscious of the bitter irony that the radar physicists like himself really helped win the war, but the bomb physicists got the credit and the support.

In any case, when I came to know him he had become obsessed with the electronic computer and with the idea that all of the important theoretical problems of solid-state physics could be reduced to mechanical computation. This was an idea that I criticized, later on, in a piece which never found a publisher in this country. It was later published only in French in the magazine *La Recherche*. There I called it the "Great Solid-State Dream Machine". It was clear that if one were to fit antiferromagnetic insulators under the general umbrella of machine computation of electronic structure, Slater's band-splitting scheme (also proposed independently by Kondo in Japan) was the only way; but this leaves one completely at sea at finite temperatures. The Slater of the 1930s, I am sure, was not capable of the single-minded determination with which the Slater of the '50s ignored finite temperature. Even worse, he ignored the enormous variety of paramagnetic insulating crystals which, we now believe, are insulating solely because of the energy gap caused by Mott's electron repulsion parameter "U". These range from hemoglobin to hematite, from blue vitriol to ruby: they are very common indeed and of great importance, as these examples show.

I earned entrée to a second magnetism meeting slightly later, at the University of Maryland, whose proceedings were later published in *Rev. Mod. Phys.* In an evening "panel" session at which Zener, Van Vleck and Slater discussed ideas on ferromagnetism, Van Vleck brought up what he called the "U-T" model, presumably a characteristically jocular reference to the University Theatre in Harvard Square. His parameter "U" was closely related to Mott's intra-atomic repulsion coefficient, U, which will reappear later; I have always suspected that the similarity of notation conceals an intellectual thread of connection. In any case, Slater quite rudely closed off the discussion by leaving the podium just as Van Vleck's student Hurwitz, whose thesis had been an early exercise in Mott-Hubbard physics, rose to comment. Slater clearly felt that the only interesting future lay in his own methods. Perhaps the most puzzling aspect of Slater's metamorphosis is the fact that the primary component of Mott's interaction "U" can be expressed in terms of Slater's own 1930's parametrization of electron-electron interactions in atoms in terms of a set of Coulomb integrals $F_0, F_1, \ldots$ and $G_0, G_1, \ldots$ It is basically simply the largest of these, $F_0 \equiv G_0$. If one read Slater's paper on antiferromagnetism carefully, one realized that he has simply and inexplicably dropped F_0 out of the calculation.

Now to return to my own story, which at this point is about the great benefits of procrastination. In 1953 or 1954 Fred Seitz began urging me to provide an article for his newly-established series of review volumes in solid state physics, based on my ideas on exchange in antiferromagnets. I accepted and promised I would have something for volume I (or 2 or 3 at least) but for 5 or 6 years my subconscious must have been aware that no matter how much guilt I accumulated, I was not really ready to write that article. Meanwhile, I was absorbing Mott's idea especially in relation to another of my great concerns at that time, shallow impurity bands in semiconductors.

Finally, in the spring of 1958 — paradoxically, it was scientifically one of the busiest periods of my life — a completely new way of looking at the problem came to me. As I began to write it up during a summer visit to Charlie Kittel in Berkeley, through pure providence I happened to hear some lectures by Leslie Orgel on something I should have known all about already, my own professor's theory of ligand fields for transition metal ions, which gave me a scheme for quantifying my idea.

But that aspect is not too important vis à vis Mott vs. Slater. What was really important is the principle, which is that antiferromagnetic exchange is entirely linked with the Mott magnetic insulating state: it is the residue of the frustrated band-forming tendency of the electrons (Figure 3).

I remember very well an incident in connection with this theory. The APS was just, in 1959, beginning to have a press-room at its meetings in those days, and a few of the speakers were invited to write a press version of their talks, and to show up and be interviewed. Rather surprisingly, one reporter showed up for me (I think from the *Boston Globe*). What seems to have really fascinated him was not what I said but the extent to which I anthropomorphized the electrons, in his view: that I saw an electron as "frustrated," for example. He seems to have thought me quite strange.

SUPEREXCHANGE (1959; and Vol. 13, Seitz-Turnbull)

ANTIFERROMAGNETIC EXCHANGE =

THE FRUSTRATED BAND-FORMING TENDENCY

COROLLARY: VERY (NEARLY) UNIVERSAL CORRELATION

ANTIFERRO ⟷ INSULATORS

FERRO ⟷ METALS

(and the exceptions do prove the rule!)

Figure 3

It is not appropriate here to go into the formal theory, but it's worthwhile mentioning two relevant corollaries:

(1) one of the most reliable generalizations of materials physics is:

> antiferromagnets are insulators
> ferromagnets are metals

It is not perfect — but the exceptions are quite well understood within the same structure;

(2) a very general quantification of strengths of exchange parameters was possible — which, in fact, explains the above correlations.

The final irony, to which my subtitle refers, is what happened next. You would suppose that Mott having explained the wide occurrence of magnetic insulators, and my theory showing how antiferromagnetism — the subject of two Nobel prizes by now — follows from Mott's idea; that these essentially simple, almost untechnical ideas — which I tried to explain tonight at the textbook level — would have become part of every curriculum in solid state physics, and featured in every textbook. Quite the contrary: with the years, they become increasingly neglected and increasingly controversial. This is so in spite of the fact that Mott's transition, and my superexchange, were a substantial part of the justification for the Nobel prize we both received in 1977.

What happened was that Slater's great Solid State Dream Machine, his vision of the computerization of physics, has indeed turned out to be the wave of the future. Not in his hands or under his leadership: although some of the most brilliant insights on which modern band theory is based came from him, the organizational and personal problems at MIT meant that not his students but centers such as Bell Labs, Chicago, the Cavendish, Berkeley, and San Diego led the succession of incremental improvements which allow us today to plug in elements, turn the crank, and get out an "electronic structure". This structure can be fantastically accurate and useful in many cases; in others, not at all. But its existence means that any requirement for rethinking of the basis on which it stands is greeted with deep suspicion and is seen as enormously complicated and difficult. "Oh, that's a many-body effect: horrors," and the one-electron-trained physicist recoils.

In blaming the textbooks' neglect of the Mott magnetic state on Slater's followers exclusively, I am not really being fair. A great deal of the fault lay with Mott, and I am not blameless myself. Mott was not at all content to simply declare victory by claiming the magnetic insulating state as his and superexchange as the proof. I did this, myself, in my much-delayed vol. 13 article in the Seitz-Turnbull series, and in another very obscurely published review about the magnetic state, in 1962, and left it at that. But Mott barely acknowledged — if ever — either his point of view, or my work, in his voluminous writings on metal-insulator transitions — for what he did was to occupy his enormous energy in the search for the *transition* that separates the metallic from the insulating case. This is a more difficult proposition, since the large scales of the energies involved means that the transition is usually not accessible; or when it is, it involves more complex situations, such as disorder, lattice distortion and polarons, and complex spin interactions.

This was just fine with Mott: he had no fear of messy, complicated situations and he loved controversy. I have, elsewhere, called him the "Happy Warrior". He had no interest in laying claim to the magnetic insulating state — declaring victory — and leaving it at that. That would have seemed very boring to him. He had written his pedagogical textbooks, in the 30s. These metal-insulator transitions afforded him what he loved most, the opportunity to be deep in the midst of a complex series of controversies, to be involved with a polyglot group of young people arguing at the tops of their voices around him. He seemed perfectly content to allow the Mott phenomenon to be partially marginalized and made controversial, because the community as a whole found it difficult to distinguish the old core from the swirl of controversy surrounding it.

Perhaps one of the most bizarre examples of these attitudes occurred as late as 1984, when after many dormant years my old friends the antiferromagnetic oxides MnO, FeO, CoO, NiO, CuO came under the scrutiny of quite a respectable group of band calculators based at IBM. They calculated with the latest frills of "unrestricted LDA theory", and claimed (falsely) to find transition temperatures in better agreement with experiment than mine. Unfortunately, they failed to mention that in their theory FeO, CoO and NiO all came out to be metals, belying the observed gap of several eV. Their response to a referee's query on this point was characteristic — they showed a firmer belief in their calculations than in the long-established historical, experimental facts. Fortunately, the paper was buried without trace by a masterly analysis of NiO by George Sawatsky in the same year.

An immensely amusing picture of a band calculator, Art Freeman (an old Slater student) is given in Hazen's book *The Breakthrough* on high T_c superconductivity. He is shown intriguing to be the first to be given the crystal structure of the new compounds, and immeasurably excited because he had "solved" it — put it in his machine and found a series of meaningless scribbles which may (they do not) tell him something about its superconductivity. Oddly enough, these compounds have brought Mott and me back to a certain raffish respectability among at least an influential fraction of our colleagues: anyone *au courant* with the field now speaks of "superexchange" and in hundreds of papers we see John Van Vleck memorialized by his subtle cinematic reference to "U–T", which are the notations used in the "Hubbard model" now universally accepted as describing at least some aspects of these materials. It is not clear that anything in this field is yet considered *truly* respectable; but that's another story.

III. Philosophy and Sociology

Introduction

This and Chapter IV represent a rather arbitrary dividing up of the miscellaneous pearls of wisdom accumulated during 6+ decades of observing scientists and practicing science. This chapter tends to deal more with broad intellectual issues, the next with practical ones of tactical approaches to the great goal of discovering the truth. But it's apparent that "pathologists", for instance, has a lot to say about both tactics and strategy; I put it in here to emphasize that under the tomfoolery there remain serious issues of what truth is.

This group of pieces more than any other represents my general outlook on the world, the universe, and all that. Of course I am not religious — I don't in fact see how any scientist who thinks at all deeply can be so, in any usual sense of the word, although apparently there are those who at least are under the illusion that they believe in some kind of God. I also have a prejudice against too extreme extrapolations, although mostly in my reviews I just gently twitted the Hawkings and the Penroses and the Smolins and the Lindes rather than attacking them head on. The word "universe" means roughly "all there is" so the dreamers about "multiverses" must not really mean what they say. But I have no logically deep objection to Everett's "many worlds" which is merely one way of describing quantum theory.

Emergence vs. Reductionism

In modern science "what you see" is very seldom "what there is". The object or fact is explained in terms of something very far from direct human perception. This fact intrigues many laymen, but may — quite properly — turn off the humanistically inclined. It may well be one of the causes of a general unease about science. Everything is analyzed in terms of indirect causes. This may have started with Copernicus' discarding the independently moving heavenly bodies (or perhaps earlier when their celestial charioteers ceased to play a role), and went on to Galileo and Newton's destruction of the intuitive Aristotelian notions of force and motion. But in the 20th century this process has accelerated: the immutable atom has become composite — and mostly empty space — and then the nucleus itself is a composite of composites, of which the sub-particles are objects which in principle can never appear in isolation.

We have an origin myth which we believe to be no myth but sober reality, yet it is more spectacular and bizarre than anything dreamed up by some primitive tribe. This origin "myth" contains (at least) two examples of emergence and of broken symmetry: the "electroweak" transition which is responsible, we think, for "inflation", and the emergence of protons and neutrons from the primeval plasma of quarks and gluons. Our biological natures, as well, are controlled by a complex and increasingly abstruse inner chemistry.

Physicists — and scientists in general — love to do two things; (a) to take apart, to analyze into simpler and simpler components; (b) to mystify: to say it's not really *this*, it's *that*; what you see is not what there is. The scientist has taken upon himself the role of the Shaman or the mullah. Everything comes from a First Cause — the First Equation —

and only he can investigate this with his very expensive equipment, and understand it with his abstruse theories.

The arrogance this attitude fosters has to be experienced to be believed. Such books as Stephen Hawking's *A Brief History of Time* covers the whole of science in six brief chapters, spending the rest of the book on personal speculations about the first millionth of a pico-second of time when, he seems to feel, all that matters happened. Much better, but still dismissive of other sciences, is Weinberg's *Dreams of a Final Theory* where he repeats the claim of 30 years ago that the only really *fundamental* science is the search for the ultimate constituents of matter — he classifies all of science by following little white arrows which inevitably lead back to the "fundamental" problem: what are the particles made of.

There is repeated reference to a hypothetical "perfect computer" which could in principle follow out all the consequences of the basic laws, making them the only input that matters. One of the elementary particle physicists even wrote a book called *The God Particle*.

It was in the first instance as a cry of anguish at this attitude of my colleagues that I wrote "More Is Different". I was perhaps, at that time, no less a reductionist, nor less willing to mystify — I merely wanted a piece of the action.

What I did was to look from the opposite point of view. The reductionist view starts from the top and searches for simpler and simpler first causes. I, on the other hand, looked at the same material facts — and accepted, in the limited sense, reductionism — but asked whether the ideal Cartesian computer that they were always assuming was really possible. What one sees from the opposite point of view is more and more complex consequences arising from simpler and simpler causes. This is also the historical order, if in fact we believe the Big Bang Theory. So what you get is the picture of, to borrow a term from 19th century biology, *emergence* at every scale.

The message is in the first instance: the world in which we actually live is the consequence primarily — or only — not of some incredibly simple but hidden "God equation" or "God particle" but of the "God principle" of emergence at every level. I first became aware of this in a very self-centered way, as the principle by which my own field of science arose from the underlying laws about particles and interactions; and it was only as I broadened my perspective that I realized how general emergence is.

The case of my own field, "condensed matter" or "solid state physics", is a crucial and instructive one. In 1932 P. A. M Dirac concluded his paper giving the theory of the electron with the words: "We have now laid down the principles underlying much of physics and all of chemistry" — which has been paraphrased as "the rest is (mere) chemistry". What he meant was that the quantum theory of the electron, the immutability of the atomic nucleus and some simple facts about electricity and magnetism could be shown to provide the substrate of laws on which all of condensed matter physics, and hence chemistry, is based.

In fact nothing we encounter in everyday life requires anything more. We did have in our hands, in 1932, the really fundamental laws for most of the sciences except astronomy and bits of geology, and within a few years progress in nuclear physics would clarify the immediately relevant parts of these fields.

Why then are we still working on condensed matter physics? The example of superconductivity is instructive. The phenomenon of disappearance of electrical resistance in some metals was discovered in 1911, and intensively studied (unsuccessfully) for 46 years thereafter. The basic underlying mechanics of electrons in metals was discovered from 1926–32 by Bloch, Peierls, and others of the early heroes of the quantum revolution. The 25 missing years (1932–57) are a crucial proof of the failure of reductionism, if one is needed.

If the strong principle of reductionism were true, the Cartesian idea that one needs merely to know the laws of motion and one should be able to compute all consequences, we should have understood superconductivity. Yet in 1956 I heard the great physicist Richard Feynman declare publicly that he had no idea what caused superconductivity; and only in 1957–63 did the problem reach a real solution. Let me add the point that it was not lack of computational power in any real sense that was the obstacle. If we had actually had the (actually impossible) ability to follow the motions of all of the electrons and ions in detail, all that the computer output could have told us would have been that the material was exhibiting all of the well-known and well-studied phenomenology of superconductivity: it could not have told us *why*, because it would not know what that question means. What it does mean is that there are certain concepts and constructs which allow enormous compression of the brute-force calculational algorithm, down to a set of ideas which the human mind can grasp as a whole.

What really was the problem is called "broken symmetry": the underlying laws have a certain symmetry and simplicity which is not manifest in the consequences of these laws. To put it crudely, a simple atom of gold cannot be shiny and yellow and conduct electricity: metallicity is a property with meaning only for a macroscopic sample. Nor can a single atom of lead be superconducting: this is also something which can only happen to a macroscopic sample. $N \to \infty$ is the jargon for it: the "thermodynamic limit". In this limit, a large collection $N \to \infty$ of objects can behave completely differently — have different symmetry from anything that the separate objects can themselves exhibit. A molecule of salt is not a cube; only a full-size salt crystal can have cubic symmetry rather than total rotation symmetry.

Another lesson from condensed matter physics: already in 1960 a number of us speculated that a phenomenon like superconductivity might occur in a liquid made up of the mass 3 isotope of helium, and indeed in 1972 this was discovered. The lesson is: the emergent property — the broken symmetry — is something distinct from the substrate on which it occurs.

In solid state physics the connection both to $N \to \infty$ and to symmetry is manifest; but perhaps more relevant intellectually is the idea of intellectual autonomy of the two levels of understanding: that one can wholly understand the laws of motion of the substrate in which a given phenomenon appears, yet have no handle whatever on how that phenomenon arises, nor even what it consists in. In condensed matter physics this tends to be the rule: the theorist very seldom is able to predict the behavior of a given material from scratch. Again and again, one finds the experimentalists calling something to our attention which is bizarre, unexpected, and new under the sun. Nature is much better at lateral thinking than the human mind, and in turn the human mind is much better at it than any mechanical calculator yet conceived.

At the time I wrote "More Is Different" it was only tentatively that I put forward the possibility that what I said about broken symmetry breaking the intellectual limits from the elementary particle substrate to the problems of condensed matter was general: that as one went up a hierarchy of intellectual endeavors at each stage there would be new concepts, in principle not computable and often not even imaginable from the simpler substrate. For instance, one cannot deduce from the simple facts of chemistry that self-replicating molecules are possible. Once one has seen them in operation, it's trivial — but would one have figured it out from scratch?

Much deeper is a question I for one have no idea how to answer — how does what Monod calls "teleonomy" arise from chemistry: how does an autonomous agent, an object which acts on its own behalf, arise out of simple chemistry? We see them, we *are* them — but what are we? What of our many properties — or those of the simplest cells we know of — is the basic essential for agenthood? And how could this concept have arisen from nothing?

By now we're into biology and we begin to have some record of the stages of emergence: sex, symbiosis (in some order), morphology, the nervous system ... and on to the greatest puzzle of all, the emergence of consciousness. Much of this we have come to "understand" in the reductionist sense — much we haven't.

. From this point of view the social sciences are not distinct in principle, only in their distance from the substrate of fundamental certainty — the number of "little white arrows" which are missing. Yet the existence of parallelism across cultures, economies and societies tempts us into the conjecture that there is some kind of understanding beyond simple phenomenology which can be achieved by analyzing the way the hierarchies of social organization arise from a simpler substrate. A new thing appears here: historicity. Already at the stage of morphology, and at many subsequent stages, we are confronted with Steve Gould's question — would it have looked at all the same if we ran the tape twice? Answer: obviously not, so in some sense nothing is predictable from first principles. When history enters, Descartes must leave; yet we still have, quite properly I feel, motivation for, at each stage, understanding the workings of the "little white arrows" of Steve Weinberg's, which point to the "explanation" of each concept in terms of something simpler.

Most powerfully, it is primarily by means of such a reduction that we can rescue ourselves from the possibility, or at least from the accusation, that science may be purely a socially-constructed consensus with no objective external reality to which it corresponds. One may find many examples in history of just such constructed "sciences" which turned out badly-psychoanalytic psychology, the phlogiston theory, Ptolemaic astronomy, are some (and not the worst) of the examples which come to mind. But when one succeeds in finding a reductionist explanation for a given phenomenon, one embeds it into the entire web of internally consistent scientific knowledge; and it becomes much harder to modify any feature of it without "tearing the web" — by which I mean falsifying a proliferating network of verified truths.

It is no longer possible to falsify evolution for instance, without abandoning most of physics and astronomy as well, not to mention the damage that would be done to the many cross-links among paleontology, plate tectonics, physics, and planetary astronomy. Our reasons for believing in the objective reality of science are no different in kind, or weaker in principle, than our reasons for believing in the objective reality of sense data, in that both rely on the uniqueness of a self-consistent schematic representation of many correlated inputs. As long as a science relies only on its own internally determined and verified body of facts and ideas, it remains a "model" in the true sense — the Mendelian model, the phlogiston model, the behavioral model — and it is subject to the doubts one may have about validation of such a model. I sometimes call this "pharmacological science": verification is seen only as a matter of brute-force statistics on ever larger populations. In my opinion, it is reduction which rescues us from such quagmires.

There are, of course, at least two further, more practical reasons for reduction, and I will close by mentioning these:

(1) In many cases, understanding means power — if we know that sickle-cell anemia is genetic, we have some hope of alleviating it; if we understand superconductivity, we will understand the practical limits on superconducting magnets for MRI.

(2) Unpredictability can be only in a specific, detailed sense: often statistically we can understand distributions and correlations: we may not be able to predict an earthquake, but we know their distribution of sizes and can build for 1 or 100 or 1000 years with some confidence.

In summary, I argue against, not the reductionist program itself but the rationale and programmatic which is often associated with it, which gives the "Theory of Everything" the status of a "God Equation" from which all knowledge follows. Rather, I see the structure of the world as a hierarchy with the levels separated by stages of emergence, which means that each is intellectually independent from its substrate. Reduction has real value in terms of unifying the sciences intellectually and strengthening their underpinnings, but not as a program for comprehending the world completely.

Is the Theory of Everything the Theory of Anything?

This talk was given first as a Green College lecture at the University of British Columbia. Green College is a delightful, literally green enclave of timbered buildings vaguely recalling the local native styles, in the center of one of the world's more beautiful campuses, and dedicated to vaguely interdisciplinary and international pursuits. Green is Cecil and Ida Green, the Canadian founders, illogically, of Texas Instruments, and great benefactors to science. But the dedication to rationality did not include rational numbering of the accommodations — I stumbled up and down a number of staircases until someone finally took pity on me and led me to my somewhat Spartan accommodations.

This article is perhaps my final say on the big questions I opened up with More Is Different. It's very much a portmanteau job — I threw in a plug for "Consilience", for instance, and some thoughts on the origin of life, that are not obviously connected to the logic. I gave the talk again at a workshop on emergence organized by Philip Stamp's PITP, also in Vancouver, but boggled at continuing this "emergent" career and went back to my beloved cuprates thereafter.

It was in 1820 that the great French scientist and mathematician Pierre Laplace staked out the ultimate claim of the physicists for intellectual hegemony over all science — and possibly a lot else —:

"We ought to regard the present state of the universe as the effect of the antecedent state and as the cause of the state that is to follow. An intelligence knowing all the forces acting in nature at a given instant, as well as the momentary positions of all things in the universe, would be able to comprehend the motions of the largest bodies as well as the lightest atoms in the world, provided that its intellect were sufficiently powerful to subject all data to analysis; to it nothing would be uncertain, the future as well as the past would be present to its eyes...."

— Laplace, 1820

Ever since, this philosophical position has been named Laplacean Determinism.

Something like this kind of hubris still remains with us in the minds of some of our most eminent physicists, and has captured the imagination of the young and the fascination of the educated lay public; and it's with something like this scenario in mind that otherwise extremely hardheaded governments continue to shell out billions for scientific instruments for probing the depths of the atom and of the universe.

Far be it from me to belittle the achievements of my colleagues in the sciences of cosmology and the elementary particles. The latter group, for instance, have mostly answered a question that Laplace didn't even mention: namely, "of what is the universe made?" We can now say that without a doubt all the perceptible parts are made up from quanta of only three families of four matter quantum fields, (most, even of these, being very rare), interacting via three other types of field; and the thing that would have flabbergasted Laplace is that these constituents are absolutely without identity or differentiation. His "intelligence"'s job was a lot harder than he knew, figuring out how to tell the "bodies" apart, whose motions his "superintelligence" would comprehend. In their turn, the cosmologists can tell us a lot about what was happening within a microsecond of when it all began — and even show us a photograph of the universe at about 100,000 years, the ultimate baby picture (relatively speaking, about the equivalent of an hour of our lives — the embryo is still undifferentiated at this relative age). I realize that we are far closer to having that "one single formula" than we had any right to expect even 50 years ago when my career started.

Yet in those 50 years, in another sense, I am going to argue that the remarkable achievements of these branches of physics have made scarcely any difference at all to any practical achievement of Laplace's dream — rather, they have made it less achievable than it seemed in Laplace's day. The more we know about how the world really works, the more we find that it depends on a set of principles which detach the actual operation from the underlying "hardware" of fundamental physics. I once invented a slogan which puts the argument in a nutshell:

MORE IS DIFFERENT!

The meaning of this catchphrase is that simple laws, rules, and mechanisms can, when applied to very large assemblages, lead to qualitatively new consequences. Thus the remarkable simplicity and universality which our colleagues find in the underlying quantum physics of the universe — and the even greater simplicity for which they are presently striving — is increasingly irrelevant to the genuine complexity of the world in which we live. The multiplicity of ways in which this truism can work itself out is the subject of this lecture.

Let me attempt to list, in outline form, in shorthand phrases and in a somewhat arbitrary order, some of the ways in which this scenario works itself out, I start with the simplest and best understood.

I. Effective Low-energy Theories and the REAL TOE: Standard Model → Nuclear Physics → non-rel QM, quantum chemistry. Role of the RNG

II. Broken Symmetry: The real disconnect. Quantum → Classical Worlds. Emergence; Space, true novelty

III. The Arrow of Time: Cosmology rears its head: the Big Bang, dissipation + classical world

IV. Transition to Life: a) the role of Work (SK); b) semirigid molecules → information

V. Mysteries: Intentionality → free will → consciousness. Role of random probing?

From there, take your pick of emergent phenomena: religion, language, money, —

VI. Finally, a pitch for consilience — Ed Wilson's phrase for the great seamless web of science

I. Effective low-energy models. I think the arguments here are well understood, even by particle physicists — in fact they invented some of them. Physics actually started the other way round: The history of physics is the history of starting from the heuristic description of what you see before you, and deducing its microscopic nature. This is what Maxwell and Gibbs and Boltzmann did with thermodynamics and hydrodynamics: they showed that these are the effective theory that results from the assumption that these systems are merely large collections of molecules, very small and fluctuating at very high frequencies.

Then, when modern physics began at the beginning of the last century, the first task was to produce a theory which accounted for all the known facts of atomic physics and chemistry; and all the great pioneers such as Dirac and Einstein and Heisenberg wrote the classic papers on what their successors now call "squalid state physics". By 1932 Dirac bragged that "we have the underlying theory for all of chemistry and most of physics", and he was right: non-relativistic quantum theory, assuming that nuclei and electrons are conserved, really does exactly that; but as he was probably aware at the time, it was only a "low-energy effective theory", which didn't worry about the internal structures of any of these things. Nonetheless, it is a very good theory, and accounts well — doesn't predict, I hastily add — for most of what concerns us in our daily life — and with the addition of a bit of low-energy nuclear physics, is the physics of literally everything between here and the nearest Black Hole. It is, in fact, the *real* TOE, and no serious modification of it will ever happen.

Nowadays we have a very sophisticated way of making the connection between micro- and macroscopic, called the "renormalization group" — which is a clever way of averaging out all of those rapid, small-scale fluctuations, and rescaling, to obtain a new "effective" theory for whatever entities are left; and it is in this sense that the particle physicists can say that our theories are "derived" from theirs — yes, the Standard Model subsumes, in this sense. NRQM — but as I said, that's just where we started!

II. The idea of an effective theory, which is in some sense just a baby version of the microscopic theory, is probably as far as most theorists have thought about these things — but it is by no means the end of the story. There is something else much more important and exciting going on when we change from the atomic to the macroscopic scale: the quantitative change becomes qualitative in a very real sense. It should be a little mysterious to you that some of the simplest properties of everyday materials have absolutely no connection to the properties of the atoms of which they are made, or of the very general laws which rule the atom. In no sense is an atom of copper red, shiny, ductile, and cool to the touch — which is to say it conducts heat and electricity amazingly well. When we put a lot of atoms of copper together they become a *metal*, something conceptually new, that wasn't contained in the separate atoms themselves.

This means that the effective theory has to contain some terms of reference which are completely new — for instance, in the case of a metal it has to have what is called a Fermi surface, something that it would have if the electrons were free from their atoms and could move as independent entities. What happens is called, generically, *broken symmetry*; the idea is that the *state* of a collection of particles can choose not to have the same symmetry as the underlying laws which govern those particles. For another example, the fundamental equations recognize neither any special origin in space, nor any orientation; yet the crystal of copper is in a definite place and has a definite orientation. And when our predecessors, who were not stupid, had to describe it they did so by first setting down the nuclei into definite positions in a regular crystal lattice, and then taking the electrons into account. So the description of the system contains several totally new parameters, which weren't there before the crystal formed — its position and orientations. In a sense which is not as bombastic as it seems, if there weren't any solids we'd have no way to measure space — so by breaking the symmetry of space we have in a sense created space as a new entity.

All of this might have seemed to the founders of the theory of solids to be pretentious nonsense. Everyone *knew* that there were crystals and that space was there and you could measure it with measuring rods. (But what if there were no rods, as during the Big Bang?) It wasn't until we discovered a new type of broken symmetry, with a totally new physical parameter, that the fact that More is really Different dawned on us. This was the phenomenon of superconductivity, the existence of metals with zero electrical resistance, discovered in 1911, pondered over by the greatest minds in physics for 25 years after Dirac's TOE, and only solved beginning in 1957. Here the key to the solution was the possibility of breaking gauge symmetry, the symmetry related to the conservation of charge; and the new physical parameter was the phase of the wave function. This isn't as portably measurable as space, but it is measurable, thanks to an inspiration of Josephson in 1962. (The impossibility of such states had been the subject of a theorem "proved" by the famous mathematical physicist Eugene Wigner — and he and a number of followers remained doubters to their dying days.)

Starting from superconductivity, the idea of broken symmetry even infected the particle guys, and today underlies their extravagant search for the Higgs particle. But that's a detour. The important lessons to be drawn are two: 1) totally new physics can *emerge* when systems get large enough to break the symmetries of the underlying laws; 2) by construction, if you like, those *emergent properties* can be completely unexpected and intellectually independent of the underlying laws, and have no referent in them.

One interesting emergent property is so simple and fundamental that we don't even notice it: identity. Every quantum particle like a copper atom is absolutely identical to every other one: you can't paint it green, or call one Jack and another Frank. But solid bodies like crystals do have identity, as does any large enough piece of matter, because it's almost infinitely unlikely ever to be in exactly the same quantum state and able to exchange identities.

III. Now I'm beginning to enter into matters which are the least bit controversial, and what is more I can hardly claim to understand them completely. But here goes!

In order to have the real world as we know it, somehow the idea of time and its sense must enter. Quantum mechanics doesn't know or care which way time goes — nor do even classical Maxwellian molecules. Here one really has to invoke the cosmos, because of a paradox which was actually noticed by a man named Olbers in the 19th century: if we really lived in a static universe, everything would be at the same temperature and nothing could ever change! What has been learned in the past half-century is that *this* symmetry is, fortunately, broken by the cosmos: everything is expanding and cooling, and we can shrug off our excess entropy into the (expansion-induced) dark of outer space.

But there goes the absolute predictability of everything, that Laplace was relying on — those photons disappearing into outer space, never to return, are irrevocably entangled with everything here on earth, so for this if for no other reason quantum uncertainty releases us from Laplace.

The fundamental lessons of these two emergences, of space and time, are these. First, that already at a very primitive level, the very existence of a classical world in which there are identifiable macroscopic objects moving about is already, from the standpoint of the fundamental laws, unbelievably complicated. It's no wonder you don't understand how an electron acts; he certainly couldn't understand you! Secondly, these two stages of emergence prepare us for the realization that even the terms of reference for our description of the world can be very different at different scales — more is really different.

IV. It is presumptuous of us to ignore all the emergences that are implicit in the complexity of the cosmos — but we do. What really interests us — and where the term "emergence" arose — is the problem of how life emerged from non-life. Here there really is no agreed answer — yet there are a few things one can say. One is that what *won't* do is vague references to "dissipative structures" such as arise in certain hydrodynamic instabilities. One sees in every living object, no matter how primitive, the necessity for permanent, stable structure. These permanent structures play two roles: the more important, but less noticed, is energetic: life — including replication — involves the use of stored energy and the doing of work by it; "waiting for Carnot" in Stu Kauffman's phrase. This can be done by a structure such as the membrane of a mitochondrion, but it does require structure. The second is the stable storage of genetic information. Given those — and that is a large order — the exponential growth mechanism of Darwinian evolution can do almost everything from then on. But how it actually started to happen is beyond knowing at this moment in any detail.

V. The really hard questions come up at the next stage: life learned, in order, three things: first, to act on its own behalf; second, and almost implicit in that action, to be free to do so, i.e., to have free will; and third, to be conscious, whatever that is. I say "almost" implicit — after all, a plant acts on its own behalf by sending out branches and roots and leaving the useless ones to die, and no one accuses a plant of having free will. I suspect that's a hint — somehow the whole process of searching for light and nutrients gradually became virtual rather than real, and the result was the evolution of brain function. But in the end one has conscious objects, relating and competing, and the rules of the game have become only marginally influenced by the underlying chemistry and genetics. From these emerge such ever more complex further worlds as language, religion, money, art, politics,... Clearly, while each of these fields could probably benefit by looking more at the underlying neurological and genetic basis, each lives in its own independent conceptual world. Also clearly, at this point the efforts of scientists in understanding the emergence of free will and consciousness have so far been totally in vain.

VI. I have been trying to bring out the hierarchical structure of the scientific enterprise, but at the same time to stay free of the implication that in any sense the underlying sciences play any determining role in the behavior of more complex ones — the process of emergence seems always to lead to a totally new conceptual structure. In this I am trying to get rid of the strong distaste which many commentators feel about the enterprise of unification of the sciences — a distaste which is exemplified in criticism of the recent book of Ed Wilson's, *Consilience*. I think many have in the back of their minds Laplace's intemperate and untenable boast with which I began this lecture — that by accepting the close relationships among all the sciences they will somehow become the slaves of determinism and of the hegemony of our expensive friends who spend their lives digging deeper and deeper into the nucleus and the cosmos. My message is, "have no fear: More is Different."

Is Measurement Itself an Emergent Property?*

How can you predict the result when you can't predict what you will be measuring?

— Philip Warren Anderson

I do not argue with either of these two basic propositions: validity of all of the laws of physics in each microscopic process, and determinism of the quantum theory: that quantum physics is deterministic, in that the state of the whole system at any given instant determines the past and the future, and that all interactions can be understood as local in space-time. But complete knowledge of the state is necessarily impossible in most measurement situations.

This is the first and perhaps the least important of four fundamental sources of unpredictability in science, or, more accurately, of the inaccessibility of the Cartesian ideal of deterministic computation by a super-duper computer from the fundamental laws: the idea that "the universe runs like clockwork". I see the difficulty in each case as, in a sense, "physical" uncomputability, an exponential explosion in the computer power necessary, rather than uncomputability in the mathematical sense.

These four fundamental sources, then are:

1. The measurement process: coupling of quantum system to dissipative classical variable.
2. Emergence of measurable quantities: space, time, phase, field strength, etc.

---

*Originally published in *Complexity*, Vol. 3, No. 1, 1977, pp. 14–16. © 1997 John Wiley & Sons.

3. Sensitivity to boundary conditions of classical variables.

4. Emergence of categories and concepts, What is relevant? What to measure?

(1) Jim Hartle [1] has given us an adequate description of the decoherence process which leads to quantum "uncertainty". For example, the apparatus in the canonical Stern-Gerlach experiment entangles one quantum variable, *orientation*, with another, *position*, which can then be encouraged to trigger a "dissipative process", i.e., to couple to a large entropic bundle of histories that rapidly decohere from the alternative bundle from the other half of the wave function. (This is what happens in the detector.) I have nothing to add to his description, except to say that even this basic description of measurement sees it as an emergent phenomenon, in that it is irrevocably tied to the possibility of irreversibility; which emerges only in a sufficiently large system. Your result only exists *after* the measurement.

(2) There is another aspect of the measurement process that is emergent in a quite different sense. The quantities that we measure on the quantum system of interest are variables which are defined not by that system but by the macroscopic apparatus with which we do the measurement: orientation, position, velocity, field strength, etc. We cannot imagine a way to set up a coordinate system for orientation or position, for instance, without a rigid body to refer to. It was not by accident that Einstein's writings on relativity were full of clocks and meter sticks.

What in fact is the role of the rigid body in the quantum measurement process? It exhibits no dynamics, acting only as a boundary condition (slits) or static Hamiltonian (magnetic field) or at most a low-frequency driving field. It is not expected to undergo a quantum transition on scattering against the system, in fact if it does so the measurement is spoiled — as in the Debye-Waller factor of x-ray scattering or the Mössbauer effect, which cause reductions in the number of measurements proportional to the number of recoil quanta. It is this kind of "zero-phonon process" in which an arbitrary unquantized amount of momentum or other quantum variable may be absorbed by the macroscopic order parameter, which is responsible for some of the puzzling features of the measurement process. As far as I can see, it is a crucial part of all measurements. The question becomes, then, are space and time, for instance, inevitably to measure, or are they a consequence of the dynamics of large systems, i.e., emergent concepts which are meaningless before condensation occurred? Did they exist in the Big Bang?

That this is not an idle question is made clear by the example of phase of a superconductor or superfluid. This is a quantity which can be deliberately switched on or off as a macroscopic thermodynamical variable by manipulating temperature. A given piece of metal can be a reference system for phase measurement in the same sense that a solid body is one for position measurement, or not. And, if there is no superconductor in the vicinity, there is no meaning to the phase variable.

A magnetic field may be measured by means of a ferromagnetic needle which has the relevant broken symmetry, by a current generated by a dissipative process, or by a charged beam, ditto; each of these is a spontaneous, emergent source of a field as a macroscopic coherent variable [2].

We cannot imagine another system of measurable variables; but then, we wouldn't be able to, would we? The example of superconductivity is enough of an argument that new measurable entities can arrive spontaneously.

(3) The third source of unpredictability is the notorious sensitivity to boundary condition. This has been discussed by David Ruelle [3] and needs no further from me. Even at this most predictable, the world is not predictable, but much can still be done about it.

(4) Emergence of concepts and structure. This to some extent melds into (2), but goes somewhat further. The question here is not only the absolute unpredictability of frozen accidents — could anyone have predicted the genetic code in detail? — but even more serious, the unpredictability of what kind of accident could happen — could anyone have predicted that a code would happen? The structure of Nature is so completely hierarchical, built up from emergence upon emergence upon emergence, that the very concepts and categories on which the next stage can take place are themselves arbitrary. A functional eye can be built upon several entirely different structural principles, and which a given organism uses depends on its evolutionary history, not on any traceable logical process. But there are much deeper examples, such as the concept of "agenthood" which underlies life: is "wetware" life the only example? The emergent concept, at the higher levels, becomes completely independent of its physical substrate: money can be wampum, stones, paper, gold,... Categories and concepts can have enormous and permanent evolutionary effects. Agriculture is a concept with provably independent substrates of domesticates, which arose and caused population explosions and ecological disasters, and enormous modifications in organism populations, wherever it occurred.

I would argue that predictability is certainly an illusion at even the "Darwinian" level of evolution and from there on. Futurology is a mug's game. (As one can easily tell by looking at its exponents.)

There is a fascinating side issue here: how well can we extrapolate, even within the narrow confines of the physical sciences? We have some remarkable successes: the black hole, the neutron star, the anisotropic superfluid ^{3}He, all of which were predicted long before observation; as was the Josephson effect, shortly before. However, in each case there were vital aspects which remain for observation to tell us. But my impression is that the dismal failures, the cases where we had to receive a very heavy hint from Nature before we caught on, are much more common. It took 15 years, from 1923 to 1938, before Bose-Einstein's idea of condensation was even mentioned in connection with superfluidity, and another 10 for superconductivity (Landau, late '40s). In 1956, 45 years after its observation, Feynman publicly declared his inability to calculate superconductivity. I won't mention how many broad hints eventually forced us to the Standard model. In the case of the Quantum Hall Effects, the discovery was wholly due to the experimentalists — and it took a shocking period of years for the theory to be constructed. The fact of the matter is, Nature is much better at lateral thinking that is the Turing machine equivalent we are supposed to have in our heads — and it is only when we leave that machine behind that we do make real discoveries (and we don't do it with microtubules and quantum gravity).

Postscript: Verification and Validation of Science

This is my response to some of the recent comments on the verifiability of scientific results from philosophers like Naomi Oreskes, quoted by Horgan in his attack on the Santa Fe Institute [4,5]. "Verification and validation of numerical models of natural systems is impossible" is the quote he misuses. Oreskes is expressing an epistomological viewpoint very current among modern philosophers, that allows validity only to tautological statements such as those of a closed mathematical system. What these philosophers miss is the false dichotomy of this distinction. Of course, solipsism is a logically acceptable view, epitomized by the figure quoted by Dennett [6] of "the brain in the vat" all of whose sense-input is controlled by a Cartesian supercomputer to correspond to natural experience, but there is no outside world actually supposed to be there.

As Dennett points out, the "brain in the vat" is requiring an impossibly difficult task of the supercomputer, not only because of the enormous density of sensory data it is having to be fed, but because it has a "schema" — a picture of the world and a map of the relationships in it — which it is constantly and actively checking and updating successfully. The enormous level of "compression" which the schema achieves is its evidence for validity; the solipsistic view may be logical but it also idiotic in the sense of ignoring this compression. We can distinguish, then, three different types of candidates for validity: (1) the closed tautological world of logic and mathematics. (2) Simple induction, where we have $O(1)$ measurements per fact: what is the length of a rod, or the infectivity of AIDS, or the carcinogenicity of rhubarb? Here, of course, there is not true validation in any sense. And yet a third form of validity could be: (3) the multiply connected scheme which lies behind our system of modern science, as well as our understanding of the real world around us, where enormous compression of a lot of correlated fact has taken place, and where we can check the validity of any new theory (which Oreskes misnames a "model") by applying Ockham's razor: does it compress the description of reality, or expand it? From the "nature spirits" of primitive man to the inept theorist adding one parameter per fact, it has always been easy to tell the real theories from the failures. This is what I like to call the "seamless web" of science, and it represents an incredibly powerful criterion for truth, one which the "philosophers" of science have not yet appreciated.

References

1. J. Hartle: Sources of predictability, *Complexity* 3(1): pp. 22–25, 1997.
2. This aspect has been partly captured by W. Zurek in his idea of the "pointer state" and the "quantum halo". But he does not remark on the emergence properties of the pointer state. See review article in *Physics Today*, Oct. 1991, p. 36.
3. D. Ruelle: Chaos, predictability, and idealization in physics, *Complexity* 3(1): pp. 26–28, 1997.
4. J. Horgan: *The End of Science*, Addison-Wesley, New York, 1996.
5. N. Oreskes, K. Shrader-Frechette, and K. Belitz: Verification, validation, and confirmation of numerical models in the earth sciences, *Science* 263: p. 641, 1994.
6. D. C. Dennett: *Conciousness Explained*, Little Brown, New York, 1991, chapter 1.

Good News and Bad News

In 1989 the massive international meeting which had grown up from a pre-"Woodstock" "M2S" (Materials and Mechanisms of Superconductivity) root, upon the discovery of the high-T_c cuprates, was held at Stanford. The banquet was in the San Francisco Exploratorium, which has seating within earshot of the banquet speaker (myself) for at most 100 of the thousand or more attendees — not to mention the noise and distraction attendant on the many fascinating exhibits. I was pretty inaudible. Perhaps this was as well, considering that the references in the speech to "pathologists" might have been much more identifiable at that time — by now I am probably safe from libel suits. But I did not expect that my speech would have had the predictive value it had: the sociology of this field has continued to be, if anything, worse than what I described here. Of course the notorious Schön-Batlogg fraud appeared 10 years later; but to my mind a worse, but unprovable, fraud was perpetrated by an eminent chemist a few months after this talk — so I had omitted a whole category of misbehavior.

The talk is fortunately devoid of any reference to my own efforts, since at the time they were in large part — not totally — off base. The view of the technological possibilities has turned out not to be even pessimistic enough.

At this point, three years into the history of high-T_c superconductivity, it is hardly a surprise to find ourselves hearing both good news and bad.

It is not hard to find the good news: it is the materials themselves, our favorite superconductors, still stellar performers after three years of investigation. Ignoring the petulant outbursts from reporters on such authoritative journals as *The New York Times* and *Newsweek* — and even our own *Science* — who seem to have expected instant miracles, the news about the technological prospects is pretty good. Item: A Japanese consortium has developed metal-filled ceramic wires with far better critical currents than we dared expect. Item: Thin films in the million to ten million A/cm^2 current range have been achieved in a variety of groups. If properly oriented, these films seem capable of sustaining very high fields indeed. Item: our estimates of upper critical fields in the appropriate orientation continue to rise, and are well over a million gauss (100 Tesla, to the modern). To my knowledge there is no fundamental barrier to building a 100 Tesla magnet, given enough financial resources. Item: normal metal-high T_c contacts with negligible contact resistance, and splices, are possible to fabricate, using gold and oriented samples. Item: unique technological niches continue to open up. An example: low temperature cryogenics in the He range are very difficult in space, but it is a triviality to reach 40–50 degrees, where the new materials are entirely satisfactory. Another: the next few decades are likely to see the end of high-voltage above-ground power transmission lines, for economic reasons even in the absence of plausible medical evidence that they do us any damage. Buried, low-voltage lines cry out for the use of high-T_c superconductors.

We ourselves are somewhat responsible for the exaggerated expectations which led to this negative publicity. Any knowledgeable scientist was aware, after six months, that while these materials were better than we had any right to expect, they were complicated, extremely anisotropic, and structure-sensitive. Looking at the very earliest data it seemed clear that good orientation or even single crystal fabrication might be a requirement for many uses. This does not seem quite as formidable as it sounds when you realize that high-temperature rotors for modern jet engines are fabricated as single crystals of gigantic size. Single crystals or highly oriented films are likely to play a big role in high T_c technology. But the development of such a technology was bound to take a lot of time, as was even the task of finding a few low technology applications to get the process started. Even the transistor took a decade to break in to the big time. I feel that in our eagerness to attract funding (and media attention) some of us overstated the possibilities for instant gratification.

From the point of view of science we are already beginning to get a lot of satisfaction out of these materials. Ong and I recently organized a small workshop at Princeton, for instance, at which a striking fact was the general agreement among speakers from different laboratories as to the actual experimental properties of single crystals of nominally the same material. This agreement did not, however, extend to the different interpretations, but that is another matter — see later.

Again, from the scientific point of view we have been presented with an unbelievable bonanza. Everywhere we turn we find new science to study: new forms of magnetism, new questions in chemistry, even all the magnetic properties of rare earth substituents, in addition to the basic question posed by the high T_c, which has turned out to be uniquely subtle and complicated. There are unique properties of the flux-containing type II state, with at least two flux lattice phases.

I believe the rough outlines of an acceptable theory are gradually emerging, but whatever one may think about that, exciting new ideas have entered theoretical physics. Perhaps still more will be required to explain the many puzzling experimental phenomena associated with high T_c.

What then can be the bad news? The right quote is: "every prospect pleases, and only man is vile". The bad news is us. None of us has behaved perfectly, and many of us have been very unprofessional. Perhaps, since it was born in the Reagan era, high T_c has been infected with a general attitude which one might call Reaganethics, which has the same relationship to real ethics that Reaganomics has to real economics. Press release publication, stonewalling (refusal to withdraw discredited results and theories), competitiveness extending to unfair practices in refereeing, backbiting gossip, and great insensitivity to questions of conflict of interest, have all been apparent in our field.

Much more serious has been a lack of scientific objectivity. Among our experimental colleagues, this often takes the form of a lack of truth in advertising: what the press release, the abstract, or the discussion says is not what the data actually mean. In their desperate haste to solve "the Problem" or just to provide the interpretation which many experimentalists feel is obligatory, the experimentalists mis- or over-interpret data which quite often may say something quite different — or may be a bad material or an artefact. Few, even now, have begun to realize that there is no single key fact, and that "the Problem" is a complex congeries of related problems. Three examples (now out of date [1997]) are: (1) early tunneling data where the major feature, the ubiquitous $|v|$ term in the conductivity, was ignored and artefactual and minuscule "phonon bumps" claimed as the major result; (2) optical data of all kinds (since much improved); and (3) the great fuss made about positron fermi surfaces, now entirely dropped out of sight — I found the whole story on these rather depressing (later examples are not lacking).

Now for the theorists! Before I say some rather cutting things, let me make a few mitigating remarks about all our behavior which may be familiar to sociologists of science but not to the average physicist.

(1) Very few scientists think very hard about the underpinnings of their science: they follow "paradigms" which they accept as given from their training and seldom look beyond them. The average solid state physicist, for instance, believes firmly that there is "a" band structure for every metal, easily available from certain expert practitioners, which represents all he needs to know about the underlying electronic structure; he doesn't think about "bonds" or "ions", while his chemist opposite number may think in just the opposite way.

(2) This kind of difference in style between different fields is all-pervasive and very confusing to the narrow mind — it exists between particle-trained theorists and solid state theorists, between both and statistical mechanicians, and there are also the materials scientists, the computational physicists, and so on.

(3) Scratch any physicist and you will find that he believes in one or several sets of irreconcilably contradictory ideas. Such are the existence of "p-bands" and "d-bands" in the face of real band theory; or the ergodic theorem in the face of condensed phases. Not all of us have all these problems, but I have often enough found that arguments run up against questions about the actual nature of scientific truth that I have come to realize that such things are real.

When a really hard and really new problem like high T_c comes up, which engages the attention of a wide variety of fields, the cognitive dissonance which arises is unbelievable — all the different kinds of scientists find their unexamined assumptions and their unconscious dichotomies clashing with each other, and confusion will continue to reign for many years.

We can add to this the fact that the volume of communication has increased tenfold and its average quality decreased almost as much, as a consequence of preprint and newspaper publication, and of the drive for funding which militates in favor of haste and against retraction and correction. To the uninitiated, it all sounds as though no consensus whatever, either theoretical or experimental, is being approached — as though it's just a game which any number can and should play, and any experimental result can be as factual as any other and can form the basis for a theory.

A final contributing factor is the relative obscurity of the community which has made, and will make, whatever fundamental advances there are. The community calls itself "many-body theorists concerned with strongly-interacting electron systems", a group of people who were mulling over such non-world-shaking mind-crushers as "heavy fermion phenomena" and the Fractional Quantum Hall effect before, at a meeting in Bangalore in January 1987, they realized en masse that they had a new problem to think about. The odds against an important theoretical development in high T_c coming from entirely outside this group of at most a few hundred people are roughly equal to those against an amateur winning the world chess championship. Yet the language these people speak is almost unintelligible even to the majority of theoretical physicists from other fields, much less to the great bulk of materials specialists who are doing experiments, and least of all to reporters. The concepts being tossed around, even before high T_c, are quite difficult and in many cases inimical to the scientists' prejudices.

Enough excuses. The late great theoretical physicist Lev Davidovich Landau was not noted for tact or for his tolerance of human limitations, characteristics which landed him in great danger of his life in Stalinist Russia, from which he was rescued by the bravery of Pyotr Kapitsa, himself not the most tolerant of humans. Landau reserved his greatest scorn for bad theoretical physics, for which he used the term "pathology" and called its practitioners "pathologists". He was very tolerant of those who sincerely wanted to learn, and was much loved by the group around him, but when he perceived that prejudice, ego, parochialism, intellectual snobbery, or the like was causing people to do bad physics, he could be merciless.

So far as I know, he never subdivided pathology into separate diagnoses, but it struck me while attending his recent 80th birthday symposium that I had a unique opportunity to make up that omission, in view of the situation in the theory of high T_c superconductivity. I have therefore developed a classification of the commonest pathologies, which in theoretical physics may well replace the well-known directory published by the American Psychiatric Association. The list is:

(1) Megalomania
(2) Brain Surgery
(3) Fetishism
(4) Homeopathy
(5) Autism

(6) and, probably the source of many of these rather than separate, something I call Eidolontia Aurigena which is a free translation of "phantasies coming from gold" — I leave it to you to figure that one out.

I deal with the symptoms of these one by one.

(1) — as I already said, it is hard for an outsider to understand the level of knowledge which exists in our field, and it seems to some as though we were just a bunch of ninnies and any number can play. We are quite happy when serious theorists come to play if they inform themselves properly and have a track record in similar fields, and such as Frank Wilczek and Tony Zee are quite welcome when they come with the appropriate willingness to listen and learn. The category I designate, on the other hand, do seem to think we are incompetent, so they might as well take a stab — in general, they are people who have been getting away with quite a bit, and may think they can get away with it in high T_c. Irrelevancies such as media fame, Academy membership, big prizes, or just plain chutzpah seem to endow the scientist with a presumption of competence which is troublesome only because reporters, students, governments, and even granting agencies often take them seriously. If there are any reporters or representatives of granting agencies in the room, please will they accept this message: we specialists have quite a track record in solving problems almost as hard and confusing as this one, and we know a lot about methodology; don't imagine that some eminent amateur is likely to come along and finesse us — especially when careful examination reveals that he actually hasn't had any great successes recently.

(2) Brain surgeons isn't a very big class, but it gives me the chance to use a story I heard from Bob Laughlin, about the brain surgeon who was the only doctor on board a cruise ship and when asked to do an appendectomy, he did it very successfully — but the incision turned out to be in the skull. That was the only way, he said, he knew how to start! All of us have a little bit of the brain surgeon in us, being in love with our own technology, but some have insisted on blasting away with totally inappropriate tools, sophisticated to be sure but irrelevant.

(3) The group of fetishists seems always to be with us; as the ground disappears from under one, another new, special experimental fact catches the eye of someone. These are the people who become fixated on the very first experimental fact that they see, and mistake that part

of the problem for the whole, a part which may be as irrelevant to the real problem as a shoe is to a woman. Early examples were the triplet EPR signal; and more lastingly, the "chain" fetishists, and those enamored of the 1/2 ev optical bump. Recent examples are legion; there are bound to be a host of phenomena in any group of materials as large and varied as these to serve as fetishes for the rest of our lives. I am afraid judgment and intellectual breadth are the only prescription available.

(4) Homeopaths — I sometimes call these the bandaid physicists — are absolutely determined that either the patient is not sick at all, or that only a tiny dose of novelty or a little mending will suffice. These range from those who believe firmly in good old phonon BCS which will somehow win through if the standard Eliashberg equation is massaged enough, to those who are looking for excitonic interactions or other small variations on the conventional pairing schemes. They have in common a determination to wish away the more disturbing experimental anomalies by careful selection from the tails of the distribution of experimental results, or by applying special tricks to individual cases and ignoring the broad similarities among all the different cuprate superconductors.

(5) I was determined to put this in when I observed a (relatively young) theorist get up at a meeting at which every other talk was excitingly imaginative — often too much so — or revealed some new experimental aspect of great interest; and the young man simply blanked all that out and gave a talk which would have been dull but appropriate thirty years before. Some of us over 50, and even the younger ones, look back with pleasure and pride to the great days of our youth when we "solved superconductivity"; we may have stretched our imaginations and our understandings about as far as they could go, and we are now fixated in that glorious period and unable to take the sickening step of throwing all that out the window and accepting that it is a new ball game. Somehow we OUGHT to be able to use all the lovely old ideas, like s-wave, d-wave, Fermi liquid theory, dynamic screening, etc. It would be so nice and so easy. All of this is understandable, especially to someone of my age. All I can say to the younger theorists is: don't trust anyone over 45, except maybe me, and I'm not so sure about me. At least I'm one of the few who isn't susceptible to eidolontia aurigena! It really is a new ball game.

Let me close then with two bits of good news. First, let me repeat that we are closer to a theory, in my opinion, than most people realize. At the very least we have a good idea what doesn't work and what is relevant. At the best we have a set of new concepts, the precise working out of which is still a matter of hard work and resolution of a number of controversies. I see more and more of the most puzzling observations fitting into the general scheme. What is needed is good will, open minds, honesty with oneself and one's colleagues, and hard, hard work.

Finally, the best news of all: we are not electrochemists, and the phenomena we study are indubitably real, even in Utah!

The Future Lies Ahead

This was an after-dinner talk for the 2004 Santa Fe meeting on "many-body theory". I remember sitting in lonely misery — as the featured speaker at such meetings often does — at an isolated table in the courtyard of la Pena while everyone else noisily absorbed the excellent wine which was provided. Eventually Moses Chan took pity on me and spent an hour and a half pounding my ears about the reality of supersolid He — a case of serendipity if ever there was one. The talk ends abruptly — I think I had more to say but this was the gist of it — and I still hold with the dark view of the state of our science which it expresses. It presumes a little more familiarity with the details of the field of strongly correlated electron systems than its actual audience was familiar with — so bear with me, you'll hear more about this unfortunate field.

It was too much of a temptation, when asked to give a talk on the future of our subject, not to use the title of an old LP by the comedian Mort Sahl. The title is at least certainly true, if nothing else I say is.

Many of you may have heard Bob Laughlin's talk at Snowmass a few weeks ago. In it he proposed that we many-body theorists may be the victims of "dark laws" which make it literally impossible to solve some of our problems — specifically he was speaking about high T_c superconductivity. He supposed that there were many possible alternative orderings, so close in energy that we can't possibly resolve their differences. He seems to have been trying to find, in our future, something as mysterious and glamorous as Dark Matter or Dark Energy.

I was reminded of quite a few occasions on which I have heard eminent scientists tolling the death knell over physics — almost always near the end of their productive lives. I won't even use the myth of

Lord Kelvin and the fifth decimal place — there was John Bardeen, even, proclaiming in an interview that all the important problems in SSP, as it then was, had been solved — Jim Phillips said wickedly that he meant that all the problems known when JB was a graduate student had been solved. There was Brian Pippard's notorious "Cat and the Cream" proclaiming the death of academic physics in 1960; and Felix Bloch in 1964, as president of the APS, proclaiming that superconductivity hadn't been and wouldn't be solved — a sentiment repeated by John Slater some ten years later. So Bob is not alone in declaiming pessimistic hogwash, though as in the other cases what it no doubt means is that HE won't make much more progress.

I suppose that what this syndrome really means is that each of us has only a finite store of the somewhat irrational optimism that leads us to believe that Nature IS rational and that there IS a solution to every problem if we can only find it.

But enough of what isn't true of the future; What is? Like Newsweek magazine, perhaps I should give you red arrows indicating the state of play in each area. I would suggest that it goes this way:

Science: red arrow up
Sociology: red arrow down
Funding: red arrow sidewise

Science

Let me start with the good news: the healthy prognosis for our science. To me, the most cheering news is the appearance of many advances in the technology of our subject, advances both in experiment and in theory and calculation methodology. Accurate ARPES, for instance, made its debut just in time to play a big role in high T_c, and the enormous advance represented by the Scienta detector brought a new level of accuracy. Just coming in are two further improvements: laser-excited photoemission with much better energy resolution and greater escape depth; and the automated Dresden facility (though the results appear not yet to be interpreted with adequate sophistication). The ARPES technique has hardly begun to be explored outside of that specialized field, but promising new results have appeared for the cobaltates, for instance.

The development of STM spectroscopy has had a surprising impact on the cuprate field, in spite of the restricted range of materials it has been possible to study; and more is surely on the way.

An exciting new thing is Sawatsky's resonant x-ray scattering, which may open up to us observations of subtle lattice motions such as will accompany the checkerboard pattern seen in STM.

In general, the explosion in nanoscale technology has hardly begun to be exploited — a good example is Ong's recent measurements of fluctuation diamagnetism in cuprates. Even without having detailed knowledge of the field I am always being staggered by what the biophysicists can do with nanotechnology.

Of course I should not omit the wonderful things that are being done with Fermion systems of atoms on optical lattices, and with the availability of a knob to control the size and sign of the interaction.

In theory, the big news is the DMFT (dynamic mean field theory) which gives us a systematic way to deal with the major effects of strong correlations. After nearly 50 years, we are finally able to really understand the Mott transition, for instance, at least in three dimensions, and to model the Kondo volume collapse in cerium. I for one also believe that the sign problem is getting to be less and less of an obstacle to realistic simulations in Fermion problems, and that one of these days we will begin to trust the results, at least in relatively simple systems like the cuprates.

Sociology

Here I am afraid that I have some fairly harsh things to say on all sides. Anyone who knows even a little about the history of scientific discovery knows that almost any major discovery is likely to encounter resistance even from the discoverers themselves, who are slow to realize what they have; and almost certainly from referees, rivals — usually the two are the same — and an inevitable coterie of holdouts, of whom those with alternative failed theories are the most stubborn. But in the past the resistance has usually died out more or less inevitably, if necessary due to death or retirement of the most tenacious opponents.

I would suggest that in modern conditions — which John Ziman has aptly named "post-Academic Science", where "academic science" is his description of the way we operated a few decades ago — the inevitability of the victory of obvious scientific truth is a lot less evident. So far, I know of only one subject that has reached the stage where this may happen — namely cuprate superconductivity — but others have come very close to it in the past — a good example being polyacetylene and other polymer conductors.

What causes the problem can be described well by another retro reference — the comic strip Pogo — which produced the immortal phrase, "We has met the enemy and they is us!" It is not in anyone's self-interest to solve the problem!

The experimentalist can ensure his funding, he hopes, and produce a steady stream of papers, by continuing to search for the key effect that is, all by itself, going to make everything clear. (A particularly reward-ing game is to search for "THE BOSON which carries the interaction", a chimera based on a naive textbook picture which certainly does not apply here). It helps to emphasize, in each of these papers, the important respects in which his or her data differ from those of that other institution which is making the same measurement — even though, under careful examination, the data appear to be equivalent, often identical. (Other-wise one or the other might appear redundant). In his interpretation, he often can enlist the services of his local theorist, or better yet, do his own theory, no matter how amateurish — referees seem to prefer "interpreted" data.

The theorist has even more leeway to further his career at the expense of clarity and simplicity. He can persist bullheadedly with a hypothesis that is plausible to the naive, but physically impossible (that phonons can cause d-wave superconductivity leaps to mind) and find all the loopholes opened up by the fact that it is very hard to prove a negative, especially if your heart isn't in it. Again, the volume of possible papers staggers the mind and enormously enhances the CV — and if lucky, one can enlist some experimentalist whose suggestive but irrelevant data can start a whole fad of imitators. He can base a theory on almost any experimental uncertainty or hypothetical state of matter, so long as it is sufficiently dif-ficult to disprove (see above, but with the added advantage that the state of matter may be experimentally meaningless or nearly so). Wonder of wonders, if he is lucky he can catch the imagination of the editors of some prestigious journal like Nature, so that only papers which refer respect-fully to "the stripe theory" (whatever that may be) can appear in that august journal.

It goes without saying that the longer these controversies can be main-tained, the more funding has gone into them, the more tenure-track jobs have been awarded, and therefore the more opportunity for new people to start out new fads based on new red-herrings. Who needs a solution to the "problem" posed by the cuprates? — certainly not most of those engaged in the field! Certainly not the funding agencies, whose raison d'être is the continued confusion; certainly not the deans, chairmen and provosts whose insatiable hunger for overheads is thereby slaked; and of course, certainly not the PI's of all those contracts.

I'm sorry if the above sounds a little bitter. It constitutes my thoughts on reading an extensive file of referee's reports on a recent paper. We had attempted to find the most elementary common ground on which we thought all sensible people could surely agree, and instead we found complete unanimity of rejection but complete diversity in the reasons given, none of which made any sense to us. We could only conclude that there was a universal determination to avoid any appearance of progress towards a real solution, but no other common agreement.

I would caution those who are engaged in pushing the most recent bandwagons for nanotechnology or quantum computing to think a little about the possible consequences of bandwagon formation and hyping the agencies. Did the hype about high T_c do good or harm? Is it really true that more money = better science? Or can one slide into a situation where an unstably growing manifold of nonsense "solutions" leaves those who are actually trying to solve something wandering in the ArXiv, starved for the oxygen of attention?

Could Modern America Have Invented Wave Mechanics?

My Harvard classmate, Thomas S. Kuhn wrote, some years ago, an influential book about scientific revolutions. Fortunately, many scientific revolutions do not follow his scenario, but the one he focused on, the discovery of quantum mechanics, is well described by his model, which is most valid for a revolution occurring in the central core of a mature science. He described the appearance of "anomalies", observations which are inexplicable with current theory or even contrary to it, first in a few special instances and then in an increasing flood. One has then the "crisis" stage, a stage in which the science involved exhibits all the symptoms of a mental or social breakdown. The most common behavior is denial of the anomalies, which takes several forms: attacks on the correctness or accuracy of the anomalous observations, convoluted reinterpretations of experimental data according to conventional ideas, or, perhaps most often, blank denial of their relevance. Theorists, meanwhile, invent increasingly arcane versions of the standard ideas: in the case of quantum mechanics, for instance, more and more sophisticated ways of calculating with the old quantum theory, ways which could explain any or all observations because of their complication and opacity. Others experiment with unlikely scenarios.

In Kuhn's story, one has the impression of a happy ending: Heisenberg, Schrödinger and Dirac uncover the true theory, which very rapidly attracts many of the old quantum theorists — Bohr, Slater, Van Vleck, Born, Kramers — to the great work of confirming it, and sweeps all before it. It embodies a new "paradigm", a conceptually new way of thinking. Of course, as Kuhn himself points out, there remain many too deeply committed to their original views to change, and some of these retain considerable power: Néel, Raman, etc., even Einstein; and I would argue that Bohr, with his "complimentarity", was a very reluctant convert. So quantum mechanics was not by any means a majority view until many years later. Nonetheless, from the start there were adequate funding and excellent media in which to present their views, the paradigm shifted, and quantum mechanics prevailed.

Let us imagine, however, that something similar were to take place in modern America, and rewrite the history accordingly:

(1) Heisenberg, in a desperate attempt to achieve tenure, would be too busy writing 10–15 papers a year on the old quantum theory to stop and think. His funding has long since ceased because many of the NSF's referees would not believe even in the *old* quantum theory, and he would always receive one or two very negative reviews.

(2) Schrödinger would never have achieved the level of productivity required for either tenure or funding. Nonetheless, Schrödinger's paper was submitted but the editor of *Phys. Rev. Letters*, refusing to take responsibility himself, sends the paper on wave mechanics to a succession of hydrodynamicsts and opticians who unanimously quash it. The assistant editor to whom Schrödinger appeals sees it as "out of the mainstream" and agrees.

(3) Dirac, who is fortunate enough to have achieved some reputation in mathematics, picks up Schrödinger's discarded ideas and gives several invited papers on them at the American Physical Society and international meetings of one sort of another. These papers are handicapped by Dirac's quiet manner, and since there are so many more popular and conventional alternate "theories" given at the same meeting, his ideas are stigmatized as "controversial".

(4) In any case, the spectroscopic data on which they are based are themselves under considerable attack, since others more in agreement with the popular versions of the old quantum theory are more widely accepted and their perpetrators are better funded, though wrong. Yet others publish right results but reword their abstracts to make it appear that they had achieved the Nirvana of "agreement with theory".

(5) At this point it is recognized that the "crystal detector" used in early radio may be of great commercial importance and seems to be a quantum effect. Hundreds of millions of dollars are poured into institutes, centers, etc., for the study of the quantum. Each such center acquires a local, competing theoretical school, whose version of the facts cannot be contradicted with impunity by the well-funded center's experimentalists. Meanwhile, Heisenberg, having finally

achieved tenure, notices some interesting experiments on electron scattering from surfaces which strikingly confirm Schrödinger and Dirac's ideas. However, since these were produced independently and serendipitously, they are easily quashed by the powerful groups centrally involved in the field, on the basis of poor resolution and the unfamiliarity of the technique. The enhanced funding leads to a many-fold expansion in the volume of publications and number of meetings at which many groups argue bitterly over priority for each unsuccessful theory.

Fantasy? Not really. Some years ago Leo Kadanoff described, in this column, the present state of physics in mordant terms, including, for special mention, among other sub-fields that of high T_c superconductivity theory. That phenomenon revealed a preexisting state of crisis (rather than causing it) in the quantum theory of metals, a crisis which cries out for a major paradigm shift. But the discovery of high T_c itself seems to make this crisis almost unresolvable because of the great financial and political resources at the disposal of the many "schools" which have grown up in the theoretical community. Most of these schools are popular just *because* they deny the existence of the crisis and the necessity of reexamining the fundamentals of the subject because of the high level of competition in the field and of the flood of complex and confusing experimental information, only some of it trustworthy and much of it containing the conscious or, worse, the unconscious biases of the experimentalists. At meetings where experts congregate one often feels almost more impatient with the audience for taking the speakers seriously that with the promulgator of yet another implausible and contorted "theory". But both speakers and audience are caught in an unfortunate sociological system for which I am at a loss to suggest a remedy.

Perhaps worst of all, the very best of the young people shy away from this difficult and controversial field and wander off in search of less populous, and above all less argumentative, areas: for instance, either the fertile new pastures opening up in "soft physics", where the phenomena are so diverse and so new there seems to be a separate problem for each theorist; or the exploitation of the accepted breakthroughs associated with the Quantum Hall Effects. In this they follow many of the best of their elders (such as Leo Kadanoff) who in their distaste for the behavior of us "high T_c'ers" do not recognize a responsibility to inform themselves of the relevant issues for the sake of the basic health of the enterprise of physics.

Loose Ends and Gordian Knots
of the String Cult*

Over four centuries ago Francis Bacon, in his *Novum Organum*, out-
lined the philosophy which came to be the distinguishing characteristic
of modern science. This philosophy held that knowledge of the nature
of things was to be gained by the acute observation of nature, not by the
study of authoritative texts or of holy books, or from imaginative flights of
human fancy. The resulting explosive growth of our understanding of the
universe and of our ability to manipulate it cannot be gainsaid; whatever
one may say about the technical ingenuity of the medieval Chinese or the
early mathematical discoveries of the Indians and the Arabs, one has to
concede that nothing remotely resembling modern systematic science
developed in those cultures.

Many of us in the physics community have become increasingly disturbed by the growing hegemony in a major subfield of our subject of
what we see as a revival of the medieval, pre-Baconian mode of thinking:
that the universe is designed on some simple basic principle which can be
discovered by the exercise of pure reason, unaided, in fact unencumbered,
by experimental study. Such would seem to be the thinking of the community of "string theorists", who in the past two decades have achieved a
dominant position in theoretical physics. Every research physics department worldwide with any pretensions looks to have its own string theory
group of at least two, since only a few talk physics to anyone else. String
theory has produced at least three media superstars, whose books and TV
interviews bring to physics departments flocks of students ambitious to

---

*Book review of *Not Even Wrong: The Failure of String Theory and the Search for Unity
in Physical Law*, by Peter Woit (Jonathan Cape, 2006). Originally published in the *Times
Higher Education Supplement*, 25 August 2006.

join this brilliant enterprise which, it has been said, reveals "the language in which God wrote the world."

Peter Woit, in a strongly-argued and serious book, has taken on the task of analyzing this situation. Dr Woit is neither a sour-grapes sorehead or a sensationalist amateur, as he makes clear by describing the complex mathematics of string theory in what may be excessive detail for many readers, without, as far as I can tell, many really serious errors. He is an admirer of Ed Witten, the fabulous mathematician-physicist who has been the guru of string theory for several decades. Woit took his degree at Princeton, learning quantum field theory from several of its most influential modern exponents, and he keeps in touch with the field through numerous contacts. In fact, after spending several chapters on the history of modern particle theory, Woit adds a long and highly technical chapter praising the many exciting mathematical discoveries which have resulted from the complications of superstring theory — mathematical developments which, *inter alia*, earned Ed Witten an unprecedented and well-deserved Fields Medal and contributed to the stellar career in mathematics of Sir Michael Atiyah, ex-President of the Royal Society of London.

How then can the enterprise on which all of these unquestionably brilliant people are engaged have gone so far off the rails as to merit Wolfgang Pauli's famous putdown "that's not even wrong!"? This needs a little history, which Woit describes in the early chapters. (The experimental bits are a bit shaky, though.) The history concludes with an unexpected and glorious success: the so-called Standard Model, which was constructed in roughly 1965–75, by many of the same theorists who carried on into strings. This marvelous structure classifies all of the bewildering array of known "elementary" particles and fields by means of three families of four "particles" (Fermions) each, and two "gauge" symmetry groups which imply two sets of "fields" (Bosons) which are responsible for interactions of a particularly simple type among the particles. The way in which this structural classification fell into place, and the great leaps of imagination which it involved, justifies a degree of hubris among the few dozen truly extraordinary individuals who discovered it. It has also the very special feature that the ostensible world picture that we see at the ordinary chemical scale has very little resemblance to the underlying structure, but instead is *emergent* from it: for instance, neither nucleons nor light quanta are fundamental in the Standard Model scheme (the electron happens to be.) Both this hubris, and the complexity of the result, fed the temptation to go on leaping, and to forget that each of these earlier leaps, without exception, had taken off from some feature of the solid experimental facts laboriously gathered over the years.

It is conventional to say that the Standard Model vitiated all of particle theory as we had come to know it, because everything measured since 1975 has "agreed" with it. This is problematical; what is true is that measurements since 1975 have left the structure intact, but with modifications in detail. Some of these modifications are very ad hoc, such as those necessary to fit the observed weak time asymmetry, and the neutrino masses and oscillations. As so modified, there remain at least twenty-five arbitrary constants which must be determined from experiment.

Woit begins the story of strings with a cautionary tale: the sad fate of the fad of "particle democracy" which had overtaken the theory world in the 60's. The hope it offered — as later with string theory — was that the mere structure of certain equations would lead to a unique theory of the world. The dream lives on in the minds of a few adherents, and in a couple of books still popular among new agers. It was, however, in the course of fumbling around with this mathematics that the defining ideas of string theory emerged and eventually began to catch the fertile imaginations of the particle theory community. These ideas are:

(1) That what we had been treating as particles (or fields defined at spacetime points) are not that, but are little wiggly strings which define a *surface* in spacetime as they move;

(2) That (for reasons related to working out a consistent theory of these objects) space is really at least ten-dimensional, with most of the dimensions curled up so tightly we can't tell;

(3) That, by similar logic, the underlying symmetry of spacetime must be *supersymmetry*, a generalization of the ordinary relativistic symmetry which requires among other things that for every fermion there must be a corresponding boson (in strict supersymmetry, of the same mass) and vice versa, since the generators of the supersymmetry group switch the two. Hence *superstrings*.

I need hardly say that none of these ideas have any experimental basis. So why then are they taken seriously? Perhaps there are several reasons — reasons which do not include the possibility, or even the hope, of experimental confirmation.

At the outset, the main job was perceived as bringing general relativity into the fold, a task at which ordinary quantum field theory has always failed. With these visionary postulates, it did seem possible to make a reasonable supergravity — in ten supersymmetric dimensions, of course. These difficulties appear when gravity becomes a quantum theory, at the "Planck scale" ten to the 15th power smaller than the scale at which the Standard Model was constructed, which gives plenty of room for inconvenient things to disappear by the time we get to any scale capable of being studied experimentally.

Gravity has always had a very special cachet among mathematically inclined theorists just because of its gorgeous mathematical expression and owing to the myth that it sprang full-blown from Einstein's brain (which it did not, he saw its outlines already in 1907 as a consequence of experimental arguments, and spent 8 years learning the math to do it right). Even though we now know that there are other ways to converge gravity, one may nonetheless concede that it is a useful exercise to create an example of a theory which does not fail at the Planck scale.

To my mind, the most valid point in favor is that the Standard Model exhibits several "internal local symmetries" which receive a natural interpretation in string theory as rotations in the extra dimensions. This idea dates back to Einstein's time.

A third line of thinking seems to be purely aesthetic. The theory (in some one of its many versions) has actually been justified as "just too beautiful not to be true". Woit is particularly dismissive of this claim, suggesting that intricacy, abstruseness and novelty are no substitutes for the simplicity of, for instance, the Standard Model. There is doubtless a fascination and excitement in being in possession of a particularly esoteric and complex body of knowledge or ritual. String theory has begun to seem an obsession, even a cult — the kind of thing which leads the young to wear T-shirts with slogans like "why be stuck with only four dimensions?"

The major source of optimism, however, was the dream that superstrings would furnish a *unique* theory, because of the constraints found on usable versions of the ideas. It is this dream which has evaporated as the mathematical understanding has increased, in fact to the point that string theorists have come to accept that there are almost no uniqueness properties at all; in other words, one may have whatever universe one pleases. It may be said that there is so much freedom in string theory that our present universe, even at the elementary particle level, would be the result of historical contingency. (A term which I prefer to calling it the "anthropic principle".)

Finally, Woit points out, a motivation for pursuing superstrings is often quoted as, "it's the only game in town", implying that if you want a job — or a position at a prestigious institute, or even a MacArthur genius award — you'd better learn string theory. He notes that we are getting perilously close to making the thesis of radical sociologists of science come true, that at least this portion of science is socially constructed.

It is time to sum up. What is Woit's argument? He is not accusing the string theorists of egregious mathematical error — of course they are superb mathematicians (he makes the unnecessary point that there is a fringe community, that being true of most fields of science.) He accuses them simply of doing pure mathematics in physics departments, i.e., of redefining the word "science". One could not possibly object to the existence of an active mathematical community pursuing such an exciting, original line of work. The objection is to the claim that this work is physics, that it possibly or even probably will tell us how the real world constructs itself. One may particularly cavil at the high degree of hype, to the point of monopolizing popular attention, and that the gigadollars of a number of philanthropists, as well as numerous physics department employment slots, are being distanced to this rather esoteric branch of algebraic geometry.

Not Even Wrong is written for the mathematically inclined. Woit seemed to feel that it was essential for this very complex subject to be covered in a serious manner. The math may have been a mistake — it will open him to nitpicking on every issue he discusses, while the central issue may not be so very complicated: the question may well just be what the emperor is actually wearing. He writes the non-mathematical parts of the book well and clearly, although not always without attitude. My personal nit-pick would be his concluding chapter, where he calls for more mathematician-type rigor, which I suspect is not the right prescription. Still, to be honest, I am pleased the book has been written and to have had the opportunity to review it.

Imaginary Friend, Who Art in Heaven*

These two books present two scientists' contrasting views of God and religion. Francis Collins, author of *The Language of God*, has had more than his fifteen minutes of fame as the successful director of the publicly-funded half of the human genome project; he presents us with a heart-felt memoir in which he evolves from an atheistic upbringing to belief. Richard Dawkins, author of *The God Delusion*, has followed up his well-known books on evolution — *The Blind Watchmaker*, *The Selfish Gene* — with a fiercely combative zinger against all forms of religion. It is roughly based on his TV series *A World Without Religion?* (for which the introductory shot was Manhattan with the twin towers intact). Per-haps his usual readers will be delighted, but personally even I found the intensity a bit extreme.

Does God exist? Already in the 13th century Thomas Aquinas con-ceived this to be a question, but it is only with the Enlightenment that it could begin to be asked in the terms it is now: as a questioning of the existence of the supernatural at all. Nonetheless, many discussions quote Aquinas' superficial "proofs" of the existence of God, as both Collins' and Dawkins' books do.

But of course the question is, what do we mean by "God" and by "exist"? Surely in the religious context Francis Collins does not really mean "exist" in the scientific sense that he would apply to his genome data. The essence of scientific reality is reproducibility, that whenever you

---

*Book review of *The Language of God: A Scientist Presents Evidence for Belief*, by Francis S. Collins (Free Press, 2006) and *The God Delusion*, by Richard Dawkins (Bantam Press, 2006). Originally published in the *Times Higher Education Supplement*, 6 October 2006.

look at it, and whoever is doing the looking, the data on the genome is the same. I don't need to belabor the point that whatever the word "God" means existentially it can't be that, nor, clearly, is that what Collins is advocating.

Many believers do in fact believe in that kind of literal existence for God, especially in the United States, where something like a quarter of all Americans seem to believe in the End of Days scenario which will have us all participating, sooner rather than later, in an Armageddon which will sort out our destinations for all eternity. But there is nothing they can point to in the way of tangible verification that would convince, for instance, a Muslim that the father and the son and the holy ghost are somewhere in the heavens judging us and watching for that moment; nor vice versa, the Muslim can't convince us that He provides 14 virgins for every suicide bomber. It is because they seem to be substantiating beliefs of this kind that Dawkins, for one, so deplores the more respected, and especially the scientific, advocates for religion like Francis Collins.

For the first hundred pages or so Collins seemed to be going to avoid the trap of committing himself to a particular choice of a tangible God, until suddenly he took me by surprise by turning out to be a literal believer in at least the New Testament Christian narrative. At the same time he argues persuasively and eloquently for completely Darwinian evolution, as of course the decipherer of the genome must. The genome, as he describes it, contains overwhelming evidence for design by chance. He would seem to believe that the Creator set the whole thing in motion thirteen billion years ago, in such a way as to lead to the evolution of modern humans about 100,000 years ago, then pretty much stood by until 2006 years ago, when he caused a sequence of miraculous events which culminated, more or less, in the accession of Jesus into that place commonly referred to as heaven. (The words heaven, hell, immortality, and soul actually don't appear in his book — I would have liked to hear his opinions on them; but in a tentative way Collins does declare in favor of selected miracles.) Actually, more correctly, He must have begun to act 100,000 years ago, when He started inserting souls with a "moral sense" into human beings, sometime after conception. In other words, Collins has designed his own religion, and he's stuck with a dilemma — is he right or is some other one of the billions of monotheists right? Or would the immense diversity of choice simply make the null hypothesis more likely?

But to return to the meaning of the word "exist". No one would ever say that syndromes don't exist — characteristic responses of the human organism to various stimuli, benign or otherwise. But increasingly we learn that these responses differ from individual to individual; what we find is that they group themselves into recognizable classifications. There is, for example, the syndrome of bipolar disorder; but no one's bipolar disorder is identical to anyone else's. Many of these syndromes are quite benign and nearly universal, such as empathy, and the imprinting that leads to familial or romantic love, and are responsible for profound beauty and great art. Dawkins likens religion to the imaginary playmate syndrome common in lonely children, but that is too dismissive. The need for God and the accompanying myths and rituals exists as a phenomenon of great power in many, if not most, of us. Its strength is attested to by the fact that for many thousands of years no one ever conceived of living without it. Dawkins can hardly expect to make it go away.

There is a third meaning of "exist" which Collins rejects and describes as the "God of the Gaps" — the explanation of the unexplained bits of creation. If one doesn't understand some step in evolution, for instance, one says "there is organization and therefore a deliberate designer" — i.e., Intelligent Design. But there are other instances. For instance, a surprising number of intelligent people believe that existence implies creation implies a Creator — a syllogism I fail to follow, especially because I know as a physicist that the decision is arbitrary as to what is cause and what effect.

Collins places great emphasis on what he calls the "Moral Sense", the more or less common ground that he finds in the ethical systems of all cultures and that he finds, by introspection, in himself. He does not, fortunately, equate religion with morality — I am sure that many of his friends are atheists, and atheists are on the whole no less moral than believers, nor do they have any expectation of forgiveness for evil they may do — but he does feel strongly that there can be no explanation for the moral sense except that it was instilled by God. To my mind, this is another case of the "God of the Gaps". The evolutionary explanation of the nearly universal tendency to moral, cooperative behavior has not yet been definitively found, but that does not mean that there is none. It seems to me that the survival value of a moral code must be immense, as one can see in contrasting life expectancy in a failed state like the Congo or Somalia, where the moral code is largely abandoned independently of the piety of the population, with that in Europe.

As Dawkins points out, it seems to anyone familiar with evolutionary biology that the religious syndrome must be a behavioral Darwinian adaptation. No other species has evolved so long under survival pressure predominantly of intraspecies competition between groups — that is, war. It may be no accident that many of the sacred texts of world religions are war stories: much of the Old Testament, the Bhagavad Gita, Homer (in effect the sacred text of Greek polytheism), much of the Koran. The religious narratives of the early American civilizations are ugly, bloody and warlike. The survival value to the tribe of a strong religious narrative enforcing social cohesion and obedience to the elders, encouraging reproduction and unreasonable optimism, and modulating the fear of death, is obviously very high in the situation of continuous intertribal warfare. How it worked itself out evolutionarily is yet to be understood.

The Language of God is Collins' personal memoir, and of course there must be adequate explication of his crowning achievement, the reading of the human genome — a fair swatch of the book is details about biology. Even this gains a lot of interest from his not-so-saintly views of his scientific competitor, Craig Venter. But more interesting to many will be the account of how he arrived at his particular version of faith, and how he tries to reconcile the enlightened skepticism which is science with its logical opposite, faith. This book is a natural for the bestseller lists, if not particularly for its literary merit.

Dawkins' book is, oddly, much more embittered and deeply emotional, although leavened by his effervescent wit. Much more than his previous books, *The God Delusion* is a direct attack on religion in all its forms as the source of much of the violence and evil in the world, as well as a strong plea for more tolerance of atheism. He makes much of the fact that polls reveal that Americans would rather vote for a candidate who is gay, or has a criminal record, than for an atheist. We atheists can, as he does, argue that, with the modern revolution in attitudes toward homosexuals, we have become the only group that may not reveal itself in normal social discourse.

Dawkins' point against even the more peaceful religions is that unquestioning faith becomes accepted as a virtue and not a defect, and that it is unquestioning faith which powers the dangerous fanatics. He is not willing to accept, as many people do, the countervailing good in the form of works and of personal satisfaction as balancing, more or less, the strictures of puritanism and the evils of fanaticism. And one must sigh at his optimism. He may be right that religion has become a poor survival strategy, leading the world in the direction of nuclear and environmental doom, but I am afraid that if so, the bottom line is that the next dominant species on planet Earth won't be us.

So here we have the pros and the cons of religion, and I'm not sure that either book told me much more about this perpetual dialogue than I had been ruminating over for many years. Both books are lively and probably, at least for those interested in hearing choir-directed preachments, worth reading. Hopefully there will also be readers who will brave the alternative choice for themselves.

(This may be my last review in this journal. If so, it seems appropriate for me to put in a final word of acknowledgement, first to the superb editing of Andrew Robinson, but most of all to the hard labors of my wife, Joyce Gothwaite Anderson, at clarifying my prose and often my thinking.)

IV. Science Tactics and Strategy

Introduction

Anyone with as long a career as mine in science begins to think about how it actually works. My first epiphany was brought on by my reading, relatively late, Tom Kuhn on revolutions; at that time I saw it as totally descriptive of what we do, and drew none of the radical conclusions that later contributed to the science wars; nor do I believe that the young Tom, with whom I was reasonably friendly at college and graduate school, really meant those implications at that time. I, at least, saw immediately that science progresses by what evolutionists call "punctuated equilibrium", giant or small jumps with very slow and steady growth in between; Tom later on began to disavow the small jumps. My next guru was my Cambridge friend John Ziman, with his anatomy of "Reliable Knowledge" which tells one why, on the whole, "academic" science works and continues to grow while all other human activities seem not really to progress. In the turbulent late sixties Volker Heine and I joined a couple of sociologists and started a course in "Science and Society" accredited for both physics and sociology, where these and other questions were discussed in a very informal way.

Later, I became intrigued by various ways in which Ziman's idealistic picture fails: the excrescence of postmodern "cognitive relativism" of the French school and the literary theorists which brought on the Science Wars, the persistence of bad science, pseudoscience, and downright superstition (creationism, Therapeutic Touch, astrology and all that); but finally, my own depressive skepticism about whether, in less cleancut fields than Alan Sokal is used to following, the sociological forces necessarily are strong enough to drive out the demons of self-interested fudging and deliberate misunderstanding. Science can survive superstitious or bombastic outsiders, but is it proof against the unenlightened self-interest of the scientists themselves?

Solid State Experimentalists

Theory Should be on Tap, Not on Top[*]

The very distinguished head of the Kamerlingh Onnes Laboratory, Cornelis Gorter, a delightful and strong personality, was editor of *Physica* during the '50s. As editor he instituted a firm rule against including theory and experiment in the same paper. This rule was old-fashioned even then, and seemed obstructive to those of us working at Bell Labs on the emerging science of solid-state physics, where such great experimental-theoretical teams as Bardeen and Pearson, Bardeen and Brattain, Yager and Kittel, Herring and Geballe were developing new systems of team-work and sophistication in the study of solids which achieved spectacular success. Later examples were the teams of Hopfield and Thomas, Rowell and McMillan, Brinkman and Osheroff, as well as many others.

A number of years later, I began to appreciate the sound logic behind Cor's apparently anti-intellectual stand. In fact, I have usually been happier with myself and with the result when I have published independently of — even if in tandem with — the experimental result. I have come to believe a theorist should appear as an author only if he has participated in the design of the experiment or done substantive non-trivial analysis of the results. The obvious pitfalls are not the worst ones, though they are bad enough: the theorist may be riding piggyback, by which I mean that he is simply adding to his publication list on the basis of some relatively minor suggestion or encouragement, better described by an acknowledgement. A second problem is that the paper has two chances to fail if it contains theory rather than one; the results may be right and the interpretations wrong. Both occur frequently — and obviously enough — to

[*]Reprinted with permission from *Physics Today*, Vol. 43, September 1990, p. 9. © 1990 American Institute of Physics.

need no detailed descriptions. A little of both, for instance, occurred in the discovery paper of phonon structure in superconducting tunneling, where the trivial, but mistaken, explanation of a second harmonic origin for a second bump at twice the frequency of the first was justification for the inclusion of a theorist in the authorship. (It turned out, actually, to be the longitudinal phonon spectrum, where the first bump was transverse, and the two were unrelated. The spectrum at still higher energies was only vaguely to be thought of as a second harmonic. Later on, the full theory and experiment were published by Schrieffer, Scalapino and Wilkins (theory) and Rowell, Anderson and Thomas (experiment); yours truly breaking his own rule because he had basically organized the collaboration — i.e., designed a lot of the experiment.)

Much more serious is the distortion of priorities, of communication, and of the refereeing process, which occurs when excessive weight is given to theoretical interpretation. We don't want to lose sight of the fundamental fact that the *most* important experimental results are precisely those which do *not* have a theoretical interpretation; the *least* important are often those which confirm theory to many significant figures. Two examples which have disturbed me, some 20 years apart, both are also drawn from the tunneling literature. Around 1965 a young visiting experimentalist at Bell Labs observed that there was an interesting logarithmic anomaly of the tunnel conductance at zero voltage in certain junctions between normal metals. He was extremely reluctant to publish results on the effect because there was then no theoretical explanation, even though the experimental data could be fitted phenomenologically very neatly by simple logarithmic expressions. It was an important, if not world-shaking discovery of some interest when the theory finally appeared: local moment-assisted tunneling with an attached Kondo singularity. If such respect for theory had been the norm when the "resistance minimum" was first discovered (fortunately in a country where Cor Gorter's rule was observed), it might have delayed the publication (by van den Berg) of the original observations of the "Kondo" effect for some 30 years.

Even more serious problems seem to be besetting solid state physics today, and especially the field of high T_c, where I am an avid, if not impartial, observer. An example is in tunneling studies both in the normal and the superconducting state. Here again there is no theory to go by for almost any fundamental measurement, not only tunneling. That is why high T_c is such an interesting problem. Yet often the experimental papers on high T_c have theoretical collaborators listed in their authorship. My own instinct is to disregard many such papers, since the data may or may not have been influenced by theoretical prejudices, and, in any case, the abstract and interpretation sections often do not reflect the experimental observations, especially where these contain — as they often do — new and unusual features.

To return to the example of tunneling in high T_c materials, one of the few consistent experimental observations is that the tunnel conductance of the normal state has a characteristic "V" shape over a wide voltage range, roughly $\sigma(V) = A + B|V|$, which persists to low temperatures as a voltage-dependent background. One always sees this behavior more or less disguised by other "bumps" or "gaps" or artifacts which are often the ostensible point of the paper as submitted, and which vary radically from specimen to specimen, lab to lab, etc. Often, in fact, the experimentalist seems to go on until he sees some simulacrum of "BCS" behavior, which, he feels, has a "theoretical" justification, and then publishes immediately. On the other hand, it took two years to see, in the published literature, a quantitative experimental study of the much more consistent, puzzling, and therefore important "V" data on the normal state. I am told that unless it was backed by enormous prestige in the form of the right author and institution, most referees were in the habit of rejecting such studies. This example appears to me to reveal a major weakness in our approach as scientists, a collective unwillingness to welcome new or anomalous results.

Of course, in urging independence of theory, one must not relieve the experimentalist of the responsibility of understanding the theoretical constraints, e.g., such things as the essential symmetries his data should satisfy, or obviously invariant branching ratios such as neutrons per fusion. Also, I am happy to see the relevant theoretical parameters presented: e.g., NMR experimentalists *should* relate their data to Korringa theory, thermal conductivity *should* be compared to the Wiedemann–Franz law, and (what is surprisingly rare) the relevant Mott minimum metallic conductivity should serve as a standard of comparison for electrical conductivity: the reader should not have to calculate the Mott number for himself. Sometimes it is even, regrettably, necessary to ask about minor experimental details such as whether contacts were interchanged *à la* Montgomery.

A number of factors such as the funding crunch and the publication explosion are working against originality and innovation in our subject of solid-state physics, and in science generally. The undue influence of theory is an unneccessary addition to this burden, since we are doing it to ourselves. The prejudice in favor of a pat "interpretation", no matter how anomalous the observed phenomena, is particularly stifling when, as in journal refereeing and grant reviewing, it is essential to get consensus, since originality and independence of mind are least likely to be found in a committee. Perhaps it is even time to return to Cor's rule and free experimentalists totally from their theoretical friends and colleagues.

Shadows of Doubt*

I am likely to be ticked off by the hubris of mathematicians. In this case it exhibits itself in the form of the Penrose Fallacy, which is to assume that all problems too difficult to be solved by the great brain of the author must be identical. Mathematicians seem to have no appreciation of the real complexities of Nature. Two of my colleagues in the local math department gave a series of lectures about Free Will which never mentioned biology or the brain, and which were even more naïve than Penrose. The review expresses my own position, based on enumerating unsolved mysteries rather than by pseudological reasoning.

About 15 years ago, Roger Penrose's former student, Stephen Hawking, devoted part of his inaugural lecture as Lucasian professor in the University of Cambridge to the prediction that by the year 2000 physicists would have been made obsolete by electronic computers. Although Penrose does not refer to this particular instance, it would seem that it is his concern about this kind of optimistic view of the capabilities of computers that motivated him to write *The Emperor's New Mind* and now, some five years later, its sequel.

In *Shadows of the Mind* he elaborates on his earlier proposals and attempts to answer his critics. I was put off reading *The Emperor's New Mind* by the many critical reviews it received, so I came to the sequel fresh, albeit prejudiced. Let me say without hesitation that my prejudices have been amply confirmed. Nonetheless, reading this new book is a

---

*Book review of *Shadows of the Mind: A Search for the Missing Science of Consciousness*, by Roger Penrose (Oxford University Press, 1994). Originally published in *Nature*, Vol. 372, 1994, pp. 288–289.

fascinating and mind-stretching exercise. I can imagine that the average scientific reader, unfamiliar with the many-faceted mental universe that Penrose inhabits, will be dazzled by his extraordinary breadth and scope. But the more extraordinary the mind, the more unfortunate it is when it is used to entertain what may well be vain speculations. Also, Penrose's great reputation, charm and skill as a writer (perhaps not as evident in this book as in the previous one) should not blind us to the fact that his professional background is not really relevant to his subject matter.

The book consists of two parts. In the first part he argues that the mind does things that are beyond the capabilities of a "mere" computing machine. (The word "mere" is a trap; in this case it is in the meaning of "mere" that the meat of this statement lies.) This is why machines are not about to replace physicists (or mathematicians). I heartily agree. But he then concludes that some novel laws of physics must be crucial to the operation of our brain, and that they might possibly relate to certain aspects of quantum gravity theory. This conclusion troubles me. In the second part of the book, Penrose goes on to discuss his view of the gaps in our understanding of physics and biology through which such radically new material could creep into the theory of the brain. I can find little here to sympathize with.

He presents four alternative propositions about the mind that are intended to cover all possibilities: A is the "mere" machine; B is the machine with an impotent "Des Cartean" observer, which as far as its method of operation is concerned is essentially A; C uses possible new laws of physics; and D is the supernatural alternative that does not obey natural law. Penrose rejects D as a cop-out, as an alternative inappropriate to a scientist. A and presumably B are excluded by very subtle and ingenious reasoning involving a restatement of Gödel's theorem as an argument about computer algorithms: that no "provably sound" algorithm operating on a Turing machine or equivalent computer could ever encompass all the correct mathematics of which the human brain is capable.

To my mind, the most likely alternative is different from all of these; it is that the operation of the mind follows the ordinary laws of physics and chemistry, without bizarre additions, but that it operates using algorithms, concepts and mechanisms that are quite outside the system of apparently rigorous "theorems" of computer theory. In computer complexity theory, for instance, the complexity classes are often meaningless categories. But by using one's knowledge of the nature of the problem to be solved, one can often do what from the theory seems to be impossible whereas nominally "equivalent" problems turn out to be inaccessibly distant from each other.

There are many ways in which computational methods using quite ordinary physics might evade the apparently "rigorous" limitations of the von Neumann/Turing architectures. What follows are just a few mainly culled from various recent books and articles about the mind.

First, the mind's hardware is by no means complete at birth. Not only its instruction kit, but its internal and external connections are constructed using knowledge of the nature of the actual world it will function in. The concepts of space-time and of objects moving continuously in space are built in; in fact, the most obvious optical illusions involve replacing discontinuous events with continuous ones. And most of all, the mind experiences objects: complexes of data that exist in space-time and show different aspects of the same entity.

Second, the brain's connections and its program are not complete until after it knows that other autonomous entities — other similar machines — constantly surround and communicate with it. There is a hint: the primitive mind animates with purpose even those objects in its surroundings that are inanimate. Communication is impossible without two factors: someone to communicate with and a common perception to communicate. Communication is a primary feature of mind, which is currently thought to be already established before the mind is complete.

Third, there is a fair amount of evidence that the mind is not a single, simple entity: it may be a number of independent, autonomous systems squabbling for a central dais. Multiple personality disorder is only an extreme form of what goes on in the mind all the time. There is no single Turing machine or single tape. It is not clear that it really is correct to model a parallel collection of semi-independent machines, that is, in some sense, wider than it is deep, in terms of a sequentially operating single algorithm. In discussing complexity, this can be a different "large-N limit", with different capabilities.

Fourth, some of Penrose's arguments, and much of computer theory, are about exact, rigorous solutions. His computers do not "halt" until they have found an exact answer. This can be crippling. In the real world it is usually adequate to "satisfice", to use Herb Simon's term. Methods directed merely at finding an acceptable way to do something can be much more efficient than exact ones. This is one way the mind can take advantage of its knowledge of the structure of the world.

As I see it, it is not really necessary to identify what particular aspects of the nervous system allow it to evade the rigid, rigorous, logical arguments with which Penrose tries to pin it down. One has merely to point to the remarkable ability of complex systems to develop emergent properties that overcome the apparent limitations of their separate constituents. Apparently rigorous "theorems" that seemed to make anti-ferromagnetism as well as superconductivity impossible turned out to be irrelevant in the face of the emergent property of broken symmetry, just as all the many kinds of argument against evolution — the thermodynamic one, for example — do not prevent its happening. What does seem clear is that the above, and other new concepts and methods using conventional physics and chemistry, are far more likely to solve the problem of mind than is quantum gravity.

It is impossible to analyze here in detail all of the arguments in the second part of the book. Let me pick out a few about which I have some independent knowledge. The long section on quantum measurement theory emphasizes the many dilemmas and queries that one encounters if one assumes, with Bohr, that there is a genuine dichotomy between the microscopic world in which quantum theory applies and the macroscopic world of measurement apparatus. These mindbending difficulties ("EPR", "entanglement", Bell's theorem and so on) are the stuff of rather boring philosophical discussions; it is hard to see how they could make consciousness easier to understand. But one seems unable to find any natural scale for this dichotomy, among other things; and many, if not most, thinking quantum physicists· reject the idea that there is any dichotomy, and assume that quantum laws hold all the way up and down. This possibility is dismissed by Penrose in two brief pages (pp. 310–312).

Penrose's primary objection to this point of view is that it is "unsatisfactory" in that it involves continual splitting of the wave function of the Universe into fragments, only one of which an observer can perceive. (This splitting is the "many-worlds" viewpoint, although there are other ways to interpret the same mathematics, among them that of M. Gell-Mann and J. Hartle.) We cannot decide for nature which of her ways are "satisfactory" or "unsatisfactory"; that is nature's call.

More seriously, Penrose makes the claim that there is no quantitative justification for the all-quantum viewpoint. In a popular book, *The Quark and the Jaguar*, published earlier this year, as well as in several articles, Gell-Mann discusses at length the rapid and complete "decoherence" between alternatives, which prevents the observation of coexistence within, for a typical case, 10^{-21} seconds, by actual and precise calculation. That this is a consistent and logically satisfactory possibility has been obvious for many years, since Fritz London first proposed it in 1938. It has now been formalized. Penrose should have been aware of this.

With regard to superconductivity, Penrose has, I think, got the implications of macroscopic quantum coherence backwards. In a superconductor, the quantum field itself becomes a macroscopic object, a perfectly measurable, rigid, thermodynamic parameter of the body on the same footing as strain, torque, entropy or magnetization, and obeying the same general laws (which derive from the general phenomenon of broken symmetry). Coherence is maintained not by an energy gap as Penrose suggests, or by some mysterious persistence of a quantum superposition, but by mundane thermal equilibrium. It has always seemed to me that for anyone in possession of the facts about superfluidity and superconductivity, it would be hard to doubt that classical behaviour is simply large-scale quantum behavior — that is, an emergent property of large quantum systems. But habits of thought die hard.

Microtubules are for Penrose the likely seat of the mysterious quantum gravitational effect that makes the mind possible. Biophysicists who specialize in their study would agree that the behavior of microtubules is indeed interesting and complex, but would see no need (nor in fact any room) for anything but the characteristic chemical control mechanisms with which we are familiar.

Penrose has written a complex, erudite and fascinating book, and my complaints about it do not mean that I did not enjoy and learn a great deal from reading it. But one should keep in mind that Penrose is a mathematician with little experience of the messy, frustrating but ultimately deeply satisfying process of checking his ideas against the experimental facts about nature. Mathematicians are used to game-playing according to a set of rules they lay down in advance, despite the fact that nature always writes her own. One acquires a great deal of humility by experiencing the real wiliness of nature.

The Reverend Thomas Bayes, Needles in Haystacks, and the Fifth Force*

In 1759 or so the Reverend Thomas Bayes first wrote down the "chain rule" for probability theory. (The date is not known; the paper was published posthumously by a "good friend" in 1763.) Bayes seems to have had no idea that his simple formula might have far-reaching consequences, but thanks to the efforts of Harold Jeffreys, earlier in this century, and many others since, "Bayesian statistics" is now taught to statistics students in advanced courses. Unfortunately, however, it is not taught to nutritionists or even to experimental physicists.

These statistics are the correct way to do inductive reasoning from necessarily imperfect experimental data. What Bayesianism does is to focus one's attention on the question one wants to ask of the data: it says, in effect, how do these data affect my previous knowledge of the situation? It's sometimes called "maximum likelihood" thinking, but the essence of it is to clearly identify the possible answers, assign reasonable *a priori* probabilities to them and then ask which answers have been made more likely by the data. It's particularly useful in testing simple "null" answers.

Consider, for instance, the question of looking for a needle in a haystack. Actually, of course, there are no needles in most haystacks, so the question doesn't come up unless I happen to suppose that at some particular source of hay there are a lot of absentminded seamstresses or bucolic crack addicts. So I might look for needles to find out if a particular set of haystacks came from that bad source and therefore shouldn't command a high price.

*Reprinted with permission from *Physics Today*, Vol. 45, January 1992, p. 9. © 1992 American Institute of Physics.

Let us set it up: there are two sources of hay, one with no needles at all and one with up to 9 needles per stack. Let's assign precisely probability $\frac{1}{2}$ for the sake of argument, to the case where I'm buying from the needle-free source. (This represents the "null hypothesis" in this example.) If I'm dealing with the potentially needly hay, let's assume that $p = \frac{1}{2} \times \frac{1}{10}$ for 0, 1, ..., 9 needles in any one stack.

I search for needles in one stack, and find none. What do I now know? I know that this outcome had $p = \frac{1}{2}$ for needle-free hay, $p = \frac{1}{20}$ for needly hay; hence the probability of this outcome is 10 times as great if the hay is needle free. The new "*a posteriori*" probability of the null hypothesis is therefore $\frac{10}{11} = (\frac{1}{2})/(\frac{1}{2} + \frac{1}{20})$ rather than $\frac{1}{2}$. Clearly I should buy this hay if that is a good enough bet.

Now suppose I was an ordinary statistician: I would simply say my expected number of needles per stack is now down to 0 ± 2.5, and to get to 90% certainty I must search at least ten more haystacks, which is ten times as boring.

Thus it's very important to focus on the question I want to ask — namely, whether I have reason to believe that there is any effect at all. In physical experiments one is often measuring something like a particle mass or a Hall effect, where we know there is some finite answer; we just don't know how big. In this case the Bayesian approach is the same as conventional rules, since we have no viable null hypothesis. But there are many very interesting measurements where we don't know whether the effect we're testing exists, and where the real question is whether or not the null hypothesis — read "simplest theory" — is right. Then Bayes can make a very large difference.

Let us take the "fifth force". If we assume from the outset that there *is* a fifth force and we need only measure its magnitude, we are assigning the bin with zero range and zero magnitude an infinitesimal probability to begin with. Actually, we should be assigning this bin, which is the null hypothesis we want to test, some *finite a priori* probability — like $\frac{1}{2}$ and sharing out the remaining $\frac{1}{2}$ among all the other strengths and ranges. We then ask the question, does a given set of statistical measurements increase or decrease this share of the probability? It turns out that when one adopts this point of view, it often takes a *much larger* deviation of the result from zero to begin to decrease the null hypothesis's share than it would in the conventional approach. The formulas are complicated, but there are a couple of rules of thumb that give some ideas of the necessary factor. For a large number N of statistically independent measurements, the probability of the null hypothesis must increase by a factor of something like $N^{1/2}$. (For a rough idea of where this factor comes from, it is the inverse of the probability of an unbiased random walk's ending up at the starting point.) For a multiparameter fit with p parameters, this becomes $N^{P/2}$. From the Bayesian point of view, it's not clear that even the very first reexamination of Roland von Eötvös' results actually supported the fifth force, and it's very likely that none of the "positive" results were outside the appropriate error limits.

Another way of putting it is that the proponent of the more complicated theory with extra *unknown* parameters is free to fix those parameters according to the facts to maximize the *a posteriori* fit, while the null hypothesis is fixed independent of the data. The Bayesian method enforces Occam's razor by penalizing this introduction of extra parameters; it even tells you when you have added one too many parameters by making your posterior probability worse not better. (A fixed theory with *no* new parameters, of course, does not have to pay any penalty.) It also turns out to be independent of tricks like data batching and stopping when you're ahead — but not of discarding "bad runs".

All of this as folklore is not news to most experienced experimentalists. Any good experimentalist will doubt a surprising result with less than 5–6σ "significance", for instance. Nevertheless, the common saying that "with three parameters I can fit an elephant" takes on a new and ominous meaning in the light of Bayesian statistics. How can we ever find and prove any new effect? Again, I think physicists' intuition operates very well: We tend to convert, as soon as possible, our unknown effect into a new or second sharp "null hypothesis". We might propose, for instance, that there is a 17-keV neutrino with some small amplitude, and test that idea and an alternative, null hypothesis on the same footing. If our further data then (hypothetically) indicate a *different* mass with a different signature, we don't take that as evidence for our new hypothesis, which is now a sharp one, but as a destruction of it. Perhaps a good rule of thumb might be that an effect cannot be taken seriously until it can be used as a null hypothesis in a test of this sort. Of course, statistics can never tell you what *causes* anything; they are not a defense against insufficiently lateral thinking (such as neglecting to ask whether both or neither of one's hypotheses is true), systematic error, or having found some effect you are not looking for — any or none of which can be operative in a case such as the 17-keV neutrino.

Still, one sees the phrase "significant at the 0.05 or at the 0.01% level" misused all over physics, astrophysics, materials science, chemistry and, worst of all, nutrition and medicine. When you read in your daily paper that pistachio fudge has been shown to have a significantly favorable effect on sufferers from piles, nine times out of ten a Bayesian would say that the experiment significantly reduces the likelihood of there being any effect of fudge on piles. While we physicists have no hope of reforming the public's fascination with meaningless nutritional pronunciamentos, we can be careful with our uses of the word "significance" and we can test our own parameter values realistically.

Emerging Physics[*]

I have always wished that I had the energy, stamina and ability to somehow present my view on the fundamental questions in popular terms. Although Bob had just published — admittedly in an obscure place, the Journal of Unreproducible Results — a rather strange, but clearly unfriendly, attack on me, I felt that it was essential to support, as far as was reasonable, his attempt to do what I hadn't done. Nothing I say in this review is slanted either way. Philosophically I think he is dead on; but it's not a message that your average reader realizes is important.

I should make my interests clear right at the start. For many years I have thought that a book such as this should be written, and have been urged to write it myself. I didn't do so, and could not possibly have written it as well as this for the audience one hopes it attracts. This is a book about what physics really is; it is not only unique, it is an almost indispensable counterbalance to the recent proliferation of books by Brian Greene, Stephen Hawking and their fellows promulgating the idea that physics is a science predominantly of deep, quasi-theological speculations about the ultimate nature of things. Thus the enterprise as a whole has perforce my strong endorsement, and any disagreements or criticisms should be read in that knowledge.

The central theme of the book is the triumph of emergence over reductionism: that large objects such as ourselves are, in myriad ways, the product of principles of organization and of collective behavior which in

[*]Book review of *A Different Universe: Reinventing Physics from the Bottom Down,* by Robert Laughlin (Basic Books, New York, 2005). Originally published in *Nature,* Vol. 434, 2005, pp. 701–702.

no meaningful sense can be reduced to the behavior of our elementary constituents. Large objects are often more constrained by those principles than by what the principles act upon. The underlying laws of physics have no sense of time, give us no clue either to measuring or locating ourselves in space, and have no clue to identity — we all are made up from nothing but waves in a nonexistent medium (as Bob aptly quotes from Christina Rosetti's "Who has seen the wind"). Our identity and perceptions are all collective behaviors of "ghosts" who borrow their reality from each other.

Laughlin gives the reader a quick tour through much of physics (without, by actual count, even one equation). There is some emphasis, but not exclusively, on the quantum theory of condensed matter, insofar as it explains such things as computers (with a skeptical side glance at quantum computation), the properties of ordinary metals, and the like. There is an enlightening discussion of the special quantum phenomena, the Hall and Josephson effects, which through the "protection" of collective behavior allow measurement of h and e to enormous accuracy. Some of this will be hard for the layman; at almost no point is it out of his reach. The pedagogy is leavened by anecdotes, occasional eloquence, and characteristically pungent diction (Laughlin once interrupted a scientific talk with "Liar, liar, pants on fire!").

There are idiosyncratic views of a wide variety of scientific topics. Laughlin reveals his view of nanotechnology with the chapter title "nanobaubles" (a view with which one can concur); relates some inside dope on Star Wars and its notoriously fraudulent x-ray laser; then continues on to the nanobaubles of life, for which he has considerable admiration. Then we hear his own ideas on biology, which surely will not be everyone's but are thought-provoking. Finally, his view of complexity science surprised and pleased me with its relative benevolence.

In spite of the above fulsome praise, not by any means is this the perfect book, even for its purpose. Laughlin is not reliably careful with facts, whether scientific or historical. For example, it had rhetorical value but is incorrect to give his great hero (and winner of two Nobel prizes) John Bardeen mythic status, and to demonize the "engineer" Bill Shockley; Bill was indeed verifiably contrary and sometimes mean, but he was also a great physicist, and he, rather than John, was responsible for creating the great research center at Bell Labs, with Fisk and Kelly, and for hiring John as well as many of the other stars who graced the place, such as Townes, Herring and Matthias. John was human, and wrong as often as he was right. It would have been instructive to point out that he had published two mistaken theories of superconductivity, 15 and 7 years before he got the right one. Laughlin's history and emphases are too much those of his generation.

Laughlin makes too much of the role of the renormalization group and other "protection" principles, as opposed to mechanism, in determining the properties of things. Was it Pierre Weiss in 1905, with his mysterious molecular field and Weiss magnetons, or Heisenberg, with the quantum theory, who explained ferromagnetism? — I think the latter. And possibly partially a personal motive (for another example) leads him to misleadingly accuse two unnamed physicists of predicting superconductors to be limited to below 30 K (the actual figure was 40), when what they said applied specifically to a particular mechanism — and is in that case true.

In my experience, which, incidentally, is longer than his, underlying causes often enlighten even our conceptual thinking as much as precise numbers do, which Laughlin seems to deny here. After condemning the astroparticle types for overemphasizing deep thoughts and broad vistas, he seems to reveal a certain measure of "particle envy" and distaste for the messy, quarrelsome, but rivetingly absorbing ways of doing the real sciences. What made John Bardeen great, as indeed he was, was stubbornness and experimental taste, and Laughlin dismisses these values.

Those who devour the work of Greene, or decorate their coffee table with Hawking, will find this book a useful antidote. It should spike the interest of those who read the physics popularizers, although in its personalized coverage and opinionated style it is *sui generis*. My message is this: buy the book.

On the Nature of Physical Laws[*]

Human beings — even physicists — are very capable of holding two totally incompatible concepts in the mind at once. Few of us have never attempted verbal communication with a pair of dice, for instance, while at other times we subscribe firmly to the laws of Newtonian dynamics.

Deterministic dynamics of macroscopic objects is just one, if possibly the best tested, of the laws of physics. It is the nature of physics that its generalizations, such as this one, are continually tested for correctness and consistency not just by careful experiments aimed directly at them but, usually much more severely, by the total consistency of the entire structure of physics. That the trajectory of a roulette ball is deterministic was tested, rather thoroughly and directly, by a group of now eminent physicists calling themselves the Eudaemonics[†], for fun and profit: but much more severe tests are made every day via the internal dynamics of our instruments and our technology. We can measure the fundamental constants to precisions of 10^{-7} to 10^{-8} using the modern quantum technologies due to Josephson and von Klitzing, and we measure time, using the wonderful technology of millisecond pulsar timing, to 6 orders of magnitude better than that, Any such precise measurement is a triumph of deterministic dynamics, as is the achievement of the almost unthinkable precision which brings beams of protons and antiprotons together at LEP. The values of $\hbar/e^2$ and $e/\hbar$ do not depend on the mood of the experimenter, and evil thoughts do not prevent those beams from

---

[*]Reprinted with permission from *Physics Today*, Vol. 43, December 1990, p. 9. © 1990 American Institute of Physics.
[†]This episode in the lives of Ralph Abrahams, Jim Crutchfield, Doyne Farmer, Norman Packard and others is very readably told in the book *The Eudaemonic Pie* by Thomas A. Bass.

colliding. Recently the entire science of deterministic dynamics — misnamed "Chaos" in the popular mind, but really, as we all know, meaning "deterministic Chaos" — has very much bolstered our understanding that what goes on in the most apparently random physical systems, such as turbulent jets and convection cells as well as dice games and roulette wheels, is simply "sensitive dependence on initial conditions" acting in a perfectly deterministic system.

It is disturbing, then, that some who call themselves physicists set out seriously to test the effect of thinking at them on sensitive electrical measurements, on card shuffling machines, or on bouncing ping-pong balls. It is much more disturbing to announce positive results on the basis of statistical deviation at the few-δ level. The problem is, of course, the question of consistency of the structure of physics: if such results are correct, we might as well turn the National Bureau of Standards into a casino and our physics classes into séances, and give back all those Nobel prizes, since the measuring apparati we think we have been achieving all this precision with can actually be bent out of shape at the behest of the first Uri Geller who comes along, and our vaunted precision is all in our heads.

It is for this kind of reason that physicists, quite properly, do not take such experiments seriously until they can be (1) reproduced (2) by independent, skeptical researchers (3) under maximum security conditions and (4) with totally incontrovertible statistics. Oddly enough, the parapsychologists who claim positive results invariably reject these conditions.

Less thoroughly entwined with the very nature of physics, but still very much subject to this important concept of the immense overdetermination of the structure of science in general, are various other laws which have been questioned recently: e.g., the equivalence principle in the weak sense that gravity and inertia do not depend on internal states of motion, or the principle of invariance of the branching ratios for nuclear reactions. With the fantastic level of confirmation of the laws of general relativity which has recently been achieved, especially by Joe Taylor's group studying the binary pulsar PSR 1916, for example, and the recent very severe tests of the equivalence principle, it is hard to see why statistically weak and physically naïve challenges to these laws deserve PRL publication and front page coverage. Equally, when cold fusion is claimed to produce heat without neutrons or neutrons without tritium it takes very little thought to realize that some very basic principles on which whole technologies have been based have been suddenly abrogated, and one is better advised to examine the challenger than his results.

My moral, finally, is that physics — in fact all of science — is a pretty seamless web. If we challenge one of its smaller generalizations, we may be successful if we replace it with something else which holds all of the strands together. It is wonderful to discover that no outside fact forbids 5-fold symmetry or high T_c superconductivity, only our prejudices: but even those prejudices were soundly based, since the prior conditions for these were explicitly and clearly stated in the classic literature, the first that it was *assumed* that a periodic structure existed, the second *assumed* the BCS-Eliashberg dynamic screening mechanism. Result which rip the fabric to shreds must, by this time, be assumed to be almost invariably wrong, and it will save everyone a lot of energy and time if we recognize that such results should be examined with a tougher mind that we physicists are used to applying — perhaps, as has been advocated elsewhere, we should call in those who are more used to dealing with film-flam, such as magicians and policemen.

On the "Unreasonable Efficacy of Mathematics" — A Proposition by Wigner

Joel Lebowitz asked me to join a panel at one of his semiannual Stat Mech meetings, I think in 2006, to discuss Wigner's famous proposition. I was way out of my mathematical depth — other members were Sinai, Jürg Fröhlich, Michael Fisher, I think — but that didn't stop me from my usual determination to put the cat among the pigeons. As a result of many decades of interaction with Eugene on many levels — and particularly of his complete refusal to take my views about physics seriously — I came to the task with a very negative attitude.

I have a feeling that I am expected to represent the negative on Wigner's proposition, but actually I am personally quite neutral about it. I do believe in the principle of emergence, that the complex tends to emerge from the simple and that therefore I guess things will tend to get simpler as one analyzes them more finely, and therefore be more susceptible to mathematical description; but I think almost all the things worth studying are irreducibly complex. To me Wigner's remark seems more or less on the level of the anthropic principle, best summarized by the silly old song "We're here because we're here because we're here …".

A way I can satisfy expectations by being a little negative is to express some cautions about the limitations of the use of mathematics in physics and the physical sciences in general, under three headings:

I. In my experience, interesting and relevant mathematics is more often stimulated by interesting experimental results or questions than vice versa.

II. Interesting mathematical ideas can misdirect you into scientific dead ends — they can become answers in search of a problem.

III. Complicated or lengthy mathematical procedures are very easy to use as a cover-up for shoddy or dishonest science.

I. Some examples in physics of experiment-stimulated mathematics are Wigner's own discovery of the random matrix eigenvalue distribution stimulated by nuclear physics, the wonderful profusion of mathematical curiosities coming from experiments on the quantum Hall effect, and even general relativity, which was an attempt to explain the experimental equivalence of gravity and inertia. From my own experience, if you will forgive me for being self-referential, I could bring up several examples of mathematical discoveries (which are surely not my speciality) arising from experiment. Perhaps the best known is localization, which is actually a purely mathematical statement about eigenvalues of certain linear problems; another is the spin glass phenomenon, which has led to a congeries of fascinating and still open mathematical questions: a third still fertile area is the subtle problem area opened up by the infrared catastrophe phenomenon and the Kondo effect. All three arose directly from surprising experimental phenomena which conventional thinking could not comprehend.

II. The answer in search of a problem. From time to time the theoretical physics community has been seduced into hieing off after fascinating mathematical ideas which turn out in the end to have had little or nothing to do with physics. The most notorious example was "particle democracy", the dream that self-consisting dispersion theory was going to tell us all about the elementary particle spectrum, which led to the learning of a lot of esoteric mathematics (I even endured a course in it) but nothing else. The present-day candidate in my field is the Quantum Critical Point, which seems so far to have answered few questions anyone wanted to ask, although it is dragged in whenever anyone sees anything he can't understand. Of course, one has to leave the allegation unproved, but neither Grand Unification nor superstring theory seem clearly to have escaped this category.

III. A very serious problem in today's literature and with today's peer review situation is the amount of shoddy, incorrect, and perhaps downright fraudulent theory that is published in prestigious peer-reviewed journals and often invited for presentation at major international congresses and discussed in the public press. Here I am specifically referring to my own field — I do not know the situation in other fields, and at least hope that other peoples' standards are better. The computer is the source of much of this mess, and perhaps that's not relevant to real mathematics — I encountered a case of this kind early in the high T_c days, where an honored professor hit the headlines by calculating the T_c of three or four compounds accurate to three significant figures using his proprietary software. He actually seemed to have no idea of the real physics, and showed it by "calculating" a value of another experimentally known parameter (but not known to him) which was off by a factor of 10. We do now know what T_c depends on but even now a 50% error bar should be attached to any such attempt, and as far as I can see this work was pure fraud, unprovable of course because neither the input nor the calculation was available to others. But the fact that generally computer work, either simulations or calculations, is not held up to even the same standards of reproducibility and relevance to reality as experimental investigations leads to a great temptation to fudge, usually of course unconsciously.

More relevant to the subject here is the individual — and there are more than one — who has built up a massive body of work, and often a minor following of devoted students and experimentalists grateful for his citations of their work. His basically erroneous or deliberately obfuscated premises are concealed behind a mass of mathematics in a series of self-referential papers which no referee can ever follow in detail. Based upon this infinitely elastic foundation, he can produce the "A–" theory of this phenomenon and that phenomenon, adding one concealed parameter per measurement, and keep his funding officer and his department head gloriously happy with his productivity.

All that suffers are the rest of us, finding the avenues for communication of genuine theoretical physics increasingly clogged with this kind of nonsense.

In conclusion, yes, mathematics can be strangely useful, but not necessarily for the purposes Wigner intended.

When Scientists Go Astray

Every Nobelist has been asked to do a number of talks for lay groups, and a few even get quite good at it — and bless them, communication of science to the lay public is a Good Thing. I am not very good at it, or at least I have not noticed an increasing demand for my services in this regard. This was one such talk which I gave at the Harvard Club of Princeton in, from internal evidence, about 1991–2, and kept around because it is on a subject which continues to interest me. This collection contains another piece on scientific integrity, "The Future Lies Ahead", which is considerably darker because of my more recent experiences in the high T_c superconductivity melee — it suggests that under some sociological circumstances the scientific truth may never be attainable. The present piece represents a more balanced view and has my views on the cold fusion mess, which I kept fairly up on for a long time, though I didn't know any of the participants well.

More recently, I did fairly personally encounter a flagrant case of scientific fraud, the Jan Hendrik Schön case at Bell Labs in 1999–2001. From the point of view I expressed in the Harvard Club article, the behavior of Schön himself was very atypical and therefore uninteresting: he is one of the rare scientists who commit deliberate, barefaced fraud with intent to deceive and no belief in his own results: he was a "mouse-painter". His modus operandi was to take the samples from Bell Labs to the apparatus on which he had done his thesis work, which was in Switzerland, where he would invent a set of imaginary data in agreement with whatever plausible theory he had last listened to. His apparatus was hypothetically capable of achieving unique — unreproducible elsewhere — experimental conditions, so no one could check up on him. It is, in fact, not clear whether there were ever actual experimental data; certainly much of what he published was purely imaginary. He was caught, eventually, basically because the agreement was too good, with theories which were not too good. I myself had been uneasy and suspicious about some of the results, but — such was the reputation of his superior Batlogg who gave all of the talks, and of my beloved Bell Labs — I failed to see the possibility of deliberate fraud until it was called to my attention in winter 2001–2 by a junior colleague, Lydia Sohn, who had worked near him at Bell.

Yet the case, at another level, is absolutely typical — it was the department head, Bertram Batlogg, and some of his other colleagues, who exhibited the overoptimism and apparently deliberate blindness to the possibility of fraud or error which characterize pathological science. Batlogg was a reasonably good scientist, originally, but had been caught up in the somewhat overblown enthusiasm about the new superconductors. This field, I learned too late and to my detriment, was where the superiority and self-confidence of the old, magnificent Bell Labs had begun to unravel; they were no longer producing the best samples and taking data which was the standard of excellence, as I had always relied on them to do. (We now know that the best crystals at first were at Princeton, the best infrared spectroscopy at IBM, the best superconductivity measurements at British Columbia, etc., etc.) Apparently, when the apparent bonanza of Schön's data was in his hands, he began to dream of a return to the glories of the old Bell Labs. Some combination of temptation, pressure, and enthusiasm overcame any natural caution Batlogg may have had, and in the late '90s every meeting had his talk about this or that spectacular new result from Schön et al. He never, ever, insisted on seeing the apparatus or the raw data. When he caught on to the scam is not known, but anecdotal evidence suggests that it was before the rest of us.

The concluding paragraphs of my Harvard Club talk are still valid, and I recommend them to you, in spite of the dated examples.

I am going here to focus on how, why, and when scientists go astray in what they ought to be good at — discovering the truth. The costs of such malfeasance are simple and straightforward: we pay scientists to discover the truth and we hope they will advise us wisely, and when they do not succeed in doing so that is a cost to society as a whole. I will argue that there are characteristic failures to carry out these responsibilities, that even a layman can identify some of them, and will discuss a few of the disturbing examples I have seen in the course of a long career in science.

Deception, fraud and misbehavior in science have become increasingly fashionable topics, catching the attention of congressional committees, serving as the subject of popular books, and, in a significant new development, producing a boom in professional chairs and departments of "biomedical ethics", "scientific ethics", and the like in many of our major universities. Let me say first that the personal and especially the financial ethics of scientists are, on the whole, somewhat better than those of the clergy — we have no Jim and Tammie Bakkers nor even any Vatican Bank scandals — and probably better than those of lawyers and bankers. (Update to 2008 — I said a mouthful there.) The number of scientists caught with their hands literally in the till is small enough that I know of only two cases, one of which was in Italy a long time ago, and the other of which is perhaps illustrative of my point: a well-respected New York biological researcher who created designer drugs in his laboratory for street sale — not for personal profit, but to keep his research funded. Moreover, more recently, when it was revealed that a very well-known scientist had been buying archaeological artifacts which had been smuggled illegally into the US, he felt impelled to return not only the questionable ones but to contribute from his own collection to the government museum of Peru — not, in contrast, to demand a million dollar ransom for them, as the family of an (American) army officer recently did in a similar case involving stolen German artifacts.

It is not that we (scientists) are saints or fools, but that, in general, our motivations, while just as strong, and often as corrupting as others', are not focused on quite the same rewards. When we cut corners, cheat, or lie, it is usually in pursuit only indirectly, if at all, of fortune: it is basically for the much headier goals of professional recognition, fame, and, if possible, immortality, within the quite remarkable professional sociology which has grown up in science. The separation of scientists into a distinct caste was marked by the formation of the first professional societies in the 17th century: the "Academy of the Lynxes", by Galileo and friends, and the Royal Society by Newton's circle. We have become more and more a group apart ever since. If some of the things scientists do seem a bit bizarre, just remember that it could be immortality which is at stake.

Let me say also that, as emphasized by Ziman, this very sociology has its self-regulating features: we get our goodies for what he calls "verifiable knowledge"; a rational being would calculate that in the long run being right is much more likely to pay off than not being right, because if one's contribution is false that is likely to be found out eventually, because the real payoff only comes when someone else uses or repeats the discovery.

By and large, much of the fuss about fraud and ethics is about biology and, particularly, biomedicine — and properly so if it is outright fraud you are after. In medical research fraud is all too easy: experiments often consist of careful statistics on effects one understands little about and on small or irreplaceable samples, such as groups of twins raised apart, or a small population of people with a given disease. Until recently there have been few overarching theories to tell one that a given effect is non-sensical, out of proportion with its supposed cause, or unrelated to it. Everyone has his list of outright fakes: Piltdown Man, painted mice, made-up IQ statistics.

Much less eye-catching, but very disturbing, is the realization that the quality of biological research in some fields such as psychology or nutrition is suspect because of the "inbreeding" effect, which I intend to discuss later for the hard sciences: the tendency of whole disciplines to encapsulate themselves and to lose touch with the quality standards of the "hard" science community, in which very little deliberate fakery has come to light. The great advantage of the hard sciences is reproducibility. One of the first and most influential demonstrations of reproducibility was by Galileo: after discovering the telescope Galileo made copies for each of the crowned heads of Europe, so that everyone could look and verify Galileo's claim to have seen moons orbiting Jupiter. On the whole, with remarkably fewer problems than you might think with unreproducible, one-of-a-kind experiments like large telescopes and particle accelerators, reproducibility as a criterion has meant that really serious error has a finite lifetime.

Actual fraud is perceived to not pay off. Most often its perpetrators are outside the establishment, like the perpetual motion machine inventors who in the end have been caught with buried electrical cables or other hidden sources of energy but who nonetheless often attract sympathetic coverage in the press as having outwitted the professionals. It is apparently part of the human animal's heredity to want to read stories about miracles accomplished by little guys in their garden sheds, but few take them seriously (and those that do lose a lot of money).

Another very valuable protection against fraud and error in science is the existence of what I call the "seamless web" — a body of firmly established theory, now extending from physics through molecular biology, which, in many situations, traps dubious observations. Already known laws like conservation of energy, quantum mechanics, relativity, and the laws of genetics, constrain the explanation of any given result in a fashion which can be unique, or nearly so, and makes errors easy to spot. Much of science is "overdetermined" in this sense.

Nonetheless, I'd like to tell you a few stories about how scientists in our supposedly rigorous, error-proof, falsifiable sciences do go astray — in some cases misleading the general public, with consequences which are usually only financial, not deadly, but which can seriously impact on the efficiency of the non-negligible fraction of the GNP which modern society spends on science, and which contribute to a general public suspicion of the reliability and validity of scientific research which in itself can do considerable damage. I'll conclude with some very *inconclusive* suggestions as to how the public — *and* scientists — might deal with these problems — suggestions which emphatically do *not* include regulation or legislation. However badly science regulates itself, it is far more efficient (at doing so) than adversarial or bureaucratic procedures (such as handing the problem to the courts).

A most provocative and instructive example of the problem of scientific misbehavior is the recent case involving Stanley Pons, Martin Fleischmann, and cold fusion. Outside Japan, and some readers of the Wall Street Journal, there can be no illusions left, and we will take this only as a example of how egregious error can originate and propagate. Some of my facts are borrowed from a book by Gary Taubes, but most of them are common knowledge: in view of the recent publication of Taubes' book I will give only the bare bones of the story, leaving the low comedy and high drama to him. What I want to pinpoint are the features which this story has in common with other examples — in this is it most helpful in that it contains essentially *all* the characteristic problems, except, possibly — but not certainly — outright fraud.

You will remember that roughly four years ago (in 1987) it was announced with great fanfare, by these two researchers from Utah, that nuclear fusion of deuterium could be produced by a simple process of electrolysis with palladium electrodes in heavy water. Very large amounts of excess energy were claimed to have been produced (and the energy problems of the world were solved). Within days, confirmations rolled in from around the world: India, Texas A&M, MIT, Frascati in Italy, Hungary, one department at Stanford. However, within a week or two, non-confirming reports from reliable sources closer to the scientific mainstream began to trickle in, and were soon to become a flood which eventually put out all but a few embers of the publicity brushfire, and left at least all technically sophisticated opinion completely convinced of the non-existence of the effect. Before treating this wave of PR, which in itself has very instructive features, we should take a look at the pre-March '88 Pons and Fleischmann and the history of what led to their announcement.

Pons is the simpler and easier. He comes from a somewhat isolated small religious community, and apparently trusts easily only very close, long-term associates, such as his lawyer, a Mr. Trigg, who has served as his major spokesperson. Through college and graduate school he seems to have remained focused on electrochemistry, a field of considerable economic importance which has had a strong intellectual component in the past, but has grown out of contact with the mainstream — a field in which he operated with reasonable success, to a great extent because, for the requisite six years to achieve tenure, most of his work was funded within a small section of the Office of Naval Research, in which a close friend from graduate school was the contract officer.

One of our valuable customs is "peer review" but in the nooks and crannies of the Department of Defense this can be not very rigorous. Pons did a certain amount of contract research for Dow Chemical, but after a couple of his claims didn't pan out — one of them almost as exotic as cold fusion — and he developed a pattern of clamming up when questioned or criticized, Dow dropped him with prejudice on their part. Here we have object lesson #1: the possibility of a man operating in three separate closed circles with no true feedback among them, in one of which he had been spotted as deceptive and unreliable, while the publications generated in the second, very narrow world assured him, without serious outside review, a rapid, unquestioned advancement in the third, the university setting, mainly because he was able to attract funding (which benefited the university because of overhead charges). Here is perhaps the most dangerous entity of all: the closed, self-referential circle, acting in concert with the pure greed not of the scientist himself but of his institutional base, a story which becomes familiar.

Martin Fleischmann, an Englishman working on leave at Utah, also comes from this somewhat backwater world of electrochemistry. One piece of his history is relevant: he is the discoverer of an unexpected and exciting spectroscopic effect known as surface-enhanced Raman scattering. This effect met with some skepticism, and I expect Fleischmann sees his history as a victory over a hidebound establishment — but in fact (after a period of controversy, quite short as these things go) his effect was quickly repeated and taken up by other groups, and resistance to his own later publications (in that field) came entirely because he insisted on coupling his observations to a manifestly incorrect theory which he clung to doggedly.

Nonetheless out of his discovery he achieved fellowship in the Royal Society; of the rest of his reputation I know nothing. Here we have a second component which is not uncommon in such cases — early success perceived as putting one over on the establishment leaves, for some character types, a predilection for climbing ever farther out on the limb despite its being in danger of being sawn off.

When Pons conceived the idea of attempting cold fusion, for the second time in his life he went outside his closed circle for funding — in this case to the department of far-out ideas in the DOE. His proposal was submitted for review to his Utah neighbor Steve Jones, at Brigham Young University, a reasonably competent physicist with a bee in his bonnet about cold fusion as a geological process. Steven returned it to Pons with suggestions for improvement — which is **o.k.** — but also began to set up to do similar work — which is not o.k. The resulting chain reaction of accidents, misunderstandings, and plain incompetence (mostly on Pons' part, but also Fleischmann's) led to totally unrealistic claims which otherwise sane university administrators inflated on the way up, rather than exerting their appropriate function of consulting a wider spectrum of colleagues in relevant fields; on the contrary, the deans and president called in their patent lawyers and apparently endeavored to conceal what was going on not only from the physics department but also from the rest of chemistry at Utah. The inflated form and premature timing of the announcement were primarily due to a paranoid misperception that Jones might forestall them with some equally spectacular claim, while in fact Jones had in the meantime, quite properly, become convinced that no useful power could ever be expected, and was quite unaware of what all the fuss was about.

The subsequent public relations mess had its amusing as well as disturbing aspects. Amusing was that copycat blunders in science can be just as pervasive as copycat crimes in response to widespread publicity. Most of these disappeared without a trace: a few (one in Texas A&M, one in the EE department at Stanford) persisted for a while, but have led subsequently to some rather unpleasant revelations of possible outright fraud, to verified tampering with the degree-granting process, and to resignations. Serious attempts to replicate the work, especially at Caltech, revealed a number of blunders it would be easy to make, but it is not even yet known whether there was any plausible basis for the original announcement in terms of mistaken calorimetry, the most likely of these. The main cause of Pons' belief in his work seems to have been an explosion witnessed by no one, but not unexpected in a system involving hydrogen and electricity.

The problem which, perhaps, disturbs me the most, and is to this day seldom discussed, is that even a moment's study of the original claims as they turn out to have been made should have and did lead knowledgeable physicists to the conclusion that they were nonsense. The process of fusion is not an unknown one, and however it is achieved the particular way in which the deuterium nuclei come together is not going to affect what happens to the resulting compound nucleus subsequently, which is very well understood. We can know certainly that for every calorie of energy the system will produce a calculable, rather large number of neutrons, and a batch of tritium, a highly radioactive nucleus. In large quantities, both are quite deadly to humans. What should have been evident to everyone was that if Stanley Pons had produced the amounts of heat through fusion that he was claiming, his appearance should have been that of a Hiroshima victim. In fact, there has never been any demonstration of the nuclear consequences of fusion significant for energy production — what claims of neutron and tritium production exist correspond to energies billions of time smaller, and are in every case just barely at the limit of detection.

And this, to my mind, is the biggest failure of all — the failure of my "seamless web". I suppose the web fails because many otherwise sound scientists are reluctant to risk their reputations for being objective, fair-minded, judicious good-guys, and hesitate to apply the marvelous apparatus of concept and theory which they possess to anything outside their own immediate fields. The version of the "scientific method" they learned early in their careers blinds them to the importance of strong logical inference, and prejudiced in favor of even most unreliable and implausible "direct" experimental evidence. This is why Stephen Jay Gould says, quite properly, that scientists are the worst, not the best, people to investigate someone like Uri Geller who may be fraudulent: science is in a sense "institutionalized gullibility" rather than, what it claims to be, institutionalized skepticism, or at least institutionalized common sense. Also, and partially for this reason, good scientists tend to be poor copy — as is not unusual or unexpected, the media reports in the cold fusion case came out initially representing a favorably biased spectrum of scientific opinion.

But in fact, at least three scientists of reputation contributed to the media feeding frenzy of those first few weeks. The "reclusive brooding genius" Hegelstein at MIT produced a (proposed) Physical Review Letter claiming to explain how it all happened, and, worse yet, MIT, without consulting *their* in-house physics department, got patent applications under way for him. Hegelstein was a protegé of Edward Teller and Lowell Wood at Livermore National Laboratory's x-ray laser project whose reputation was enhanced by a journalistically questionable article in *The New York Times Magazine* (possibly emanating from Livermore management) from which I quoted above. His Letter never saw the light of day. Hegelstein has subsequently achieved tenure at MIT, for no obvious reason, and possibly due to intervention from above. Edward Teller, a name we all recognize, in the first days predictably reacted enthusiastically as he would to any fusion project — and, saddest of all, a true "reclusive brooding genius" Julian Schwinger, a brilliant prodigy of my Harvard graduate school days and a Nobel laureate, also seems briefly to have been persuaded, and to have spoken to the media and granting agencies to that effect. The media had found its stable of high-profile advocates; dissenting opinion, regardless of the reputation for reliability of the source, was not sought out. And the scientific community is reluctant to speak out in a negative mode.

Here is a personal example of the latter problem. Just recently I was listening as politely as I could to a very eminent theoretical physicist who was doing a Tarzan — about which more later — in my particular field, expounding a strange stew of nonsense and old hat at great length to a confused audience. I knew the same material had been presented in his native country previously where it was taken very seriously indeed, since he is in good repute with the regime. There is in-group funding and media naivete here, but the fact is that here and now I am avoiding naming the scientist and fingering the error, although I have already gone further in identifying him than most of my colleagues would be willing to do. No one is going to openly attack a colleague who is probably a long associate, possibly a personal friend, and very likely — and is in this case — public-spirited in service to the field. (Note added in 2008: T. D. Lee.)

Now we come to the Tarzan complex — an expression I picked up at Bell laboratories. This is a problem to which physicists are particularly subject. The Tarzan thinks that because he has been extraordinarily competent in one subject he can come surging through the trees and become an instant expert on any other subject, especially those which a reporter asks him about. And, parenthetically, before I get into trouble, let me add that here the exception proves the rule: some eminent scientists the like of Hans Bethe or Luis Alvarez, for instance, seem to be able to become serious, professional-level experts on a very wide variety of subjects. By the same token, this type of scientists are careful not to speak when they *don't* know.

Returning to cold fusion, Taubes in his book suggests that a great deal of the responsibility for the mess must be ascribed to the mixed message which came across from the media, in that the natural bias of the media for sensation fed on the eagerness of the copycat blunderers and the Tarzans to rush into prominence, and this in turn inflated the self-confidence of the original Utah group and their followers, who in the first months actually became more and more sure of themselves.

Among the many sad consequences of the affair was that, for a while, institutional pride affected even the professional societies involved: the electrochemical society sponsored a session at which only favorable opinions were scheduled, and serious chemical societies favored Pons with standing ovations; there were editorials praising him for showing up the arrogant establishment physicists; who were charged with wasting millions on magnetic fusion when it could be done in a little bottle.

To go on with this epitome of a case, much more serious problems occur when an eminent scientist who has lost touch with any community but his own constituency gains the ear of politicians at the highest level. It is not generally recognized how extensive is the damage that has been done by the once-great scientist, Edward Teller, whose persuasions and prejudices have made him useful politically for ultra-conservatives and the military. He was instrumental in leading us into the unrealistic, costly, and even dangerous Star Wars program.

That Teller's scientific judgment had atrophied over time can readily be demonstrated in various well-publicized cases such as his initial support of cold fusion and the infamous x-ray laser, and is reinforced by tale after tale from visitors to his — until recently — virtually private preserve of Livermore National Laboratory, tales from consultants whose strong cautions about one program or another were ignored, from speakers who found their words twisted or misunderstood, from portentiously announced theories which came to naught. To elaborate one example: the x-ray laser project is a matter of public documentation. The laser was a scientifically dubious component of the Star Wars program which, since it was powered by fusion bombs, may have been Teller's incentive for pushing that program. As its tests became less and less favorable, pressure from Teller led to increasingly skewed evaluations, to the point that eventually Livermore's director resigned in protest at what was going on in his own laboratory. By the way, it was during this period that the "reclusive brooding genius" Hegelstein slipped away to MIT.

Teller is not a unique example, by any means — he was simply acting as our Lysenko of physics. The damage done to the Soviet Union by the mistaken views on genetics of Lysenko hardly needs to be retold; but it is forgotten that in the immediate postwar years a strong campaign in favor of Lysenkoism was mounted by France's influential leftist intellectuals, a campaign which wounded French biology noticeably but temporarily.

Some four decades ago the physical chemist I. Langmuir identified a syndrome which he called "pathological science", which played only a small role in the cases I have discussed so far, but which is common enough in the hard sciences as well as in pseudo-science. Langmuir's essay is very amusing and provides considerable insight into the nature of human gullibility. The characteristic of pathological science is the employment of extremely elaborate, sensitive, and unnecessarily complicated apparatus to observe an effect which is just on the edge of measurability. In fact, often the effect is seen only when "specially trained observers" (read mediums or "sensitives") are taking the data, or after skilled manipulations with the statistics. It has long been recognized that this was the case with J. B. Rhine, the famous researcher into "paranormal" phenomenon, whose technique was to discard runs from his statistical series which were not "successes", on the basis that his subjects were not "sensitive". Regrettably this problem still crops up, even in the hard sciences; code words for some examples are polywater, N-rays, illinium and alabamine.

A famous case was the astrophysicist Joseph Weber of the University of Maryland, whose ingenious and very expensive scheme for detecting gravitational waves coming from explosive events in the cosmos is still widely imitated. Fortunately, most other groups do not use Weber's statistical methods, which were shown by a colleague of mine at Bell laboratories, Loren Miller, to involve essentially the equivalent method to Rhine's in terms of discarding unfavorable data. Weber was somewhat discredited when he announced a coincident event from two antennas, one near Chicago and one in Maryland, but failed to note that the event occurred at one antenna in Eastern Standard time and the other "simultaneously" in Central Standard time.

What tells us a great deal about institutional pride — and in this case the corrupting effect of institutional funding — is that the administration of the University of Maryland effectively blocked the dissemination of Loren Miller's report by exerting pressure on AT&T, possibly without consultation with Weber himself, who is a reasonable guy.

I have left myself a little time to talk about another type of scientific pathology, the closed self-referential circle and the non-falsifiable truth. This was a major part of the cold fusion debacle, namely the narrow and inbred atmosphere of the electrochemical community. It is said that Pons, and particularly Fleischmann, are still widely respected and that the community still cannot quite see what they have done wrong.

This overlaps with an observation of Karl Popper, that a theoretical paradigm and a field of science can become so closed that there can be no way for a theory to be what he calls "falsified" — that is, tested against real reality instead of internal reality. Popper argued that this had become true of Freudian psychology, but I see the same things happening in other places — the "behaviorist" school of psychology was a prime example. Even in my own field, I run into scientists who have receded into the details of their own mechanical calculation schemes so far that they reject and resent being forced to think about the actual physical entities they are simulating. I tried and failed, a few weeks ago, to convince a young calculator that her repeated assertions that the calculation was "fully internally self-consistent" was not the right answer to my question about why it did not correspond to reality. In greater or lesser form, the narrow specialization which leads to this kind of problem is an increasingly serious brake on scientific enterprise.

Well, finally I have to come to a set of conclusions. In the first place, I don't want to leave you with a completely downside impression. Science is great fun and full of wonderful discoveries for those who try to keep their eyes a little open for what the fellow next door is growing in his garden. And, happily, that "seamless web" of overarching theoretical understanding is becoming firmer and tighter every year. Some of the wiser science funders — the MacArthur Foundation, the National Science Foundation to an extent — are trying to do something about excessive specialization, and in its glacial way even the NAS is thinking about inter-disciplinarity. One must take into account that historically science has always fumbled its way along, starting hand in hand with alchemy and astrology and gradually and progressively down the centuries forswearing the false and discarding the errors, using its marvelously adaptive institutions to improve and grow. The process is not one-way and inevitable — good science can and does get lost or forgotten — but it is mostly so.

What I have said is not meant to encourage you to go out and campaign against the research budget; what should be done to rationalize the way research is funded is a whole other story I can't hope to treat here. Neither too much nor too little money is being spent overall, (though I think that in some fields excessive funding can also slow progress rather than help it).

Basically I have three messages — one about journalism, one for the public, and one for whose of us who are scientists.

Responsible scientific journalism is indispensable (and certainly worth enduring the other kind for.) Journalism is one of the few ways in which the countless little cliques and closed self-referential circles can be exposed to the light of day. In the end, cold fusion was debunked by the more responsible journalists learning which sources to trust and what was really going on. Unfortunately most scientific journalists do not know their turf as well as their corresponding numbers in politics, business, or sports; they do not often develop "deep throats" within Livermore National Laboratory or the SDI office, for instance. It is a constant source of amazement with what complete surprise and ignorance even major news media greet each new Nobel prize, even when the scientific community is perfectly aware of the likely candidates. One knowledgeable scientist, Bob Park, working part-time for the American Physical Society, has been able to scoop the professionals of *Science* magazine, *Nature*, and the *New York Times* again and again.

Irresponsible scientific journalism is everywhere — sadly, often appearing in otherwise reputable publications such as the *Wall Street Journal*, which is the *National Enquirer* equivalent for science. Another recent British example is *The Sunday Times'* campaign to discredit responsible action on AIDS. Even *The New Yorker* fluctuates wildly. The reporters' natural sympathy with the underdog often benefits your local astrologer, homeopath, earthquake predictor, or perpetual motion inventor — so be it.

For the public — whenever science reporting touches you, look hard at the source and also at what is being said. Is the effect barely measurable or very sensitive? Can only special people or special apparatus see it? Did the last study say the opposite? Who is talking? Is it confirmed by major people in reputable departments? Is the announcement sensationalized? Is the person quoted defined simply as a Nobel Prize winner or MacArthur fellow, or is this his field of expertise? But — if a responsible majority is overwhelmingly saying one thing … if people who aren't usually public scientists — the Kissinger equivalents — are talking about it ... you *better believe*, as in the cases of cigarettes and health, SDI, global warming and CO_2. Perhaps it's even a good bet that if the WSJ (or *The Sunday Times*) is for it, it's wrong!

Finally, there is a very important set of messages for scientists. If I had to condense it all to one plea, it would be for us to reexamine our prejudice in favor of organized gullibility. Outside evidence is *not* irrelevant. It is *not* equally probable that some observation is true or that it is not true, if it requires us to discard the whole apparatus of physical law. Good probability experts will tell us that it never makes sense to estimate probabilities in the absence of prejudice — which is what we often pretend we can do. Do not deconstruct the laws of physics — they have a lot to tell us, and what's more, it's all quite true, even though some of it was said by dead white males. Science and scientists may be fallible, but they are all the truth we have.

Further Investigations[*]

I have been puzzling over the question of how life got started since the late '70s. For over a decade I have had constant discussions on this topic with SK, and for the past five or so years I have read over and commented on successive versions of his book *Investigations*. Aside from the early papers I wrote, I have collected the resulting thoughts only in a couple of sets of unpublished lectures, at Princeton for my course on "origins and beginnings" for nonspecialist undergraduates in 1993–5, and at Gustavus Adolphus in a complexity seminar in 1995. I think it is time that I got some of this material written down, even though it might appear that to some extent it is only a commentary on Kauffman's book. I can put some of his speculations on a relatively firm footing, and change the emphasis in a number of places. But please do not take this piece as in any way critical of the very important and basic ideas in that book.

Topics:

(1) The role (or lack thereof) of "dissipative structures".
(2) The role of "work" and of cyclical operation.
(3) Is there a "Fourth Law" for open systems, and what is it?
(4) The role of synergy — recombining of the branches, tree structure is not the model.
(5) What else do we need to get up the scale from molecules to Einstein? Where do free will and consciousness come from?

(1) My first real interest in this subject was sparked by the claims made for Ilya Prigogine's ideas about "dissipative structures" in systems far from

---

[*]Thoughts about the questions raised in Stuart Kauffman's book, *Investigations* (Oxford University Press, New York, 2000).

equilibrium (IP happened to win the Nobel Prize in the same "class" as myself, and I listened to his Nobel address with considerable disbelief. This disbelief was justified, since the version of a "fourth law" that he was proposing at the time turns out to be incorrect.) It was a time when pattern formation and broken symmetry in dissipative systems was rather a fad. Convective rolls in Bénard cells, spiral patterns in reaction-diffusion systems like the B-Z reaction, and so on were the kind of phenomenon he adduced to demonstrate that dissipative systems could produce a kind of order spontaneously, and "therefore" — the reasoning has missing links — could lead to life.

The trouble is that the more one knows of biology, the less one sees anything that looks like a "dissipative structure" playing a crucial role. The important "structures" are stable, "condensed" structures like double helices of DNA, proteins which fold into stable or metastable configurations, membranes made of lipid bilayers, microtubules made of condensed rods of tubulin, and the like. The nearest thing to a dissipative structure may occur in the early stages of morphology for multicellular organisms, where a reaction-diffusion system sets up a gradient in the egg cell which determines segmentation — but the more we know about that process, the more clearly it seems under tight genetic regulation. (See recent work of Leibler.) Most of the real structures of the cell follow the rules determined by local energy minimization, the phenomenology of condensed objects with spontaneously broken symmetry. This was the fallacy of Prigogine's theory: he tried to find a variational principle which would similarly account for the observed structures in dissipative systems, ending up with the hypothesis of "maximum entropy production" which is just not in any sense true, as shown in example after example (the late Rolf Landauer was the first to point this out).

In my original paper, given at a Solvay Congress and then at Gene Yates' memorable meeting in Dubrovnik in 1980, I emphasized that such structures would not have enough permanence to allow faithful replication, or have any ability to act on the environment.

(2) I think Stu has really put his finger on it when he emphasizes the vital role of work — the idea that a basic requisite for being alive is the ability to convert energy into work, and that all known ways of doing that involve some form of constraint. (Work is energy without entropy, essentially.) Two simple examples of ways to produce work are his example of the Carnot cycle, and one which is even simpler, the hydroelectric dam. (Actually, the close analog is to an old-fashioned miller's dam where you could use the water power directly for any of a variety of processes.)

The primitive process used by the cell is like the latter. There is a semi-permeable membrane which allows it to accumulate a chemical potential difference, which then allows the cell to use work for all kinds of purposes: proofreading its own transcription and replication, motion of cilia, and so on. I do not see how a "dissipative structure" can play the same role. I was alerted to the necessary role that a source of work has in making information processes sufficiently accurate for life by Hopfield's marvelous paper on the subject.

The role of work seems so fundamental that one is tempted to suggest that the prime requisite to get evolution started may have been the formation of liposomes, which are an equilibrium structure under some circumstances. This could provide two of the essentials: individuation, without which natural selection can't operate, and a possible source of work in a chemical potential difference between inside and outside — which might in the first instance be simply used to maintain the individual's integrity.

Looking at actual life processes at a fundamental level, when they finally got started, it seems that there is inevitably a cyclical progression, as Stu emphasizes (following, to some extent, Iberall's point of view of a couple of decades ago — though I think Iberall overmechanizes and over-generalizes, trying to reduce the true complexity of life to a Marxist simplicity). The result of a process which is driven out of equilibrium, as for example replication, is captured in an equilibrium structure, such as the DNA double helix of the genome, which then controls nonequilibrium processes which act on the environment to generate the next stage — and so on. Even when patterns are formed in nonequilibrium processes, as in the example given above of reaction-diffusion gradients, or in the cellular automaton which patterns certain shellfish shells, these patterns are soon captured in a permanent, equilibrium structure.

At quite high levels of organization, animals use the temporal patterns that are set up by quasiperiodic attractors (for more on attractors, see below) for life processes — our hearts seem to be controlled by one, and speech seems to be a sequence of dynamical attractors which are modulated via control parameters — but in each case the organism depends very much on careful feedback control, and disaster happens when the nonlinear oscillator is left to run free.

(3) The inklings of a general law for open systems came to me in the period when people had all become fascinated by the "chaotic attractor" and the whole concept of attractors in dissipative, dynamic, driven systems. It also has a component which comes from John Wheeler's

discussion of why the arrows of thermodynamical time as well as physiological time have to be the same as that of cosmic time.

The thing which most puzzled me about dynamical system attractors was just that they were always attractors. When I asked why, I was pityingly assured that they were dissipative systems, and that took care of it. Of course it does; but that fact reveals that to decay down to an attractor, we must have an open system, which is shedding entropy to some microscopic system of degrees of freedom which in turn are passing it off to some external reservoir. If some degrees of freedom are becoming more restricted — the attractor has fewer dimensions than the space — by Liouville's theorem the system must have expanded in some other dimension of phase space, which is the fundamental nature of dissipation.

When you really analyze what kind of system one is thinking of, you realize that it is being driven by applying work — that concept again! — which is energy without entropy — energy at the absolute zero, really. One assumes a more or less steady state, so the energy has to be radiated away, otherwise the temperature of the bounded system would rise indefinitely. But of course that energy goes out at the temperature of the system, so carries entropy with it. The "system", therefore, has to lose entropy so it will approach the least entropy state available, the attractor, and get ever closer to it.

The simplest attractor is a point attractor, an equilibrium condensed state or, for a driven system, a steady state such as laminar flow. In the former, there is no dissipation so there is no need for driving work. The system just cools down to the temperature of space, in the end. The next simplest attractor is a limit cycle, and finally there is the chaotic attractor, which happens depending on whether the dissipation can balance the input of work, and more complex motions usually expedite dissipative transport so tend to occur in strongly driven systems (not, however, with the force of a variational theorem *à la* Prigogine.)

Thus, what seems the essential feature about open systems is that the local system — the part which concerns us — will in general go toward a less probable state, an attractor in the case where the phase space of the local variables is restricted, something more complicated if we are dealing with a more complicated system — such as a turbulent fluid, or on the cosmic scale the formation of galaxies and stars. This point of view has resemblances to Bak's idea of self-organized criticality, but is not as specific. The idea of Wolfram, paraphrased as that complexity is the inevitable outcome of the operation of natural processes, is also related in that it is a somewhat oversimplified version of the same insight. (Oversimplified because he stacks the cards by limiting himself to automaton models which are not necessarily equivalent to real physics.)

The relationship to Wheeler's arrow of time is that of course this only works in an expanding universe — or perhaps the only universe we could inhabit is expanding — because the entropy has to have available the final sink of radiation into space.

(4) This makes room for a remark. I do not see how life can have developed on the basis purely of the selection of variations alone, at least in the simple, straightforward sense of a pure evolutionary tree structure. Of course, this is to an extent a truism, in that Margulis has essentially proven that the eukaryotic cell is a synergy of at least two more primitive organisms. But I think that synergy — which can be merely a polite name for parasitism, of course — must have happened repeatedly in the course of evolution. The near universality of the genetic code, on the basis of the pure evolutionary tree, would imply that all of life arose from a single organism — one would have to ascribe any of the known variations, in mitochondria for instance, to later mutations. This seems to me implausible.

In particular, it is hard to see how replication and the work mechanism can have been evolved together. But very soon they have to be coupled, because Hopfield has shown that nucleic acid chemistry cannot be accurate to much better than 1% without coupling to a reservoir of work, and other types of catalysis can be hardly better. Thus no organism containing more than 100 bits of information could evolve by chemistry alone. I have conjectured that one of the components of the original mix was a kind of "chemical factory" producing a rather arbitrary mix of nucleic acids, short protein chains, phospholipids etc. more or less by means of the autocatalytic mechanism Stu emphasizes in his books; then some of these must synergize with our work reservoirs.

(5) One of the deepest and most puzzling questions about life has to do with how various forms of autonomy and teleonomy (in SK's phrase, the ability to act on one's own behalf) arose. It is essential that there be autonomy — separation into individuals — for any meaningful evolution to take place. As I pointed out above, the universal cell wall of lipid bilayer is a good model of what got things started, in terms of both individuation as well as of the possibility for energy storage. In, possibly, some synergistic event, such cells could acquire molecular information and the ability to replicate, which might have developed separately via some "algorithmic chemistry" scheme. Then how do they begin to act on their own behalf?

One point which is rather neglected is that there are a number of ways in which they can do this, and not all of them are as complicated as human selfishness or greed. In particular, there is a whole kingdom full of organisms which simply grow: the plants. Plants basically try out a selection of the alternative paths which are open to them, and the shoots which succeed become the future plant, the shoots which fail die off. Of course, modern plants are a little more sophisticated than that — they bias their search by various sensing methods, and have a number of automata-like mechanisms for distributing their effort well, but basically that's it.

This method is remarkably close to simple Darwinian selection among a reproducing population, like mildew growing on a wall; the only added sophistication comes from cooperation among the cells of a given organism, with for instance branches and roots seeking different goodies for the benefit of the whole. So in that sense teleonomy can be a simple consequence of selection — plus complexification, however, the idea of a multicell organism as a unit.

Interestingly, the cell itself has developed a more sophisticated scheme, which also resembles that used by certain bacteria. Microtubules, which are the structural skeleton of the cell and help organize vital processes such as cell division, are molecular cylinders which grow from certain centers implanted in the cell membrane. They seem to start out growing in arbitrary directions, and extend quite rapidly. But if the extension encounters no positive reinforcement such as another cell wall, the tubule decides to ungrow, equally rapidly and by an active process, and start out in another direction. This is again a testing of alternatives by going in arbitrary directions, but with the testing probe retractable. There is an analogy with the behavior of ciliated bacteria, whose mode of finding food sources is to start out swimming in an arbitrary direction, but to stop if the food gradient is unfavorable, spin around to another random direction, and start again. I think it is really an important matter that the organism is exploring a whole pathway rather than exhibiting a tropism toward a preferable environment, which is more plantlike.

None of these cases are exhibiting what we might think of as free will. Clearly, the behavior is computationally so sophisticated that it couldn't be reproduced except statistically without going through the same process — so that disposes of Stephen Wolfram's conjecture about free will — but somehow something further is needed.

The next step seems a long jump. When in the fossil record it occurred could be the subject of a delightful amount of controversy. This step is the virtual exploration of alternative pathways. Of course, as evolution acted the sense organs for discerning which pathways were good and which were bad improved, but that does not involve a different principle. The only new principle comes when we're not actually there, but only imagining possible courses of action. (Actually, Lakoff tells us that we often activate the actual neuron systems, even at the human level, when we speak — or presumably think — about an action without carrying it out.)

We know that this is something which we do, and in doing it we feel that we are exercising free will in making our choice after considering all the consequences that we may see. And, probably, that's the end to it. In a sense, Wolfram's conjecture is more or less right — nothing can work out what your decision will be except the same computer programmed in the same way, namely you.

It is reasonable to me that the same kind of choice is made by at least the higher animals, as anyone who has owned a dog must testify, watching it go through agonies of indecision as to whether to obey or not. Does it come along with the cortex of the brain? — I am hardly the person to answer this.

Conclusion

I offer my apologies to those who take a much more professional view of evolution, and for the last section to the neuroscientists. The point I am trying to make with most force is that there are definite physical and logical constraints on the course of early evolution and the nature of life and that many conjectures in the past have seemed to me unrealistic.

V. Genius

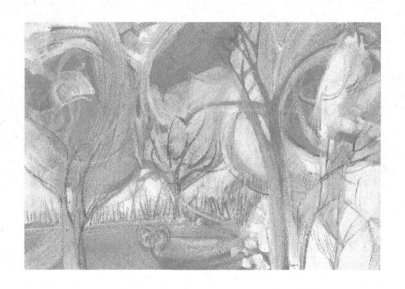

Introduction

One of the privileges of being a theoretical physicist is that one gets to know numbers of certified geniuses. (Not that all theoretical physicists are geniuses, or even very many of them.) I have had the opportunity to review many biographies of my colleagues and friends, even several titled "genius": just *Genius* in the case of Richard Feynman, for instance, and also *True Genius* about John Bardeen, and *Broken Genius*, about Bill Shockley. There were two books about Murray Gell-Mann — his own, *The Quark and the Jaguar*, and George Johnson's superb *Strange Beauty* — who surely fits in this category if only because of his long-term relationship to Feynman. The only other individual to whom I have publicly awarded this soubriquet is Francis Crick, and so I will include my review of his autobiography, *What Mad Pursuit*. I have reviewed books about or by a number of other theorists, and these may be included in other sections of this collection; and, Lord knows, I make no invidious comparisons; but these are the particular cases in which the "genius" question has willy-nilly arisen.

Dick Feynman was simply overwhelmingly competent at whatever he chose to do — safecracking, or bongo drums, or — a story I never have seen publicized — he trained himself to recover primitive man's sense of smell, and as a party trick would identify by smell objects — say books — handled by specific individuals in the room. Or leading a group at Los Alamos, or establishing instant rapport with the space engineers who really knew about Challenger. If he had wanted to become a world icon like Einstein or Hawking, he could have done that too — but he loved his anonymity, that allowed him, for instance, to join — and charm — my daughter's drawing class during a summer at MIT, with no one in the class thinking of him as anything but "the geezer". He managed not to be an "intellectual" but to be in fact an overwhelming intellect — to, for instance, go deeper into the theory of computation and of information in the quantum theory than anyone before him. I knew him personally only late in his life, and we argued a bit, in a friendly way, about these aspects of theory — and I have an uneasy feeling that he may have been right.

Murray, in his way, was much more of an intellectual, in that he did not really pick and choose what to be competent in: he felt he had to really know everything worth knowing, and he came as close to that impossibility as anyone I've ever heard of. One of the most unexpected thrills of my life was when, at one of the SFI workshops — I think it was a Sloan-sponsored one on "The Limits of Knowledge" — someone asked a question to which the obvious answer was a description of my concept of emergence — i.e. the "more is different" spiel. I was about to open my mouth when suddenly Murray launched into a description of my point of view which was certainly superior to anything I could have composed on the spur of the moment. He gave no indication that this was what he personally believed — nor did he give me any credit for it. Yet such a performance would have been beyond any of the defenders of strong reductionism such as David Gross or Steve Weinberg, otherwise brilliant individuals who to this day seem never to have understood that particular message.

When the discussion came up of personnel for a new workshop at SFI on some topic, far from Murray's presumed expertise, he would drop his copy of *The New York Times,* which he enjoyed putting us down by reading in steering committee meetings, and suggest four or five relevant scholars; the unfortunate organizer would say — oh, I didn't think of them — what a great idea!

Yet woe betide he who does not pay the appropriate homage to Lord Murray. I made that mistake — at least once. I organized my own work-shop in an area that I could presume to be competent in, and which vaguely had to do with physics — I believe it was pattern formation, non-linear dynamics, and self-organized criticality. Somehow this workshop, though quite memorable *per se,* never received Murray's imprimatur — and has consequently virtually disappeared from the proud history of SFI. (M. G-M, I believe, later ran an imitation to which I wasn't invited.) But yes — his total contribution to science — including his hobby-horse at SFI, the "complex adaptive system" at which straitjacket I sometimes pub-licly bridled — was greater than Feynman's.

"True" vs "Broken"? Broken or not, one must come to the defense of Bill Shockley who has been accused, for instance, of being a "mere engineer" (Laughlin). The Bell Laboratories in its forty-year heyday before the 1984 breakup was an unparalleled scientific institution and the physical sciences part of that was largely due to Bill Shockley. At the time he hired me, responding to special pleading from my professor, his biography tells us that he was deeply involved in advising the military as well as running an enormous group focused around semiconductors and the physics opened up by the transistor discovery, and at the same time writing the textbook for that field; but it does not tell you that he was full of exciting new ideas in unrelated fields of solid state physics: magnetism, ferroelectricity, cyclotron resonance,... for which he was hiring people like myself, Jack Galt, Bernd Matthias, Harry Suhl, who became the nucleus of Bell's years of greatness.

Yet Bill's hubris, which was to be his nemesis, was already evident, in hindsight. When individuals of this quality took on a project the full time exercise of their talents was almost inevitably going to transform it into something probably better than, but certainly different from, the original brilliant insight of a few moments' thought by Bill. But Bill could not love someone else's idea — from that moment he seemed determined to get rid of the offender, who at Bell could be rescued as I was by his co-managers, but in normal industry went off to found another competitor for Shockley Semiconductors.

John Bardeen: it is almost a cliché to remark that the quality of his genius is evident to everyone who has a connection to UIUC (University of Illinois, Urbana-Champaign), but not necessarily to everyone else. He fled to UIUC as the first — though possibly not even the most prominent — victim of Bill's hubris, and promptly earned for them the unique NPII — no wonder they loved him. On the other side were especially the Russians — his modus operandi in theoretical physics was so utterly anti-Landau that they could never believe he was anything but lucky. To them he was a "mere engineer", exactly what Laughlin accused Bill Shockley of being.

Yet that negative characterization is not true either. I had lots of occasion to use the details contained in the BCS paper, and concealed in these humdrum-appearing calculations is brilliant insight after insight. (Of course, one of John's quirks which must have annoyed the Landau gang was that he always referred to theories as "calculations".) And the Landauians' nose was out of joint — not only had John scooped them, Bogoliubov, given a hint, reproduced the whole theory, with improvements, in a few months, while the Landaus were snubbing it, possibly because of their low opinion of Bardeen.

What that paper reveals, really, is what John was best at: worrying and worrying the experimental data surrounding a puzzle until there was no doubt in his mind what were the requirements of a true theory. If he was a genius, he was a special variety; uninterested in abstractions like the marvelous explosion of theory which ensued, the "coherent states" of which this was really the first exemplar, and the extensions to other fields; he liked most the "coherence factors" with which you could calculate real things and get them *right*.

But there is a sense in which the Russians were right, too: John's two prizes were for discoveries the implications of which transformed all of technology and all of physics. He may be excused for not having a big effect on semiconductor technology — Shockley turfed him out of that very firmly and fast. But it was clear at the recent gathering for the 50th anniversary of BCS, for instance, that John had had little direct influence on the intellectual ferment caused by BCS. The emissaries to the wider world of physics were Schrieffer, Pines, myself, and the Landau group, and not a single substantive further development has his name on it.

What Mad Pursuit*

Francis Crick is known for his occasionally acerbic wit and his breezy manner, which has been interpreted on occasion as arrogant or frivolous. This reputation did not lead one to anticipate this graceful, pleasantly modest, serious autobiography, written for the Alfred P. Sloan Foundation series that seeks to promote public understanding of science. In my small experience with Crick, I have always perceived science as totally central to his being, as it is to any great scientist. His autobiography can be read as a manual (or in his case a record) of how to change the world through science. He brings in enough personal detail to give one a sense of a life fully lived, but the focus is on the science.

Crick tells us how at ago 30, after a somewhat undistinguished undergraduate degree in physics, a start on obviously boring research and a stint of building and designing mines during the war, he found his vocation by means of the "gossip test": "What do I most enjoy gossiping about?" Crick's answer: how to determine the chemical basis of life.

He seems to have set out his basic program by about 1948–49, after intensive reading and some biophysical research experience. Four years later, still an overage graduate student in the Medical Research Council unit at the Cavendish Laboratory, Crick, with James Watson, proposed the double helix for the structure of the genetic material DNA — the first of the several breakthroughs associated with Crick that did, in fact, change the world. One is reminded of Einstein's years in the patent office, forming underlying viewpoints that led to the extraordinary burst of creativity that created modern physics. Crick is not Einstein — for one thing,

*Book review of *What Mad Pursuit: A Personal View of Scientific Discovery*, by Francis Crick (Basic Books, New York, 1990). Reprinted with permission from *Physics Today*, Vol. 42, July 1989, p. 68. © 1989 American Institute of Physics.

most of Crick's achievements were collaborative — but one can argue that the 13 years that followed the revelation of DNA's structure had at least as much effect on the world as Einstein's great decade, both in a practical and in an intellectual sense: Biology became firmly rooted in chemistry and physics, and much of the mystery was removed from evolution. Crick's penetrating, critical intelligence illuminated the whole of that remarkable "classic" period of molecular biology, and in his book he runs us through several of the discoveries of that era as they happened.

Crick places the greatest emphasis on matters of scientific style, intellectual — even philosophical — content, and personal and psychological attitude. Do not be afraid of making mistakes, he says: No single idea, no matter how brilliant, is going to solve a hard problem; persistence is all; evolution seldom chooses the elegant solution. He stresses that "professionals know that they have to produce theory after theory before they hit the jackpot." But, he also stumps for another very important type of professionalism: knowing *all* the relevant facts, ideas, constraints. In his case this meant knowing the biological evidence, the stereochemistry, the way nature usually works and so on. To my mind, the greatest weakness of young scientists in *any* field is a reluctance to go outside their narrow professional specialty for evidence and ideas. Crick emphasizes these rules as pertaining to biological research, but I think he underestimates the complexity of some of nature's other manifestations. Mathematical elegance is sometimes useful in physics, for instance, but much less often than he seems to think, relative to a knowledge of empirical fact. I can hardly imagine a field of natural science where his advice would be irrelevant, and I would strongly recommend this book to a young theorist in any discipline.

Do not be misled into thinking the book has no fun in it. There is, for example, the chapter in which he reviews books and movies based on the story of the discovery of the double helix rather than retelling the real story himself, and the chapter called "How to Live with a Golden Helix", which contains the following charming passage: "Finally one should perhaps ask … am I glad that it happened as it did? I can only answer that I enjoyed every moment of it, the downs as well as the ups.… But to convey my own feelings, I cannot do better than quote from a brilliant and perceptive lecture I heard years ago in Cambridge by the painter John Minton in which he said of his own artistic creations, 'The important thing is to be there when the picture is painted.'"

And this: "There was in the early fifties a small, somewhat exclusive biophysics club at Cambridge, called the Hardy Club.… The list of those early members now has an illustrious ring ... but in those days we were all fairly young.... Jim was asked to give an evening talk to this select gathering. The speaker was customarily given dinner first at Peterhouse. The food there was always good but the speaker was also plied with sherry before dinner, wine with it, and, if he was so rash as to accept them, drinks after dinner as well. I have seen more than one speaker struggling to find his way into his topic through a haze of alcohol. Jim was no exception. In spite of it all he managed to give a fairly adequate description ... but when he came to sum up he was quite overcome and at a loss for words. He gazed at the model, slightly bleary-eyed. All he could manage to say was 'It's so beautiful, you see, so beautiful!' But then, of course, it was."

Complexities of Feynman[*]

"Complexities of Feynman" is Science's non-descriptive title for my review of James Gleick's Genius, a biography of Feynman. This book is a classic in every sense, in the quality of the writing, the depth of the research and analysis, and the importance of the subject — perhaps the only man since Einstein who could unquestionably justify the unqualified title.

Richard Feynman, the subject of this extraordinary, complex, but very readable book, was, of course, not just the joker and self-created legend of *Surely You're Joking, Mr. Feynman!*, nor was the celebrated incident of the Challenger Commission, described in its sequel, *What Do You Care What Other People Think?*, more than a footnote to his real career. Feynman was undoubtedly one of the great geniuses of modern physics, one of the finest among an extraordinary generation who built, on the basic framework of quantum theory and relativity, the accurate, detailed, wide-ranging, and beautiful edifice that is modern physics. His personal aura was such that, as Gleick relates, his mere presence changed the sound level in an auditorium or caused a visible stir in a student cafeteria. And for all his determined pose as a "natural man" with the unmistakable accents of Far Rockaway in his voice, Feynman was one of the greatest expositors of the methodology and of the beauty of science.

In Gleick's account we follow Feynman from Far Rockaway, MIT, and Princeton to his extraordinary two years of leading the computational group that made the crucial calculations for the atomic bombs

[*]Book review of *Genius: The Life and Science of Richard Feynman*, by James Gleick (Pantheon, New York, 1992). Originally published in *Science*, Vol. 259, 1993, pp. 537–539. Reprinted with permission from AAAS.

(calculations that, incidentally, were unprecedented in method and scope). Feynman's wartime boss, Hans Bethe, then persuaded him to come to Cornell, where after a brief sterile period of decompression he solved the difficulties of quantum electro-dynamics in a burst of renewed energy, using a revolutionary methodology that remains at the core of all of quantum physics (which is of course all of physics, in a sense). Then to Caltech, via an episode in Brazil that included a stretch in a samba band, and to a long career of further research that alone would have been distinguished enough to win him the Nobel Prize — say, for his quantized vortices in liquid helium or for the theory of the weak interactions, for which Murray Gell-Mann and he invented the crucial apparatus of current algebra.

Gleick resists the temptation to simplify his main character. Was he admirable for his unswerving honesty and refusal to sanction any kind of bombast or irrational cant — even refusing to say the Kaddish over his father's grave — or reprehensible for his attitude toward women and his rejection of conventional aesthetics? Admirable for his eagerness to talk to undergraduates and his democratic associations with samba bands, bar girls, and the Esalen Institute, or reprehensible for his behavior toward the many colleagues whom he almost deliberately insulted or intimidated and for his refusal to take on the social responsibilities — refereeing, hiring, memberships, recommendations, and all that — of his profession? We are left free to decide for ourselves.

Feynman's relationships with women are, also, left deliberately ambiguous. He had two touching love affairs and successful marriages, the first with his boyhood sweetheart, who was dying of tuberculosis as he labored his 16-hour days at Los Alamos, and the second with the wife of his last three decades, who started as his au pair, imported by legal legerdemain, but whom he came to praise loudly and publicly as "a great English lady". In between were a disastrous marriage, in itself almost an expression of contempt for women, and a succession of predatory affairs with women who sometimes bitterly reproached him; not to mention the bar girls. Yet, in the notes to the book it is revealed that one of these women friends was, much later, instrumental in his being chosen for the Challenger Commission. On the whole Feynman gets by on the Henry Higgins defense: male or female, obscure or influential, he treated everyone with equal rudeness.

Gleick sets his hero off against capsule sketches of three other extraordinary men: Julian Schwinger, Freeman Dyson, and Murray Gell-Mann.

The competitor in Feynman's greatest work was Schwinger, a mathematical juggernaut, suave and elegant, "speaking in full paragraphs", a prodigy recognized by Rabi at age 14, while Feynman seemed rough, crude, and inelegant. Schwinger was the same age as Feynman, and his advantages seemed so close to what Feynman's mother felt she might have been able to provide for him had the family's fortunes not declined that she kept track of their relative progress throughout her life. Ironically, though Schwinger disdained low-energy physics, his methods (when combined with Dyson's and Feynman's) and his students found, in the end, their greatest influence there. Already in 1956, when most of theoretical physics met at the Seattle international congress at which Feynman gave his famous talk on superfluidity and superconductivity, Schwinger seemed to be dropping away from the leading edge. Feynman, by contrast, to his great credit, disavowed the arrogant snobbishness of the particle theorist, paralleling Lev Landau in his eclecticism and excelling in his feel for experimental details and for engineering.

The one of the three who is most clearly also a genius is Gell-Mann, an equally complex and difficult character who, like the later Feynman, seemed to have "created himself as a mask and then became indistinguishable from it". Their relationship is one of the fascinating points of the book; a long-term association with another genius on the level of equality must have been very difficult to maintain on either side. Gleick resists the temptation in describing the physics to oversell Feynman's wonder-working and gives Gell-Mann, as well as Dyson (and also Lars Onsager in the theory of superfluidity), some measure of appreciation.

The core theme of the book, as proclaimed in the title and as emphasized by the inclusion of these alternative models, is the nature of genius in general and of Feynman's in particular. Extensive quotations from other writers on the subject as well as from Feynman himself end with Feynman's cryptic remark, "no one is that much smarter". This, to me, implies something Feynman kept emphasizing: that the key to his achievements was not anything "magical" but the right attitude, the focus on nature's reality, the focus on asking the right questions, the willingness to try (and to discard) unconventional answers, the sensitive ear for phoniness, self-deception, bombast, and conventional but unproven assumptions. Somehow a mind utterly devoid of all the standard distractions is rare enough — if it has adequate quickness — to be called "genius" and to produce results as if by magic. It seems to me that it may have been because of this knowledge of his own finiteness that Feynman could not bear to see minds as good as his own hobbled by unnecessary nonsense, and this in turn led to some of his rudest outbursts.

The review of this book in *The New York Times*, which seemed to treat this as a popular physics book, on the model of Hawking's *A Brief History of Time*, was wide off the mark. This is not intended to be a book to learn physics from, though in fact Feynman's work lends itself well to verbal description rather than to a cursory mathematical one. On the whole the book is not scientifically inaccurate; it gives the non-specialist a good overview of the essentials and even gives some credit to others, as I remarked above; at worst, it glamorizes some of Feynman's minor works a bit. It deserves to be a bestseller as a good, well-written biography of one of the century's outstanding geniuses, not as coffee-table science.

Coffee-Table Complexities[*]

To describe how theoretical physicists feel about Murray Gell-Mann, I think back to the era when he was inventing the quark and various other essential components of the "Standard Model" of the elementary particles. It was then widely agreed that the truth about the elementary particles would be quite hard to understand even after Murray revealed it to us.

In this remarkable book, Gell-Mann has taken it upon himself to "reveal" an even wider slice of the cosmos, from cosmic beginnings, the nature and interpretation of fundamental physical laws, through the origin of life and evolution, to culture, ecology and economics; and, for his peroration, he switches from dispassionate observer-scientist to environmental activist.

There can be no doubt in the mind of anyone who has worked with him, as I have, that if anyone could have accomplished the task he has set himself it could be Murray, and that he has a more profound knowledge of a wider range of subjects than anyone living. But he has taken on a nearly impossible task, and he is handicapped as far as the layman is concerned by his unwillingness to resort to the more superficial and available level at which Stephen Hawking and others cover a small portion of the same material. Nonetheless, before going any further let me say that, because of the reach and scope of what is attempted, this is certainly the ultimate book that should grace your coffee-table.

There were several books I thought Murray might be writing: an autobiography, which might give one some insight into a remarkable person

---

[*]Book review of *The Quark and the Jaguar: Adventures in the Simple and the Complex*, by Murray Gell-Mann (Henry Holt, New York, 1994). Originally published in *Physics World*, Vol. 7, August 1994, pp. 47–48.

and how he got that way; a personalized stroll down the "Eightfold Way" (his title for one of his major contributions to the Standard Model); or a description of his crucial role in two important institutional innovations during his later years — the MacArthur Foundation and the Santa Fe Institute. These are each touched upon but relegated to the later personal memoir; certainly the account of his youth is inadequate, and the SFI is described in terms of Murray's own interests rather than as a fascinating, complicated interaction of many people.

He has attempted, rather, to lead off from books such as Hawking's *A Brief History of Time* or Weinberg's *Dreams of a Final Theory*, but with the added dimension that almost half of the book deals with matters of complexity. Here he selects for special emphasis ideas on measures of complexity based on information theory, as extended by Bennett and others, and his own preoccupation with the "complex adaptive system", a generalization which attempt to encompass all objects or systems that can learn from the environment and act accordingly — such as ourselves, ecologies, or primitive living organisms. His idea of the "schema" or compressed, internally generated description of the environment is crucial.

To organize a description of the scope of the book we can use the table of contents. The first section, "The Simple and the Complex", is a fascinating mélange, including a few personal details as well as general remarks on the nature and power of theory and the scientific enterprise, and on measures of complexity and on complex adaptive systems. The second, "The Quantum Universe", encompasses not only the fundamental laws as now known but also the philosophical basis of quantum mechanics, cosmology and "Time's Arrows", and possible future shapes of theories. "Selection and Fitness" is a view of biological evolution but also contains discussions of superstition, "maladaptive schemata" and lateral thinking. Finally, "Diversity and Sustainability" is Murray as environmental activist and policy advocate.

This is a book, in sum, with so much intellectual meat to be tasted and digested that one must heartily recommend it to anyone who wants to think about the nature and philosophy of science or of the most important modem ideas. It is written in a lively style with a good eye for humorous, illustrative anecdote. Finally, as a member of the Santa Fe Institute myself, I am heartily in agreement with the emphasis on complexity and on the broad, general view.

I do see weaknesses in the book. The first it has in common with Hawking and others: in the end, even a mind as brilliant as Gell-Mann's, when it gets to the end of what is generally agreed, begins to follow the well-worn trail into its own special research programme, without a clear warning that what follows is not completely accepted. This is most certainly true of the Hartle–Gell-Mann theory of measurement in quantum theory; this is particularly unfortunate when we are entering the era where the questions it raises are becoming subject to experimental test. Furthermore, the complex adaptive system and its "schemata", as well as Bennett's measures of complexity, are only one of several promising ways into the understanding of complex systems.

More serious still are a few spots where even Homer nods. In a passage on p. 151 he refers to the creeping helium film as a "bizarre quantum effect ... associated with low inertia". This lack of the appreciation of broken gauge symmetry is crucial to the issues in measurement theory which his ideas finesse.

The chapter on "Time's Arrows" is a confusing blend of speculation and possibly wrong ideas. Missing entirely in this context is the crucial role of cosmic expansion, which in fact is explicitly denied in favor of a mysterious (equally to the physicist, as to the layman) "boundary condition". The universal cooling and increase in order of condensed bits of the Universe, which leads in time to astronomy, geology, life and us, is a consequence not of the closed-system equilibrium that is so lovingly described, but of the loss of entropy to radiation, which causes successive breakings of the primordial symmetry, which constitute cosmic history. What he calls "signals" and "traces" would evaporate in a Universe that was heating up.

Finally, there is a comment I feel I must make: the "missing acknowledgement". The SFI in which all of this book was written, and which much of it describes, was not created by Murray Gell-Mann alone or even primarily. I would have liked to see acknowledgment of the four people (at least) who kept the machinery going for its first formative five years, and instituted many of its programmes and practices; George Cowan, president, L. M. Simmons, vice-president, and Ginger Richardson and Andi Sutherland, administrative facilitators. Other names are noticeably missing, but this is, of course, not a scientific treatise in which all sources of ideas must be credited.

But in conclusion, however, I may carp at details, a book like this, from an intellect like Murray's, is an event!

Search for Polymath's Elementary Particles*

A biography of Murray Gell-Mann is long overdue, and this book by George Johnson, a science writer and correspondent for *The New York Times*, successfully fills the gap. Those of us who had hoped that Gell-Mann's own book, *The Quark and the Jaguar*, would give us insight into the man and his role in the triumphs of 20th-century physics, were disappointed for reasons that are nicely laid out in this book. Murray (often referred to here as "MGM") is to be congratulated for his generous cooperation with what is not an "authorized" biography, as he himself has assured me.

The literate public will hardly have heard of Gell-Mann, and few who have heard of him will have much idea of what he has done. Physics in the 20th century has undergone not one but a series of revolutions, and the physics of today differs from that of Einstein, Heisenberg and Dirac almost as much as their physics differed from its "classical" antecedents. Gell-Mann dominated the collective of remarkable scientists who made the most recent revolution, as Einstein did the earlier one. But unlike Einstein or, later, Richard Feynman, MGM has refused to settle down to become an even superficially relaxed old character.

In addition to his involvement with the notorious (to some) "Jason", a division within the government's Institute of Defence Analysis, and with the president's Science Advisory Committee, and then his work in the world of foundations, he has, over the past decade and longer, joined in founding an institute at Santa Fe with the aim of fomenting intellectual revolutions across the board. Meanwhile, he continues avidly to pursue

---

*Book review of *Strange Beauty: Murray Gell-Mann and the Revolution in 20th-Century Physics*, by George Johnson (Knopf, 1999). Originally published in the *Times Higher Education Supplement*, 14 April 2000.

archeological collecting, bird-watching and obscure languages and to work substantively on his favorite cause, wilderness conservation.

The story starts in New York with a lonely prodigy's childhood, his family displaced gentry from the far reaches of the Austro-Hungarian Empire. His pedantic and perfectionist father was reduced to language teaching, inserting the hyphen into Gellman to enhance the name of his ultimately unsuccessful Manhattan language school.

Helped into Columbia Grammar School by relatives, he graduated from school at 14 and from Yale University at 18. He attended both on full scholarships: even then it was obvious he was uniquely bright, if compulsive about setting others right (a habit he retains to this day). Two-and-a-half years later (six months over his private projection), he got his PhD from Victor Weisskopf at the Massachusetts Institute of Technology, who sent him off to Robert Oppenheimer's Institute for Advanced Study at Princeton, where, rooming with Francis Low, fresh from his PhD and ten years MGM's senior, they together wrote the first of the remarkable series of papers that made MGM's reputation. As Johnson notes, this first paper was no slouch: later it was metamorphosed into one of the basic methodologies of modern physics in the hands of Ken Wilson and others.

The following chapters will divide readers into two categories. There will be those like myself, for whom they constitute a compelling page-turner — for example, those who have followed the Standard Model of the elementary particles or have read one or more of the popular accounts of its genesis. But for the general reader, this will be a challenging initiation into what is one of the great intellectual achievements of the human race, no less than the discovery of the underlying structure of the physical universe. I suspect this general reader will be rewarded with some sense of the intense struggle that led to this synthesis. It is an account that nowhere falls back into mystification or solecism and is a masterpiece of scientific explication for the layman.

Overall, *Strange Beauty* is a remarkably accurate study of the psychology and sociology of competitive scientific research at the highest level, interspersed with incidents in MGM's personal life as the story unfolds, all in the course of traversing the world from one physics meeting, or centre, to another.

At the time, it seemed like a succession of long intervals of bewilderment between short bursts of discovery, the latter mostly involving the proving in of a new superaccelerator and, usually, MGM and someone else sorting out the results. Actually, it was not more than 15 to 20 years, from the first burst of new quantum numbers and particles appearing in about 1953–54 just as MGM came into full possession of his abilities, through the discoveries of the electro-weak theory in 1967 and of quantum chromodynamics in the late 1960s to early 1970s, culminating in the discovery of asymptotic freedom in 1973.

As the 1970s wore on, we all came to realize that the Standard Model, as the result had come to be known, would not endure further revision without its being replaced by some higher-level theory of completely different form, such as "superstring theory". Even with string theory, MGM played an essential leadership role by setting up a "cocoon" at Caltech, where the early workers on string theory could labor undisturbed in the decade 1975–85.

"And now for something completely different." MGM has always had wide-ranging curiosity about the world, including roughing it in areas of great wilderness. He tended to become almost professionally knowledgeable about any field that interested him, be it history, archaeology, psychology, or literature. I think of him as a true intellectual, as opposed to the literary intellectuals of the "chattering classes". This breadth stood him in good stead on the board of the MacArthur Foundation, which led to a wide acquaintance among leaders in many fields. (It is one of the few disappointments of the book that it passes over the MacArthur experience very lightly.) MGM was, therefore, the ideal person to provide intellectual leadership for the interdisciplinary initiative that became the Santa Fe Institute, an establishment that has had an influence totally out of proportion to its minute budget in fields as far apart as immunology, archaeology, economics and quantum epistemology — the aspect of his life that I have, to an extent, shared.

For those readers who have little interest in modern physics, there remains the elaboration of MGM's complex personality. The enormous intelligence, coupled with an eidetic memory, which brought him through what could have been a disabling childhood, still could not always gain him empathy. There are a considerable number of solid, long-term friendships and relationships, such as those with Murph Goldberger, the first mentor whom (I deduce) he could genuinely learn from, with his first collaborator Low, with Harold Morowitz from his Yale days, as well as with many students such as Seth Lloyd, and his family's closeness to David and Suzy Pines. His first marriage was a very close partnership; after Margaret's death he found partners and, eventually, a new wife, the poet Marcia Southwick. On the other side, there are a number of instances of collaborations that broke up with real rancour, notoriously that with Abraham Pais, involving MGM's first venture into new quantum numbers.

It is not atypical for him to have dubbed Pais the "evil dwarf" and maintained the nickname and the rancor for decades. Johnson tells this story even-handedly, and also that of the period late in the game when MGM's historical talks seemed to arrogate to himself even more of the credit for the Standard Model than he actually deserved. We all tend, perhaps, to feel that the notebook or the casual remark deserves credit as the moment of the discovery, but it does not. To keep one's equanimity one learns that "what you lose on the swings you gain on the roundabouts", and through his enormous reputation and his facility with names ("quark", "strangeness", the "Eightfold Way"), Murray is here shown to have gained more than he lost.

Johnson makes the point that Murray's perfectionism, perhaps owed to his father, has dogged him all his life. Again and again, in the course of creating the Standard Model, he backed off from publishing correct insights until compelled to by competitors, simply because of a reluctance to publish something that was not exactly right. For instance, where his work on quarks was, in the end, the ultimate vindication of quantum field theory, he was almost the last to stop treating them as a semi-fictitious aid to calculation. That perfectionism can lead to troubled relations with others is not surprising.

MGM's life had its share of tragedies, notably the early death of his wife Margaret and the estrangement for more than a decade from his daughter Lisa, while she became involved in a hyper-Maoist splinter party to a level of irrationality that must have been most painful to Murray. There were embarrassments, such as his failure to produce *The Quark and the Jaguar* in time to keep his record-breaking dollar advance (or to match the sales of Stephen Hawking, a scientist not of Gell-Mann's stature). Johnson relates how a succession of ghost writers found it impossible to deal with his perfectionism, his procrastination and his eagerness to be anywhere but at his desk — and, ultimately, his unwillingness to step out from behind his own persona. His book turned out to be a superb exposition, but lifeless.

Very amusingly told by Johnson is the tale of MGM's missing Richard Feynman's birthday as a result of a customs raid on his collection of pre-Columbian artefacts, some of which turned out to have been smuggled — the resolution being a substantial gift to the Peruvian national museum.

Finally, there is a lot about the fascinating relationship with Feynman, who brought MGM to Caltech and thus began a connection that lasted much longer than anyone could have expected but was hardly cosy. For these decades, these two geniuses, each eccentric and difficult in his own way, argued, collaborated occasionally and somehow managed to live across the hall from each other. MGM thought of Feynman (correctly) as a relative ignoramus outside of physics and a bit irresponsible. Yet it was Feynman who had the popular touch and became the best-selling author, which turned an already chancy relationship quite sour.

I got much enjoyment and enlightenment from this book. For anyone interested in physics and physicists or in the make-up of a certified genius, I consider it a "must-read". For anyone else, *Strange Beauty* has much to recommend it, even if it may be rather tough going in parts. Science writing doesn't get much better than this.

Giant Who Started the Silicon Age[*]

Genius? In the community of theoretical physicists all are geniuses — just ask a mathematician. But Bill Shockley, who invented the transistor, was something else again.

Joel Shurkin's biography follows the career of this remarkable, disturbingly flawed man from his lonely childhood to and through the 20 years during which he built Bell Labs' research up to the level it maintained for 50 years, then through providing the impetus for the phenomenon of Silicon Valley, but losing his company there, and his moorings as well; and finally, the self-constructed tragedy of his final campaign as the major domo of the eugenics movement — the campaign which will forever color his perception in the eyes of the public, and even his colleagues and followers.

I was hired by Shockley in 1949, at a time when, from the histories of the transistor breakthrough, he should reasonably have been completely consumed by the role he played in that history. In fact, he was involved simultaneously in exciting research in several other fields, recruiting spectacularly (for instance, four of us who went on to substantial careers of our own arrived at Bell on the same afternoon). Whatever scientific interest he picked up, he seemed well on the way to a solution before others grasped the problems. Almost incidentally, he was writing the definitive "bible" for semiconductor engineers — and at the same time, we learn, he was carrying a full load advising the military. It was an extraordinary time — physics was full of excitement and Shockley was in the thick of it — sometimes for better, sometimes for worse.

---

[*]Book review of *Broken Genius: The Rise and Fall of William Shockley, Creator of the Electronic Age*, by Joel N. Shurkin (Palgrave Macmillan, 2006). Originally published in the *Times Higher Education Supplement*, 16 June 2006.

Shurkin's access to detailed records made available by the Shockley family, including some previously unopened safes, as well as to oral history interviews available from many of the key figures, makes this a revealing book about Shockley's personal life as well as his scientific achievements. To his credit, he does not presume to come to any pat psychoanalytical conclusions, but allows us our own view of the man based on voluminous data.

Shockley, born in 1910, was the only child of an already mature and highly intelligent woman, married to a considerably older mining engineer whose money-making talents did not match his charm and erudition. The result was a somewhat itinerant childhood with few friends of his own age. Fortunately, the family exchequer extended to sending him to the then young, already vibrant Caltech as an undergraduate, where he could learn firsthand about the revolutions then taking place in modern physics.

In 1932 he and classmate Fred Seitz drove across the country together in his somewhat battered DeSoto, to graduate school at MIT and Princeton respectively. These were two centers where the new field of the quantum theory of solids was being born, and both young men were full, even key, participants as students (Seitz wrote the first textbook so titled, in 1940). Their journey together would not be repeated, though both returned to California for the next summer — in 1933 Shockley had a new wife, Jean, after a brief and somewhat lusty (according to the materials in those safes) summer courtship.

In graduate school began his lifelong near-obsession with physical fitness — he even earned the odd sum modeling, though he was no Schwarzenegger. His sport was rock climbing — he had routes in the Shawangunks and elsewhere that were named for him as pioneer. Also in this period he somehow became an accomplished magician, a skill he relied on in many — not necessarily appropriate — social situations, and, entirely appropriately, to charm children. In both avocations one felt the edge of a certain competitiveness.

By 1936 the Bell Labs, recovering from the Depression, had begun to recruit. The young scientist-executive, Mervin Kelly, on the advice of Shockley's mentor at MIT, Phil Morse, brought in Shockley and his friend Jim Fisk. Others hired at the same time included stars-to-be like Charles Townes and John Pierce. The idea was to see what the new quantum physics could do to replace the awkward technology then in use in the telephone system, in particular the ubiquitous vacuum tube. A number of veteran Bell scientists such as Walter Brattain joined these "Young Turks" in a research department which was to be the fore-runner of the fabulous postwar Bell Labs. This was an environment in which Shockley immediately assumed a leadership role.

An odd interlude — Shockley and Fisk (who later rose to be president of Bell Labs) were, for a month in 1940, withdrawn from their lab benches and given the ultrasecret assignment of assessing the consequences of uranium fission. The resulting report so resembled the design of the atomic pile later built by Fermi that it appears to have trumped all later patent claims, though it is not clear whether it ever broke through the fog of secrecy enough to be useful.

WW II broke up this unique research group, some staying on at Bell and, like Fisk, contributing importantly to radar. Shockley was recruited, again by Phil Morse, to join the nascent operations research effort (the Weapons Systems Evaluation Group — WSEG), then desperately trying to help resolve the U-boat threat in the Atlantic. Shockley became second in command. The achievements of this group in improving the performance of the Navy boggle the mind — for this alone the country owes Shockley a great debt of gratitude. The commitment of time and energy, however, led to increasing strains in his marriage — there was a young family to which he was able to give only a small fraction of his time, and the vacations at Lake George which had included many of their friends, such as the Seitz' and the Brattains, were a thing of the past.

After the war Kelly, then the president of Bell, and Shockley set out in earnest to create a group which could revolutionize technology and in particular return to the problem of creating the semiconductor substitute for the vacuum tube of which they had dreamed. Their stellar cast included recruit John Bardeen and a number who had continued doing war work at Bell, such as Brattain and other experts on semiconductor materials. Other noteworthy recruits were some of Shockley's acquaintances from the WSEG like Conyers Herring and Charles Kittel. It's worth emphasizing that semiconductors were not the only scientific interest; this is a misleading impression fed by the natural *ex post facto* bias of historians, and repeated in this book to some extent.

By Christmas 1947 Bardeen and Brattain had created the first semiconductor amplifier — and from that point on the triumphal story of Bill Shockley the wonder worker began to unravel. What Shurkin describes here is that the Bardeen–Brattain phone call to Shockley elicited not joyful triumph but blinding anger that the amplifier, which he had expected to be the crowning achievement of his life, had been stolen from him by scientists he saw as subordinate, even inferior. Shockley moved quickly to fill, and have witnessed, notebooks with extensions and generalizations of the original Bardeen–Brattain device in order to ensure a separate patent claim for his (undoubtedly superior) version. He soon isolated Bardeen and Brattain from the group he set up to exploit the breakthrough, and saw to it that the publicity still which was distributed showed himself at the bench experimenting, with Brattain and Bardeen in the background looking on.

It is ridiculous to argue, as some have, that because Bardeen and Brattain did not predict some details of the device they created, Shockley's behavior was partially justified. Their device by its very existence revealed the basic principle of "transistor action", the use of a forward-bias junction for injection, and started the semiconductor industry on its merry way. Bardeen's depth of understanding of semiconductor physics is manifested by a review he wrote in the next year with Pearson, as much a basic text for physicists as Shockley's book was for engineers.

At first things went swimmingly for Shockley. His residual group after the excision of Bardeen and Brattain (and a few misfits like Kittel and myself), augmented by well-chosen recruits, soon achieved steadily increasing control of their materials, and created his improved design, the junction transistor. By 1951, Bell Labs hosted a noteworthy symposium at which the transistor technology was thrown open to the world, the media began to notice — and in all this the visible leader was Shockley. Within Bell Labs, however, he had violated the cardinal rule of management that enjoined competing against a subordinate, and it was beginning to look as though his future was limited. He still had influence: I learned from this book that it was Shockley who pushed through the expansion of the salary range at Bell which took place in 1954–6, a significant and valuable move — but he would never rise to a higher level of management, and he was spending more and more time advising at the Pentagon and, ominously, developing his theories of "human quality".

Some of his colleagues had made profitable moves to the outside competitors such as Texas Instruments which were springing up in the now lucrative transistor industry, but only Shockley saw the enormous potential of what they had begun. In 1955, he began the process of forming Shockley Semiconductors as an independent company (funded by Arnold Beckman) in a small building in Mountain View, California, just south of Stanford. This was Silicon Valley's first startup.

At the same time, Shockley's marriage finally broke down, as recounted by Shurkin in rather excruciating detail. Enormous solicitude for Jean's uterine cancer was followed by abandonment. There was a brush with suicide by Shockley, and some episodic flirtations; but in the end Jean was replaced during the same busy year by the surprising Emmy, a plain, no-nonsense, bright nurse-educator who went with him to Stockholm in 1956 for his Nobel Prize, and stuck with him through the whole unraveling of his final years.

True to form, Shockley's first California team was brilliant: it included such legendary figures as Bob Noyce and Gordon Moore. But within little more than a year, the boss's management style had alienated and challenged the top eight of his new staff, who left him to form Fairchild Semiconductors — which was bought out from under them, providing funds to found Intel, from which defectors founded the next startup… . Meanwhile Shockley's firm languished and slowly died. In 1965 he returned to a much diminished role at Bell Labs.

The slow demise of his company left him time to think about his ideas on "human quality" — basically, that it could be quantified and that in the absence of some better measure one might as well use IQ, a measure with which he was somewhat familiar because of his mother's early acquaintance with the Terman program at Stanford. (Oddly enough, he was twice rejected for Terman's "gifted children" program, a selection, by measured IQ, of high school students from the area who were treated to special educational opportunities. Terman was the son of the creator of the Stanford–Binet test and also president of Stanford.)

In a sense, his philosophical approach came in a direct line from his wartime work with WSEG, much of which amounted to "find a quantitative measure of success and act on the obvious consequences". To him, the obvious consequence of quantifying human quality was eugenics, in all its easy fallaciousness: opposition to remedial programs such as affirmative action and Head Start, support for sterilization of the "unfit", and the lot. He spent the remaining 25 years of his life pursuing such a program with characteristic singlemindedness, exploiting every opportunity for publicity afforded by his Nobel Prize. The story of those 25 years at the barricades is not pretty — my own memories include the disruptive sessions Shurkin describes at the National Academy, at which he broke for life with his old friend Fred Seitz (now president of the NAS). As Shurkin describes, it was difficult to spend time with him, but one could still sense that there was somewhere hidden a reasonable, likable, even admirable man if one could avoid his obsession.

Well how can one summarize such a man? Perhaps his physics is the clearest clue. He called his method of thinking "try simplest cases", but it's really what many good theoretical physicists do when confronted with a complex problem, to try to find a simple model — only he elevated it to a mantra. He could see his way through the first few stages of any problem very quickly, but rarely employed his talents to look further below the surface or to check his models against reality. People of the quality for which he had an undoubted eye — like Bardeen and Noyce — would almost inevitably go off on their own, following their own instincts, and he could never allow that. He felt it as competition — he was intensely competitive, even in his recreation — and a loss of his authority. I know: it happened to me, in a small way, in 1950, and, as others, I survived and may have been the better for it.

How about the book? On a mundane level it needs better copy-editing, and even the spellcheck program nods. His mother is described as a "mathematical protégé" on the very first page. Not all the facts are reliable: the organization of Bell research is dicey: for instance, John Bardeen never ran a theoretical group, one did not exist as such until 1956, and Bill's secretary was not yet Betty Sparks in 1947. It is perhaps inevitable that a historian is not the best explicator of science; I would have liked a lot less talk about the mystery of quantum theory, replaced by at least a mention of energy bands — the term band never appears. The band theory behind the transistor is not very abstruse.

My quibbles are perhaps unfair — few biographers are reviewed by someone who was literally on the scene much of the time. Shurkin is a good storyteller and an even better researcher of the personal facts. After all, Bill Shockley arguably saved Britain from the U-boats during the battle of the Atlantic, and he is certainly the true father of the age of silicon; he is even the inventor of the graphite-moderated nuclear reactor. Yet the public will remember his name as that of the nutty Nobelist who donated his sperm to a genius bank.

Who would not want to hear the whole story?

The Quiet Man of Physics*

John Bardeen was an extremely quiet man. An anecdote in this book avers that when he was selling his house in Summit, New Jersey, the prospective buyer was so disconcerted by Bardeen's silence that he raised his bid by many thousands of dollars while waiting for him to speak. As an old friend of Bardeen's, I don't find that at all implausible.

Nonetheless, this quiet man led the way in two earth-shattering developments: with Walter Brattain he devised the first working semiconductor amplifier, jump-starting the information revolution; and with two young associates, he solved the 46-year-old puzzle of superconductivity, with repercussions not just in that field but in fundamental aspects of nuclear and elementary particle high-energy physics. He also helped to plan the research laboratories of the giant Xerox Corporation, and was a friend and consultant to the founder of Sony. By the way, he is the only person ever to have won two Nobel prizes in physics.

Still, as Lillian Hoddeson and Vicki Daitch bewail in this book, to most of the world he is "John who?". They intend to start remedying this situation. The second theme of the book is also the source of the title: they want to emphasize that it is possible to be a "true genius" without being bohemian, neurotic or particularly difficult to get along with — and that perhaps most geniuses do not fit the popular stereotype.

Of course, the biographer's task is much harder if the subject is not eccentric in any spectacular way. Even Bardeen's precocity — he entered college at 15 — did not seem to impair his ability to get on well with his

•

---

*Book review of *True Genius: The Life and Science of John Bardeen*, by Lillian Hoddeson and Vicki Daitch (Joseph Henry Press, 2002). Originally published in *Nature*, Vol. 420, 2002, pp. 463–464.

peers. A few years of indecision as to his natural calling ended happily in 1933, at the age of 25, with admission to Princeton University's graduate school, where he joined the fortunate little band of extraordinary students of Eugene Wigner who founded the modern quantitative theory of solids.

From Princeton he went on to be admitted into the elite Society of Fellows at Harvard, which gave him time to attempt several of the toughest problems in his field. He finally achieved a professorship at Minnesota University, and married Jane Maxwell after three years of courtship — he had the old fashioned determination to have a positive cashflow before marrying. Jane was attractive, universally beloved, patient and, happily, chatty. Soon called into war work, Bardeen rose to head a group of 90 engineers and scientists at the uncomfortable and crowded Naval Ordinance Lab in Washington.

The drama in his life was largely compressed into the next dozen years. A generous offer drew him to Bell Labs, where a group was being assembled under Bill Shockley to attempt to create an amplifier based on wartime advances in semiconductor science. The story of that achievement (by December 1947) is told more fully in Hoddeson's previous book *Crystal Fire*, written with Michael Riordan (Norton, 1997). But in *True Genius*, Hoddeson and Daitch recount the full story of the personal conflict with Shockley that ensued when the latter used his position to isolate and overwhelm the two original inventors. Perhaps, it must be said, this had lasting benefits for later developments in semiconductors, because it drove Bardeen to work on advancing the science generally, while Shockley pursued a line that eventually petered out.

By 1951, Bardeen had been lured away to the University of Illinois by old friend and fellow Wigner student Fred Seitz. There he not only fostered an independent semiconductor-engineering group with Nick Holonyak, but built a theoretical physics group within the already thriving condensed-matter physics enterprise.

He was at last able to indulge his obsession with the puzzle of superconductivity, left over from his Harvard days and reactivated in his final months at Bell. Using again his obstinacy and the careful assembling and critical interpretation of experimental data that had carried him so quickly to the transistor, Bardeen and David Pines soon formalized the crucial interaction. By the end of 1956 (with a slight interruption to pick up the Nobel Prize for the transistor), Bardeen, his student Bob Schrieffer and postdoc Leon Cooper had solved that problem as well .The resulting paper *(Physical Review* 108, 1175; 1957) deserves a place among the all-time classics of science. The intellectual ramifications of their central hypothesis — known as broken gauge symmetry — pervade all of physics, and have underpinned at least four other Nobel prizes to nine individuals. There will doubtless be more to come. (note: indeed there were, Nambu and Abrikosov, for example.)

After that, the rest of Bardeen's life may seem like an anticlimax, even though it included key roles in the creation of the Sony Corporation and the Xerox Palo Alto Research Center laboratories, and high office and influential advisory roles in the scientific establishment. Two well-written and well researched chapters are devoted to scientific controversies in which Bardeen became embroiled in later life, in several of which it is hard in retrospect to support his side of the matter. He may have been quiet but he was not self-effacing, and was quite capable of throwing his weight around in such controversies. But when the dust settled he could be quietly generous to his opponents, and was instrumental in the later Nobel Prize awards to Brian Josephson and probably to me, with both of whom he had had differences.

To return to the genius thing. Is it possible to be a genius when you are sometimes wrong? (Personal glamour seems to have little to do with it one way or the other, in science as in the arts.) Actually, few scientific geniuses are even close to infallible, and some can be wrongheaded indeed — look at Linus Pauling and vitamin C, or Julian Schwinger and cold fusion. Bardeen cracked a problem that dozens of the greatest minds of physics had failed at — isn't that enough?

There are, almost inevitably, glitches in the details in this book, but the authors' admiration and affection for their subject illuminates the biography. At the same time, they bring readers with varied levels of expertise to a real understanding of the complex workings of science as they are actually experienced by those of us who do it. Will the book appeal to the mass market and thus make a dent in the "John who?" problem? I'm not convinced, but perhaps it should.

A Theoretical Physicist*

Clearly both the book and the review are very dated. Landau's group and its offspring have spread round the world, starting well before 1989. I don't think I would make the comparison with Oppenheimer quite so invidious now. But the great strength of the Russian school, especially in condensed matter, and Landau's influence on its origins, are still true.

Anna Livanova quotes the poet Mayakovsky as saying, "I am a poet. That is what makes me interesting." Lev Davidovich Landau was a physicist, and, as Livanova says, that is what makes him interesting. This little book, although it is written with a bit too much naive adulation and Russian schmaltz, is interesting because it presents a true picture of Landau and the attitude toward him in his school.

Landau was as extraordinary a man as he was a physicist. He claimed with characteristic immodesty, ignoring his friends Peierls and Weisskopf among others, to be "the last universalist". He was one of the giants and had the charm, quickness, and fire — and ability to advertise himself — that evoked the kind of adulation that surrounded that other "swami" of physics, J. Robert Oppenheimer. Landau is often compared to Pauli because of their quickness, cocksureness, and stinging tongues, but it is the charismatic Oppy I find myself thinking of. There is even a physical resemblance between Landau and Oppenheimer, and again and again in the book the quality of Landau's personal friendship and his warmth and gaiety in social situations are attested to, as is the case with Oppenheimer but not with Pauli.

*Book review of *Landau: A Great Physicist and Teacher*, by Anna Livanova and J. B. Sykes (Pergamon, 1980). Originally published in *Science*, Vol. 211, 1981, p. 158. Reprinted with permission from AAAS.

These two "swamis" of physics, however, exerted two different forms of moral leadership and had very different effects on their worlds. Oppenheimer tasted power, built the Bomb, and by his own testimony "knew sin," returning to the seminar room only late in his career. In the Soviet Union, the bright, irreverent Jew Landau barely managed to stay in the seminar room but nonetheless built a great school and a far greater edifice of scientific achievement; beside Landau, who is responsible among many other things for modern condensed matter physics, Oppenheimer is merely a footnote in theoretical physics. The kind of leadership image Landau represented is expressed in a key quotation from Livanova's admiring introduction:

"Though seeming to some a person aloof from ethical questions, he became something of a moral paragon by his purely scientific and professional work.... In all his actions Landau was essentially defending physics against ... debasement.... Genuine science is essentially moral."

We will never know how much the differences between Landau and Oppenheimer were conditioned by the differences between the two countries.

Livanova's charming book is full of Landau stories; as with those of Bohr, Onsager, or Pauli, Landau's legend was maintained by characteristic anecdotes. But the book concentrates primarily on the physics: half of its all too few pages are a very good popular but historically accurate account of Landau's theory of liquid helium, for which he received the Nobel Prize. Of Landau's parents and childhood, we get one page; of his wife and son, three sentences. His troubles with the regime are all conveyed between the lines; we note the statement that at Kharkov his closest and dearest friends were experimentalists — Lev Shubnikov and his wife ... Shubnikov was "Fat Lev", Landau "Thin Lev".

A page later: The conflicts in which Landau and some of his friends and pupils became involved led to considerable unpleasantness that was a serious matter. In the end, it was necessary to think of moving to another city.

To interpret these passages properly, one has to remember that Shubnikov was killed by the Stalinists and that Landau had to move and spent some time under arrest. It is perhaps amazing that in a Russian book of the Soviet period the evidence is there at all. There is also an easily interpretable account of the kidnapping of Kapitza and his helium liquefier from Cambridge by the regime and Kapitza's later protection of Landau.

Much fuller is the description of Landau's teaching methods, of the informal atmosphere of his seminars, at which the audience joked and roared and the speakers quaked in fear of being dismissed with "What next?" Reading of the atmosphere around him, one begins to find almost plausible the incredibly great school of pupils and associates he built up in the most miserable conditions, surrounded by incompetence and cut off from the world of physics by barriers erected by a barbarous regime.

Livanova is open enough to document a few of the weaknesses of Landau's cocksure, ad personam, open-and-shut judgments, but she fails to mention the most important: his refusal to accept the relevance of Bose condensation to superfluidity, probably because it was suggested by minds he could not tolerate — notably that of Laszlo Tisza. Perhaps this was behind the famous incident, documented in Abrikosov's London prize lecture, of Landau's advice to Abrikosov which led him to put away his theory of Type II superconductivity for six years.

Nonetheless, this is a minor fault when compared to the admiration evoked by a man who could say things like the following:

> "You ask ... which branches of theory are the most important....
> One must have a rather ridiculous immodesty to regard only "the most important" problems of science as worthy of one's interest. [A] physicist should ... not embark on his scientific work from considerations of vanity."

In the end, I am as committed a member of the Landau cult as any of his pupils and as sensible to the loss of him, far too early, from an auto accident. I am also sensible to the loss, caused by the Soviet regime's policies, of the West's communications with the superb Landau school, which was just beginning to recover from the death of its leader.

Some Thoughtful Words (Not Mine)
on Research Strategy for Theorists*

I quote from one of the greatest theoretical physicists of the postwar era:

"The principal error I see in most current theoretical work is that of imagining that a theory is really a good model for ... nature rather than being merely a demonstration (of possibility) — a 'don't worry' theory. Theorists almost always become too fond of their own ideas ... It is difficult to believe that one's cherished theory, which really works rather nicely, may be completely false. The basic trouble is that many quite different theories can go some way to explaining the facts. If elegance and simplicity are ... dangerous guides, what constraints can be used as a guide through the jungle of possible theories? ... The only useful constraints are contained in the experimental evidence. Even this information is not without its hazards, since experiment 'facts' are often misleading or even plain wrong. It is thus not sufficient to have a rough acquaintance with the evidence, but rather a deep and critical knowledge of many different types, since one never knows what type of fact is likely to give the game away ...

Theorists ... should realize that it is extremely unlikely that they will produce a useful theory just by having a bright idea distantly related to what they imagine to be the facts. Even more unlikely is that they will produce a good theory at their first attempt ... they have to produce theory after theory ... The very process of abandoning theories gives them a degree of critical detachment which is almost essential."

---

*Reprinted with permission from *Physics Today*, Vol. 43, February 1990, p. 9. © 1990 American Institute of Physics.

The missing words indicated by dots would give the game away, that this is Sir Frances Crick talking about theory in biology, at the conclusion of his autobiography, *What Mad Pursuit*. He, in fact, distinguishes biological theory from physical theory on the basis that the mechanisms arise from the complex process of evolution. But in the absence of definitive advice on this matter from such other successful theorists as Crick's contemporaries, Richard Feynman and Murray Gell-Mann, it seems to me that one should, perhaps, take him more seriously as a guide to how theory is actually done than he may himself do. After all, in physical theory, we now know that whether or not the original cosmic egg was as scrambled as some astrophysicists such as Linde seem to think it was, almost all the phenomena we study, both in condensed matter and in particle theory, are the result of emergent processes and broken symmetries nearly as complex and evolutionary as biology.

My own experience has certainly been that most successful theories are the result of successive corrections to errors that may verge on the ludicrous, corrections normally dictated by a careful look at experiment. The long and tortuous tale I have told elsewhere of spin glass is one example; another is localization — who could have guessed, even in 1978 after certain prizes had been given out, that potential scattering, spin-orbit scattering, and magnetic scattering would turn out to give qualitatively different localization phenomena? Localization, in the presence of a magnetic field, seemed simple at first — until the experimentalists showed us that it led to the utterly unlikely phenomenon of Hall resistance quantization, leaving us theorists scurrying to catch up. In another example familiar to me, at least, the right *A* phase of superfluid helium-3 was predicted by solving the wrong Hamiltonian in the wrong way. Yet that is, too, a delightful example of Crick's "demonstration" theory: that paper demonstrated that phases of different symmetry were possible, which, in the end, turned out to be the really useful and important conceptual result.

Young theorists in my field, especially, would do well also to take Crick's words about experiment to heart. They often seem to believe that there is some kind of "Miranda rule" about what kind of evidence is admissible. Theorists discuss theory either in an experimental vacuum, or in relation to experiments endorsed by some previous paper or produced by the most fashionable experimental methods, rather than searching out the anomalies which are the real guide to the truth.

As I see it, even the "standard model" of particle theory — like it or not — was arrived at by the same kind of random walk guided at every stage by experiment, and many of its features still seem to have been as unpredictable on the basis of general principles of elegance or simplicity as the convolutions of biological evolution.

In conclusion, it appears that in all its branches physics is still an experimental science. Its basic goal is not mathematical elegance or the achievement of tenure, but learning the truth about the world around us, and Sir Francis Crick's words are as good a guide to that end as I have seen.

VI. Science Wars

Introduction

For a time I became quite exercised about the "Science Wars", and even did a freshman seminar on them, after reading "The Flight from Science and Reason", a well-written polemic on the subject. Much earlier, I had begun a collection of nonsensical items from my mail which I called "psychoceramics"; after a couple of years and with my increasing reputation it became too bulky to keep on with and is gathering dust in a far corner of my office. But eventually I came to feel that not all of the nonsense is from harmless crackpots. The varieties of nonsense that originate from political and financial self-interest I have dealt with separately under "politics and science", and the varieties that involve scientists themselves shooting science in the foot show up throughout this collection. The science wars are specifically about those in the academic community — mostly not professional scientists themselves — who claim to have found crucial flaws in the processes of science itself, invalidating our supposed epistemological basis, and who build considerable reputations within their own specialties thereby — in the process adding their bit to the public's negative attitudes about science. It is ironic to remember with what clear consciences we used, in the Cavendish, to pass on the weaker second class physics students to HPS (history and philosophy of science) — no sin goes unpunished.

They Think It's All Over[*]

In a review in *The New York Times* of Bob Woodward's latest book about American politics, I found the following: "Mr Woodward's victims know they are faced with a choice — either they can refuse to cooperate, in which case they will be described — as they were described by their enemies — or they can cave in and tell Mr Woodward their version. Either way — they know that they are at his mercy because ... the book is expected to be a bestseller before it is written."

John Horgan's book represents the arrival of such political reporting styles on the serious scientific scene — with the difference that we interviewees did not realise we were at his mercy until too late, having been pampered by respectful reporters lacking an agenda or platform of their own, hence obligingly furthering ours. Much of the fury this book has aroused in the scientific community is caused by the resulting sudden irruption of our sometimes imperfect personalities, our mannerisms, our casual backbiting, our unguarded boasts, and the like, into the public eye. Insofar as the scientists have eagerly angled for public notice, and in many cases for our own bestseller status, one might feel we have reaped a whirlwind of our own sowing; nonetheless where our remarks are shoehorned into favouring Horgan's personal agenda, we have a legitimate complaint.

Horgan, a staff writer for *Scientific American*, does occasionally catch our idiosyncrasies remarkably well. I laughed aloud at his interview with Tom Kuhn, taken back 50 years by his description of phrasing and manner to wartime Harvard (where Tom was junior marshal of Phi Beta Kappa and editor of the *Crimson*, and I a green scholarship kid from

---

[*]Book review of *The End of Science: Facing the Limits of Knowledge in the Twilight of the Scientific Age*, by John Horgan (Addison-Wesley, 1996). Originally published in the *Times Higher Education Supplement*, 27 September 1996.

the Midwest). That said, often Horgan goes too far to put his victims at a disadvantage. It seems completely irrelevant to his thesis that the late Karl Popper's housekeeper believed erroneously that he was famous even among the local London cabbies.

The book's thesis is that science is the vitim of its own success: that all the really interesting and soluble problems have been solved, or will be in a few years, so there is nothing for us to do but either accept our role as footnote writers to scientific history or to adopt an "ironic, postempirical" style. These are two useful terms adapted from the humanities: "ironic" from literature, the mode of echoing, reinterpreting or even mocking the Great Masters one cannot hope to surpass; and "postempirical", the mode of theorising about matters one cannot hope to approach experimentally: in both cases the enterprise may well be creative and exciting, but shares with philosophy and literature the possibility of endless reworking and reinterpretation — exactly the opposite of the clean, falsifiable truth which science claims to seek. The great masters to which Horgan refers in the case of science are the pioneers of quantum mechanics, relativity (which he broadens to include the hot big bang cosmology) and Darwinian evolution. Indeed, it is not possible to think intelligently about modern science without "playing off" from these great ideas, which inform everything we do — this is also true of the calculus, logic, the real numbers. Science intrinsically goes on from, rather than ironically negates, the great discoveries of the past.

There seem to be three more-or-less cogent arguments for the thesis Horgan advances, as well as some not so cogent. The first is in the words of the scientists themselves: the unfortunate propensity of senior scientists to declare that their field is now a wasteland of solved problems. Jim Phillips, who is often psychologically astute, remarked long ago of a column to this effect published in about 1959 by John Bardeen (coinventor of the transistor): "Of course, he has solved all the problems that were problems when he was young." Notorious in this mode was Brian Pippard's essay, "The cat and the cream", averring that academic research in condensed matter physics was futile, just at the time that two Nobel prizes were hatching in his own department, one for his boss, Nevill Mott, and one for his student, Brian Josephson. Horgan gets considerable mileage out of a correspondingly depressed book about biology by Gunther Stent called *The Coming of the Golden Age* and of a symposium organised by Stent titled "The End of Science?" Second is Horgan's very clever use of the philosophical naivety of some of his subjects — he even remarks on it in the case of Ed Witten. These are people who really believe that the goal of science is the complete answer to some all-encompassing question; people who, like Witten and others in physics, Stephen Hawking in cosmology or Richard Dawkins (and possibly even Stuart Kauffman) in biology, believe there is some unifying, underlying principle or law whose consequences then need only to be explicated in order to understand everything. The position is called "naive reductionism" and is typified by Witten's curious remark, that "every exciting discovery in physics follows from string theory".

This naivety is reinforced by great reputations fed by the enormous appetite of the general public for speculative musings on the "mind of God", the ultimate theory of everything, or the fate of the universe, and seems to be shared by Horgan himself, who, let's face it, is not visibly deep scientifically. It is a regrettable fact that the same naivety is shared by generation after generation of able students who follow these pied pipers into the morasses of string theory, cosmology and other postempirical subjects. Horgan is, in a way, quite right in his description of this kind of work as ironic and postempirical, but wrong in seeing it as the essence of science; science itself is still an empirical subject. Kuhn's "normal science", in my mind, can be described as a search for answers, great science as a search for questions, the greatest science as a search for the form the answers may take. These last two types of search are sometimes hard to distinguish from postempirical science until someone has invented the apparatus or the type of argument necessary to check them out, but the scientific community itself is often able to tell the difference. One good test (which several of Horgan's interviewees fail to pass) is whether the scientist involved has any history of dealing successfully with empirical facts.

The least plausible of Horgan's three arguments is that further progress in some fields, notably neurophysiology and the social sciences, is blocked by the sheer complexity and difficulty of the subject. He trots out, for instance, old and new speculations that he calls "mysterian" (not his neologism), to the effect that the human mind is incapable of understanding itself, but without giving any cogent argument except the opinions of various of his postempirical interviewees.

If we were to accept that all still-open questions are either too hard to solve or trivial, it would indeed appear we are approaching the end of science. But do we? Let us imagine ourselves back in the 18th century, contemplating the work of Newton and Descartes. If we were Horgan-minded, we should be in great despair: the continuum of space-time had been revealed and all the mathematics for discussing it had been discovered, in the shape of the calculus. Nobody has yet invented a way of expressing truths about space that does not use Newton's boring old calculus. All the rest is trivial. Well, the same role is played by quantum mechanics. Once one realises that it is necessary to use the symmetries of space and time to understand physics, one also realises that the quantum theory is probably unique as the mathematics we would have had to invent to express our physical ideas, and that we should be in desperate straits — reduced to the kind of strategem proposed by David Bohm — if we did not have it. But the physics expressed during the past 75 years by this simple mathematical framework contains many new theoretical insights and many exciting and enlightening empirical discoveries that are not in any sense just a working-out of the consequences of quantum mechanics. This is true obviously when we consider the particle side, where such new ideas as the gauge principle that interactions follow from symmetries, and broken symmetry, are by no means mere consequences of quantum theory. What Horgan is missing in his simplified, layman's view of science is that "the devil is in the details". Darwinism does not solve all the problems of biology any more than the quantum theory solved chemistry or relativity solved cosmology. These theories are tools, just as mathematics, the telescope or statistical mechanics are tools. We would be crazy to try to be astronomers without a telescope or biologists without Darwinian concepts; each is a liberation, not a straitjacket.

This, then, is the general problem with the book. I shall pick out four instances of it, where Horgan nods. First, he accepts the word of Linus Pauling that "I solved all of chemistry except maybe the sulfides" by 1931 (using, Pauling could have mentioned, the method of Heitler and London). But he does not mention that Pauling was misguided about metals (I cut my scientific teeth on his paper); and that the quantum chemistry of metals — the solved part, that is — was understood by another group of people entirely, including Eugene Wigner.

Horgan says that everything about cosmology is solved, the rest is dull. It hardly takes a sharp eye for instructive anomalies to realise that the two defects he finds in cosmological theory, the Hubble constant discrepancy and the fractal distribution of visible matter, are precisely the kind of anomaly that signals and informs the next step in science.

His interview with Per Bak reveals a similar failure of comprehension of science in his seemingly deliberate misunderstanding that a distribution of probabilities is one of the classic forms a theoretical result can take.

And his quote from Naomi Oreskes to the effect that numerical models are not "verifiable", which he applies by implication to the whole fabric of science, is a misunderstanding of the origin and thrust of Oreskes's thinking and of the nature of science. This essentially solipsistic argument (dating back to Bishop Berkeley) would deprive us of any "verifiable" knowledge of an external world whatever, even through our own senses. No sane person really accepts such solipsism. But worse is Horgan's implicit assumption that the goal of science is the Cartesian computer, a numerical prediction of everything by detailed calculation. Science (see above) is the search for understanding, and only in a fraction of cases do we have any need or desire (or capability) for a detailed numerical prediction. "Why?" and "How can I find out?" are at least as characteristically scientific questions as "How much?".

Finally, I cannot close without strongly objecting to Horgan's allegation that the some of the Santa Fe Institute's ideas derive from those of Ilya Prigogine, a statement that persuades me that Horgan's grasp of concept and of detail is just not up to the job he has set himself. Never mind the originality or the validity of any of Prigogine's ideas; they have nothing to do with concepts such as rugged landscapes, adaptive agents, complex adaptivity or self-organised criticality, which are current at SFI. The devil is in the details again. Horgan describes the serious reservations many of us have about Prigogine's ideas in one place, and in another accuses SFI of borrowing them without credit.

Despite all this I, like most of the reviewers I have seen, cannot consign *The End of Science* to the dustbin. Horgan's book is a good read, and we scientists very much need to have a few of the popular illusions about us dispelled. Some of us, few it is to be hoped, are cranks or charlatans, others may be visionaries, pointing the way to the future; but you can never be quite sure who are the visionaries and who are bewildering themselves and misleading us. Too many of those who come to represent us to the public are mere publicists, with little record of empirical or mathematical achievement. Horgan could, in a better book, have exposed some of these pretensions to good effect without writing what, on rereading, leaves a lasting impression of deliberate destructiveness. If this review sounds somewhat harsh, it is because, in spite of all the author's protestations to the contrary, it seems to me he has most mischievously provided ammunition for the wave of antiscientism we are experiencing.

Science

A 'Dappled World' or a 'Seamless Web'?*

In their much discussed recent book, Alan Sokal and Jean Bricmont (1998) deride the French deconstructionists by quoting repeatedly from passages in which it is evident even to the non-specialist that the jargon of science is being outrageously misused and being given meanings to which it is not remotely relevant. Their task of 'deconstructing the deconstructors' is made far easier by the evident scientific illiteracy of their subjects.

Nancy Cartwright is a tougher nut to crack. Her apparent competence with the actual process of science, and even with the terminology and some of the mathematical language, may lead some of her colleagues in the philosophy of science and even some scientists astray. Yet on a deeper level of real understanding it is clear that she just does not get it.

Her thesis here is not quite the deconstruction of science, although she seems to quote with approval from some of the deconstructionist literature. She seems no longer to hold to the thesis of her earlier book (Cartwright, 1983) that 'the laws of physics lie'. But I sense that the present book is almost equally subversive, in that it will be useful to the creationists and to many less extreme anti-science polemicists with agendas likely to benefit from her rather solipsistic views. While allowing science some very limited measure of truth — which she defines as truth of each 'model' in its individual subject area and within its 'shield' of carefully defined conditions set by the scientists themselves — she insists that there is no unity, no sense in which there are general laws of physics

*Essay review of *The Dappled World: Essays on the Perimeter of Science*, by Nancy Cartwright (Cambridge University Press, 1999). Originally published in *Stud. Hist. Phil. Mod. Phys.*, Vol. 32, No. 3, 2001, pp. 487–494.

with which all must be compatible. She sees all of the laws of physics as applying only *celeris paribus,* all other things being equal. Whenever other things are not equal, she seems to abandon the laws as any guide. In this sense, she advocates a 'dappled world' in which each aspect of reality has its separate truth and its separate 'model'.

Reading further, one senses that the problem may be that she is bogged down in 18th- and 19th-century (or even older) epistemology while dealing with 20th-century science. The central chapters seem to depend heavily on such outdated, anthropomorphic notions as cause and effect, recurrent regularities, capacities, etc. To me, the epistemology of modern science seems to be basically Bayesian induction with a very great emphasis on its Ockham's razor consequences, rather than old-fashioned deductive logic. One is searching for the simplest schematic structure which will explain all the observations. In particular, what seems to be missing in the thinking of many philosophers of science — even the great Tom Kuhn, according to Steven Weinberg — is the realization that the logical structure of modern scientific knowledge is not an evolutionary tree or a pyramid but a multiply-connected web.

The failure to recognize this interconnectedness becomes obvious when we are presented with 'classical Newtonian mechanics, quantum mechanics, quantum field theory, quantum electrodynamics, Maxwell's electromagnetic theory' and, in a separate place, 'fluid dynamics', as logically independent and separate rather than as, what they are, different aspects of the same physical theory, the deep interconnections among them long since solidly cemented.

Another part of the problem with this book is that the two dominant examples chosen are physics and economics, the rationale being that both sciences have 'imperialistic ambitions', the physicists aiming to provide a 'theory of everything' in the physical world, and some economists claiming universal validity in the social sphere. These two sciences, however, are on different levels of the epistemological ladder. Physicists search for their 'theory of everything', acknowledging that it will in effect be a theory of almost nothing, because it would in the end have to leave all of our present theories in place. We already have a perfectly satisfactory 'theory of everything' in the everyday physical world, which only crazies such as those who believe in alien abductions (and perhaps Bas van Fraassen) seriously doubt. The problem is that the detailed consequences of our theories are often extraordinarily hard to work out, or even in principle impossible to, so that we have to 'cheat' at various intermediate stages and look in the back of the book of Nature for hints about the answer. For instance, there is nothing in the quantum mechanics of the chemical bond which implies the genetic code in its detailed form, yet there is equally nothing in the operations of molecular biology which is incompatible with our quantum-mechanical understanding of the chemical bond, intermolecular forces, and so on. In fact in the defining discovery of the field, the double helix, that understanding played a crucial role. Thus the consequences often imply the laws without the laws implying a particular set of consequences. Physics is well embedded in the seamless web of cross-relationships which is modern physical science.

Economics, on the other hand, is an example of a field which has not yet achieved interconnection of enough related information to have real objective validity. It resembles medicine before the germ theory, or biology before genetics: there are a lot of facts and relationships, but there are probably some unifying principles and connections to other sciences which are yet to be found. Yes, the Chicago school makes ambitious claims — so did the Marxists in the past. Both, to my mind, qualify as ideologies rather than theories and belong in the political sphere. There are also serious economists — with several of whom I happen to have worked, including one of the creators of the mainstream Arrow–Debreu theory — who are doing their best to discover the deeper realities that may be there, and are in conflict with the dominant school. In science as in every other field of human endeavor the rule must be *caveat emptor*: science as a whole cannot be responsible for the temporary gullibility of the public to the claims of cold fusion, Freudianism or monetarism: these are just bad, or at best incomplete, science. In sum, whenever a school of scientists creates an intellectually isolated structure which claims validation only within its own area and on its own terms — that is, does exactly what Cartwright is claiming *all* scientists do — that science no longer has the force of dynamic, self-correcting growth which is characteristic of modern science. Cartwright's 'cocoons' are an excellent description of Freudianism or behaviorism in psychology, or of the response of electrochemists to cold fusion, but do not describe healthy science.

I have some particular reason for unhappiness about the message of the book: in a very early chapter she quotes me as being opposed to my clearly stated position. My best-known work on these subjects begins with these words: "The reductionist hypothesis may still be a topic for controversy among philosophers, but among the great majority of scientists it is accepted without question. The workings of our minds and bodies, and of all matter [...], are assumed to be controlled by the same set of fundamental laws, which [...] we know pretty well". Since it is clear that I was and am one of that 'great majority', it is disingenuous of Cartwright, who is one of those 'controversial philosophers', to quote succeeding paragraphs in such a way as to arrogate me to the opposite side.

There was a second place where I can fairly competently fault her understanding. In Chapter 8 she states that she will 'take as the central example' the BCS theory of superconductivity, an area which has been extensively studied by the 'London School of Economics Modelling Project'. I have been involved with the theory (and practice) of superconductivity for 43 years — for instance, I supplied a crucial proof which is referred to in the original BCS paper as a personal communication. In 1987 I gave a lecture studying an important and neglected part of the history of this theory, which was written up and published by Lillian Hoddeson as part of the American Institute of Physics' history of solid-state physics (Hoddeson, 1992).

My contribution was called 'It Isn't Over Till the Fat Lady Sings'. I used that crude American metaphor from the world of sport to characterize the somewhat confused period (1957–1965) between the original BCS paper, which indeed proposed a 'model', and the approximate end of the verification and validation process which made the model into what physicists properly call the 'theory' of phonon-mediated (ordinary, or BCS) superconductivity. (My usage of 'model' may be rather different from that of the LSE modeling project. What I, and physicists in general, usually mean by the term is a simplified example which is in the right universality class — for the meaning of this term, read on.) At the end of that period we were in possession (i) of a microscopic theory controlled by a small parameter, hence described by a convergent perturbation expansion about a mean-field solution which is rather like BCS, and (ii) of a detailed understanding of a much wider range of phenomenology than the Ginsburg–Landau equations could provide. This is such that the theory is no longer confined to its 'cocoon' but deals well with all kinds of messy dirt effects. (The best books describing this outcome may be Parks' two volume compendium (1969) and de Gennes' slim book (1966), both published in the late 1960s.)

Apparently, the LSE project accepts, for much of its account of BCS and G-L, a late pedagogical description (Orlando and Delin, 1990), by two engineering-oriented authors who had no part in the above history. It is known to many historians of science that textbooks tend to caricature the real process of discovery and validation, and this is an error I regret finding here. The only original literature quoted (except for BCS and for

Gor'kov's early, model-based derivation of G-L from BCS) are unsuccessful previous attempts at a theory by Bardeen himself and by Fröhlich, as well as others by such luminaries as Heisenberg, Salam, Wentzel and Tomonaga. (In the process, she renames my old friend Herbert Fröhlich 'Hans'.)

So: in 1957 BCS may have been describable as a *ceteris paribus* model, with no adequate account of a wide range of phenomena, or of its own limitations. It was made, by 1965, into an enormously flexible instrument with a high degree of *a priori* predictive power, and even more explanatory power. In fact, one of the deep difficulties of theorists of the new high T_c superconductors is persuading the community that, flexible as BCS is, new principles of physics must be invoked to explain the new phenomena. But as is almost always the case, the new ideas do not destroy, but instead supplement, the old. Just as the discovery of quantum chromodynamics left quantum electrodynamics firmly in place, no sensible theory of high T_c will displace 'BCS' — now meant as a shorthand for the full theory as of 1965 — from its validity in ordinary metals.

The story that Cartwright misses entirely, however, is the unifying and interleaving effect the theory of superconductivity had on very widely separated areas of physics. Far from being an isolated 'model' applying only in its shielded cocoon (as in the misfit metaphor she uses of the SQUID magnetoencephalograph in its shielded room) it was an explosive, unifying, cocoon-breaking event. First, in its own field: it showed us solid-state physicists that we could no longer safely ignore the general structure of physical theory: our familiar electrons were no longer little particles but unquestionably were quanta of a quantum field. True, in accepting the exclusion principle we should long since have realized how implausible it would be for 'particles' to be *absolutely* identical, but we had come to make casual assumptions about them. Then our particle physics friends began speculating how the vacuum itself might be a BCS state, a speculation ending in the electroweak theory. Finally, the nuclear physicists realized that we might have found the explanation for a series of puzzling phenomena observed in nuclei, and made the nucleus into a paired state. Yet another epiphany came when we predicted, and found, that the rare isotope of He would be a BCS superfluid of a new kind. So even though, in terms of the fundamental, unifying, microscopic laws, BCS made not the slightest change, it taught us a new way in which quantum fields could act, and also called our attention to the very general phenomenon of broken symmetry which is one of the key ways in which complexity can emerge from those laws.

Let us get back to the book. One of the basic epistemological points on which I differ radically from Cartwright is a very common misconception. Like many others, she maintains that the primary goal of science is prediction, prediction in the sense of being able — or at least wishing — to exactly calculate the outcome of some determinate set of initial conditions. But that is not, for instance, what an archaeologist is doing when he measures a carbon date, or a fluid dynamicist when he studies the chaotic outcome of convection in a Benard cell. Rather, each is searching for understanding. In the one case, he wishes to correlate different measurements to get some idea of the sequence of past events, which surely could never have been predicted in the quantitative sense but may enlighten him as to fundamental human behaviors. In the second case, he knows to a gnat's eyelash the equations of motion of his fluid but also knows, through the operation of those equations of motion, that the details of the outcome are fundamentally unpredictable, although he hopes to get to understand the gross behavior. This aspect is an example of a very general reality: the existence of universal law does not, in general, produce deterministic, cause-and-effect behavior.

Of course, in some sense there is always an aspect of prediction, in the sense that correct predictions — but by no means detailed ones — are very strong validations of the theory. If the archaeologist sees a particular kind of pottery, he may predict and then verify the carbon date; then next time he will not have to check the date. But in the epistemology which describes at least the natural sciences, I believe that the goal is exactly what Cartwright is trying to convince us is impossible: to achieve an accurate, rational, objective, and unified view of external reality. In the final section of her Chapter 2, asking 'Where Do Laws of Nature Come From?', she gives as her answer, 'always the source must be the books of human authors and not the book of Nature'. On crucial matters she is a social constructionist. I have argued elsewhere that this is not a tenable position, at least unless one is willing to accept total solipsism. Our perception of the everyday world is based on fragmentary, unreliable data which we only put together by creating a 'schema' or theory about the actual nature and objective existence of the various objects — chairs, mothers-in-law, teddy bears, or whatever — which we hypothesize to be in it. Then we correct, verify and validate the theory by making predictions from it (if I reach out and touch that brown thing, it will be fuzzy). Or I ask someone else to confirm my idea. Thus if we reject the inductive methods of science, we reject our only way of dealing with reality. In order to maintain our daily lives we have to accept the objective reality of the world and that it is the same world for everyone.

Why is this necessarily the case? Because we have so many cross-checks, so many consistency conditions. In the end, the schema contains many, many fewer bits of information than the data our senses gather, so we can be sure that no other theory could possibly fit. Now, we see that we can think of science as simply a somewhat more abstract, somewhat more comprehensive extension of our schema, describing the external world and compressing the enormous variety of our observations into a comprehensible whole.

The process of deconstructing the rest of the book in detail is beyond my budget of patience. The last chapter, in which she deals with the quantum measurement problem, for instance, seems to advocate one of the thousands of alternative incorrect ways of thinking about this problem that retain the quantum-classical dichotomy. My main test, allowing me to bypass the extensive discussion, was a quick, unsuccessful search in the index for the word 'decoherence' which describes the process that used to be called 'collapse of the wave function'. The concept is now experimentally verified by beautiful atomic beam techniques quantifying the whole process.

Another key word missing from the index and from the book — I checked — is renormalization. This is not just a technical trick but a central concept in the philosophy of physics, underpinning the physicists' use of model Hamiltonians, the passage to the limit of continuum equations, and even the modern view of statistical mechanics. A 'modeling project' which has anything to do with physics should hardly ignore the way in which we build and justify our models. The renormalization group is a way to expand the scale from the atomic to the macroscopic which shows that often the result is an enormous simplification, a division of systems into 'universality classes' all of which behave the same way in this limit; hence we may pick the simplest model for further study.

Returning to Cartwright's other exemplary subject, there is another contrast here. It is a great advantage of physics over economics that we physicists can often actually justify our use of models in this way, whereas use of the same idea in economics is almost never justifiable. An economy cannot be sorted out into macroscopic versus microscopic, with the former constructed by simply aggregating the latter: the individual agents have foresight and are of such widely different sizes and characteristics that averaging is meaningless, even if they behaved in any mechanistic or even rational way.

There is an attack on the entire science of molecular biology in the Introduction, making the hardly very philosophical plea that the allocation of funds for genetics should be slashed in favor of preventive medicine, childcare, and other worthy causes. I could agree that a very bad glitch in the patent laws — based on not very good science — has led to a frantically accelerated search for the 'gene for this and that disease', where almost all phenomena involve the collective contributions of many genes and perhaps even of the entire genome. But while we are being feminist, are we willing to give up DNA testing? Or the heavily molecular and surprisingly successful research programme on AIDS? These are political and moral questions and have no place in a book about epistemology. Science advances by looking under the streetlight where the light is, not by 'crusades' against socially acceptable targets. The political direction of scientific strategy which she appears to advocate here has a very bad historical record in which Lysenko is only the worst recent disaster.

In summary, this book seems to show that what may have happened in the philosophy of science — or at least in this corner of the field — is precisely the kind of intellectual isolation from outside sources which elsewhere leads to bad science. There is a reluctance to accept the fact that science has become a dynamic, growing web of interrelationships both within and across fields, and that even philosophers can no longer make do without taking into account modern insights into the workings of nature and of our own mentalities. The description in this book of the process of scientific discovery, in the chapter called 'Where Do Laws of Nature Come From?', is just false from the point of view of one who has participated in it. Scientists have increasingly, and in some cases consciously, had to invent for themselves the epistemology they actually use. Scientists are not particularly able philosophers, as the case of Bohr demonstrates, but at least they are in touch with reality at first hand, and their insights into the matter have profoundly changed our understanding of how we make discoveries. In the modern state of science, no discovery lives in a cocoon, rather it is built within and upon the entire interconnected structure of what we already know.

In a sense, this is a valuable book, in that it serves as a wake-up call telling me that it is time scientists themselves examined epistemology in the light of their experience of the reality of scientific discovery. When challenged on these subjects, many of us cite Popper's ideas. Though basically right as far as they go, these now seem out of date and naïve. Two scientists who have addressed these matters are Murray Gell-Mann and E. O. Wilson, and my remarks above are strongly influenced by what they have had to say. Gell-Mann, in *The Quark and the Jaguar* (1994), and even more in remarks at various workshops of the Santa Fe Institute, has emphasized the role of 'compression', while Wilson (1998) proposes the term 'consilience' for the web of interrelationships. But it is time to take a more definitive look at why — and, of course, when, science is right.

References

Cartwright, N. (1993) *How the Laws of Physics Lie* (Oxford: Oxford University Press).

Gell-Mann, M. (1994) *The Quark and the Jaguar* (New York: W. H. Freeman).

de Gennes, P.-G. (1966) *Superconductivity of Metals and Alloys* (New York: Benjamin).

Hoddeson, L. *et al.* (eds.) (1992) *Out of the Crystal Maze: Chapters from the History of Solid-State Physics* (Oxford: Oxford University Press).

Orlando, T. and Delin, K. (1990) *Foundations of Applied Superconductivity* (Reading, MA: Addison-Wesley).

Parks, R. (1969) *Superconductivity,* Vol. 2 (New York: Marcel Dekker).

Sokal, A. and Bricmont, J. (1999) *Fashionable Nonsense* (New York: Picador).

Wilson, E. O. (1998) *Consilience. The Unity of Knowledge* (New York: Knopf).

Reply to Cartwright[*]

I am afraid that Nancy Cartwright and I will have to agree to disagree, on the whole. If my review comes through as harsh, it is perhaps the natural response of a quantum theorist who has worked in economies to a book in which physics and economics are treated as epistemically identical.

One complaint is justified. I had no adequate evidence to extrapolate her concerns about the agenda in the case of breast cancer to a wider arena — 'childcare etc.' But I think she overplays the role which the scientists' intellectual drive towards unification plays vis-à-vis the unfortunate effect of commercial concerns.

My quote from her social constructionist alter ego followed her approving statement that '[n]owadays the SC's provide us with powerful arguments against taking the laws of physics as mirrors of nature', and was followed by a passage in which to my reading she explicitly denies the general applicability of the laws of physics. I think I stated her position correctly.

A few points of terminology may have obscured my meaning for her and may also for other readers of this journal. I used the words 'Bayesian induction' for want of a more widely accepted description of the pragmatic epistemology of modern science. I certainly did not mean to imply assigning explicit 'priors' in the space of all possible theories, merely that Bayesian mathematics is the best way I know of to estimate a suitably heavy penalty for introducing unnecessary complication, as well as telling us the enormous epistemic value of cross-connections between isolated subject 'cocoons'. I agree with Cartwright that the reliability of

---

[*]Originally published in *Stud. Hist. Phil. Mod. Phys.*, Vol. 32, No. 3, 2001, pp. 499–500.

the methods of hard science is a suitable subject, but feel that she has not contributed much to the answer.

Our difference about 'decoherence' is real. I find this and the cluster of ideas around it much preferable to more traditional ways of treating the quantum paradoxes just because the 'classical' apparatus is treated as a quantum system as well (an idea with a long history going back perhaps to F. London, the only philosopher among the fathers of quantum theory); and as I remarked, recent experiments have verified this approach. The difference between our two versions of 'scientific realism' comes out in this discussion: in mine, there is only one correct answer to any well-posed physics question, but we may not know yet what it is. There is a real world out there with one set of real laws, and by now we have a very good idea what they are.

This is also a good example for my Bayesian — or Ockham's razor — position. For 75 years physicists — perhaps overwhelmed by the prestige of the authors of the Copenhagen interpretation — have adhered to the idea that there is some mysterious scale at which the quantum world changes into the classical. Decoherence is simply the code word for the null hypothesis that there is no such scale: that we are quantum all the way up. I have for long felt that the question was settled by the many examples of macroscopic quantum coherence, but I am notoriously impatient and perhaps the delicate experimental work which has gone into pinning these questions down is worthwhile. There is absolutely no experimental evidence for such a scale.

I referred to the 'renormalization group', which is not the same as 'renormalizability', technically, though I believe she understood my thought. The will-o'-the-wisp of universality (I use the word in the technical sense), provable in statistical mechanics by this method, is a dream rather than a reality in the human sciences, indeed, especially economics.

Postmodernism, Politics and Religion*

Alan Sokal really likes footnotes, which may have made him uniquely qualified as a hoaxer of "science studies". The original hoax, a purposely and wonderfully nonsensical paper about the epistemology of quantum gravity, appeared in 1996 in the cultural-studies journal *Social Text*, with the enthusiastic endorsement of its editorship of eminent postmodernists. By actual count there were 107 footnotes. In *Beyond the Hoax*, the first chapter is a reprise of the original *Social Text* paper, with an additional 142 "annotations" in which he explains at length much of the complex fabric of in-jokes and bouleversements which made it so exquisitely wacky to anyone with even a modest knowledge of physics.

The remainder of the first part of the book contains well-footnoted essays on his reason for this exercise in foolery and on the various responses he has received. His inspiration was that he maintains that there was under way a serious assault against rationality by postmodernists, led at the time by a relatively small number of left-leaning academics in "humanities" departments. He felt that this would be self-defeating for the Left, while opening up great opportunities for the obfuscatory tactics of the Right. Indeed, as the recent book by Chris Mooney, *The Republican War on Science*, amply testifies, the "faith-based" administration of George W. Bush has done its best to obscure a variety of "inconvenient" scientific truths, but Sokal has found little confirmation that they have borrowed much of this obfuscation from the postmodern relativists and deconstructionists in the leftist fringes of academia.

---

*Book review of *Beyond the Hoax: Science, Philosophy and Culture*, by Alan Sokal (Oxford University Press, New York, 2008). Originally published in *Physics World*, Vol. 21, August 2008, pp. 40–41.

Part II of this book, coauthored with his collaborator Jean Bricmont, is a serious philosophical discussion of epistemology, its first chapter condemning the cognitive relativism of the postmodernists — the idea that fact A (for instance, the Big Bang) may really be true for person A, but not for person B — and the second chapter making a trial run at a reasonable epistemology for science. I was delighted to find as part of their vision "the renormalization group view of the world" in which one sees every level of the hierarchical structure of science as an "effective theory" valid at a particular scale and for particular kinds of phenomena, but never in contradiction with lower-level laws. This leads them to an emphasis on emergence as well as reductionism. I have seen few better expositions of how the thoughtful theoretical scientist actually builds up his picture of reality.

On the other hand, the Sokal/Bricmont view of science as a whole may be a bit idealized, perhaps best suited to relatively clean fields like relativistic field theory. In the murkier and more controversial field of materials science, for example, reality is not so cleanly revealed, particularly when it contradicts the personal interests of the investigators. There is a picture in the sociologists' minds of a hegemony of "old white males" which however is pretty ridiculous: youth, race and gender observably present no bar to acting in your own interests rather than those of science.

Part III encompasses more general subjects. One very long chapter explores the close relationships between pseudoscience and the postmodernists. It is easy enough to find ignominious stories about pseudoscience; some striking and important ones that Sokal picked, for example, were the widespread teaching of Therapeutic Touch (a practice with its roots in Theosophy, and not involving actual touch) in many estimable schools of nursing; and, going farther afield, the close ties between conservative Hindu nationalism and the teaching of Vedic astrology and folk medicine in state schools in India. Whether or not postmodernism has any causal relation to pseudoscience, when attacked proponents of such pseudosciences are seen to defend themselves by referring to the postmodernist philosophers. The postmodernists have been known to be sympathetic and supportive of such views — often, for example, favoring Vedic myths or tribal creation stories over the verifiable truths of modern science.

Finally, Sokal enters into the currently very much discussed, intertwined fields of religion, politics and ethics. His essay takes the form of a long, discursive review of two recent books on religion, Sam Harris'

The End of Faith and Michael Lerner's *Spirit Matters*. He promises it to be a critical review, but I found it to be rather more critical of Lerner than of Harris. He supports Harris in considering Gould's "non-overlapping magisteria" to be a copout. Sokal is an implacable enemy of fuzzy-mindedness, and makes the point that religion cannot avoid inventing factual but unlikely claims about actual events — even if one abandons young-earth creationism or reincarnation, or those fascinating inventions heaven and hell, the idea that there is an actual personal God listening to one's prayers and responding is not that far from "He is talking to me in Morse code via the raindrops tapping on my windowsill." Lerner's book addresses the conundrum of religion as "spirituality" — there is, for instance, the sense of wonder we scientists feel at the marvels which are revealed to us. (I think a more interesting book in this vein is soon to appear by Stuart Kauffman, *Reinventing the Sacred*.) Sokal, though he lets Lerner get away with dubious claims about studies of the efficacy of prayer, rather dismisses this view.

But then Sokal moves on into the political — if we want the voter to actually vote his true economic and social interests, do we hesitate to take away from him what are correctly known as the "consolations of religion"? Do we not risk his perception of the political left as condescending and elitist; how do we attempt to break through misperceptions of the true values of the conservative elite? This is not a problem to which anyone, Sokal included, has a good answer.

I too cherish long explanatory footnotes, crammed with extra ideas. But I do need to warn the prospective reader that, even skipping all the footnotes (which would be a great loss) this book is not a page-turner. The author is not one to drop a line of argument just because it wanders "off-message". Nonetheless Sokal writes lucidly; and one must not forget that his main targets, the postmodern theorists in English, philosophy, sociology, or "science studies" departments, are still doing well in our universities, even the most respected, and command enough respect for election into such august rolls as the American Academy of Science (I count two of Sokal's prime targets in as many years). They aim to persuade the elite among our students that scientific rationality is just the invention of a few white males eager to hang on to positions of power; where Sokal (and he may be right) sees it as our main hope.

VII. Politics and Science

Introduction

This is becoming ever more of a relevant topic. The last book reviewed, *The Republican War on Science*, was only the first of a series by a variety of authors lamenting the demonstrable fact that in this country, the validity of scientific conclusions has become a political issue. After the recent primary and election season it has become career-threatening for any Republican politician to show respect for the unanimous opinion of the major scientific societies on human effects on climate change, and dangerous for him to admit the fact of human evolution, the antiquity of the earth, or to support open-ended biological research. The blatantly misnamed "sound science movement", born as a tool of the tobacco, oil, and military-industrial lobbies, has arisen to apply the methods of McCarthyite redbaiting to climate and environmental scientists, and has obedient state attorneys general and likely Congressional committee chairmen ready to go into action.

The first and longest of these pieces was given as one of three Bethe lectures at Cornell, with Hans sitting prominently in the front row, making me feel like the proverbial child trying to teach his grandmother to suck eggs. As I stated there, over the century or more of relevant history the two parties have been more or less equally culpable and equally supportive in regard to science, and it has only been since the Reagan revolution consolidated the Southern and rural reactionaries under a single party label that the threat to mainstream science has become real.

Politics and Science

This lecture was one of a series of three funded by a foundation at Cornell. The other two were more technical. The reader may wonder at my hubris, talking literally in front of Hans Bethe, whom I enormously admire, about his own lifetime specialty — so do I, now, but I was braver then. The date was also noteworthy in that the Challenger accident happened during the afternoon, and the audience was rather subdued. The relationship of science to Republican administrations doesn't change much.

The problem of politics and science has been with us a long time — certainly since Archimedes was clobbered by that Roman soldier who thought — probably correctly — that he'd been helping the tyrant of Syracuse with his defenses. In fact it may be that the first politicians were scientists, namely the priests or witch doctors who doped out enough astronomical fact or herbal lore or both to make themselves indispensable and thereby became the leaders of the first organized polities. The idea of king-priest as life-giver and magician is found in most primitive cultures and religions. Frankly, I think that basically throughout history the scientist has somewhat relished his role as magician — a concept which Jerome Ravetz has emphasized — and has used it to do relatively well out of most societies and most governments, compared to his actual tangible political clout (coming either from direct immediate usefulness or numerical or economic strength). We can of course regret that the other, priestly function was usurped by various upstarts and that an even higher fraction of GNP is spent on them.

We tend nowadays to think of the relationship of science and politics in purely military terms. A hoary myth about this is the attractive idea, most recently propagated by the otherwise respectable historian

W. H. McNeill, that weaponry is the leading edge of technology and that scientific developments in weaponry are the driving engine for the wheels of world history. A famous thesis here is that the destabilizing force for the Roman Empire — its MIRV — was the breeding of horses capable of carrying knights in armor, and that the breakup of the resulting feudal system awaited the development of infantry weapons usable by the yeoman armies of the emerging nation-states. Aside from being very culture-centered, this neglects for instance the 1000-year survival of the Eastern empire, the assaults of the nomads and the rise of Islam, and the rottenness of the inappropriate and unsuitable politico-economic system of the Western empire based on slave agriculture. In the present half-century we find the USA and then the USSR helpless politically and almost as much so culturally in possession of the greatest and most scientific military power the world has known — the USSR losing many or most of its most viable clients in Asia and Europe, for instance China, and Yugoslavia, and now for practical purposes Hungary, while the USA is being led around by the nose by its supposed clients such as Israel, the Philippines, and South Africa, and bullied by Iran. Where technology clearly impacts on politics must surely be the economic sphere with its consequent cultural and military effects: for example the early industrialization of the British economy leading to world military hegemony almost in spite of themselves, and to a brief period in which, like Hellenism after Alexander, the corresponding imitative culture spread far beyond the political boundaries of the Empire as that empire collapsed in the face of assimilation of its cultural and economic message.

Actually, the relationship of politics and science in America clearly started on the very economic nonmilitary level of agricultural research and technology. State and federal subsidies to predominantly agricultural schools such as the land grant universities and Cornell's own path-breaking agriculture school, and for agricultural experiment stations, long predate any substantial government investment in the hard sciences and/or their military uses. We live with the incredible consequences of that investment: the essential evasion of the Malthusian dilemma whenever political and economic stability allows modern agriculture to take hold.

Perhaps this success story is a good place to examine what makes a successful scientific-technological mix. Clearly steady, healthy, but not palatial, support of the fundamental sciences such as genetics, population biology, biochemistry, etc. was important. This was done by a diversity of institutions supported by a diversity of means — Rockefellers were very important, the plowing back of patent resources in the case of the University of Wisconsin biochemistry program, state support, department of agriculture support, corresponding governmental mechanisms in Europe in particular, and now the NSF and NIH governmental grant systems. Clearly the educational job and the technical personnel on the extension agent level are vital. Corporate research support at the fundamental level has been hardly of much value at all except in development of specific chemicals; and the more we learn about antibiotics, Agent Orange, and other horror stories, the more we may be excused for taking a jaundiced view of the thoughtfulness and responsibility of the purveyors of agrichemicals. Under the American system of bottom-line emphasis, industry is not a trustworthy source of basic research funding in most sciences.

There have been fiascos — DDT for instance and the other persistent insecticides, and of course the casual way in which agricultural researchers have carried pests like the gypsy moth from place to place — but the field has been marked more by little, manageable pork-barrels and not in general by the gigantic multi-billion-dollar waste on demonstration projects and harebrained ideas which our harder sciences have led us into. Perhaps the most famous was the British West African "groundnut scheme" aimed at clearing thousands of square miles of jungle and replacing it with peanuts — it was a scandal at the time, having all the usual identifying characteristics of overoptimistic projections of technology, overly pessimistic projections of possible alternative technologies, and crazy market estimates — but it is happily forgotten now.

Agriculture also contains the most unmistakable and characteristic example of the consequences of political meddling at the fundamental science level: Lysenkoism, which essentially set back Soviet biological science by 30 years.

In general, the moral is: governmental support can work and is vital at the fundamental science level; politically motivated meddling is disastrous; and science can be a force for some sort of narrowly defined good.

How is the situation now? I am most unhappy that Keyworth's Science Advisory Council has no working biological scientist and only one biologist. Small as its powers are, this underlines the fact, obvious to anyone who has spoken with them, that this (Reagan) administration views science in military terms almost exclusively. Under it military spending has risen from 50% to 70% of all government R&D. Thus we have passed the British, in this regard. They have suffered for 40 years under a quixotic attempt to maintain a Great Power military in a minipower economy. Like the British, severe cuts are becoming necessary and since these will be applied "evenhandedly" we can expect the biological side to suffer most severely from them — the biological side, where any dispassionate observer must admit that by far the greatest intellectual as well as technological challenges remain to us, and which was for the British their main point of world leadership.

A few years after the war I remember reading a little story in *The New Yorker* about the first faculty Senate meeting in Fall 1945 in some small Midwestern university. There was a big, extrovert, polyprogenitive professor of agriculture and a wispy ectodermic professor of physics who until that meeting had never got a word in edgewise. At this Senate meeting for the first time in his life the little physics professor actually rose to speak — and was recognized over the ag professor, and listened to in awed silence.

That was the way it was — the physicists were the Dean Witters of the postwar years and when they spoke people listened. But while they had gained an audience, they had of course lost their innocence.

For us physicists this was the heyday of government support. In countless aspects of the war of the black boxes we had shown ourselves to be the best engineers — magnetic mines, proximity fuses, bombs, operations, of course radar, guided bombs and countermeasures, etc., and the nuclear physicists, our somewhat self-elected elite, had, if a little late for the fourth of July, given us an appropriately gaudy finale.

There are some science and politics lessons in this one too. One tends to forget that the Germans had superb engineers, who in fact produced, for instance, the first guided missiles already in 1943 — the glide bombs used in the Mediterranean. I saw the inside of one and they were beautifully engineered. There was a corps of dedicated rocket scientists who caused no end of trouble toward the end. But our overall technical superiority or equality was very much enhanced by the fact that Hitler had severely damaged fundamental science and particularly physics, which had a big role in many of the above developments. There were not only the campaign against "Jewish science" but the crackpot intellectual climate which pervaded the Nazi regime from top to bottom, with high officials believing in the "Hollow earth" fantasy and other equally peculiar cults and nostrums. There is a chilling similarity to the atmosphere in Washington today. In any case, there was no way in which reliable technical or strategic advice could reach the top levels of that regime. It is a cheap shot to liken Hitler's "intuition" to Reagan's famous "luck", since there is no resemblance in personality otherwise except for their equally deep appeal to the character of their respective populaces.

Physics, and to a great extent other sciences, had a magical appeal which led to a rapid, exponential rise in the national support, involving a series of bureaucratic manoeuvres which finally led to our present system. It is not my purpose to detail this history here, though of course it has many fascinating twists and lessons. Highlights on the positive side were the early and enlightened support by the armed services of many branches of fundamental science; the battle over civilian control of atomic energy which led us to the relatively well-managed DOE large facility programs and national laboratories; and the creation of the peer-reviewed science agencies, the NSF and the NIH, the former by political miracle and the latter based on the almost equally magical prestige of modern science-based medicine which arose in the postwar years.

A qualitatively important contribution to the science-technology mix in this period and somewhat earlier was the rise of science-based industry as a major contributor to fundamental science and basic technological research. Beginning with electric power and chemicals — G. E. and Dupont, and corresponding and earlier European companies, and then with the first high-tech industries, telecommunications and radio, the postwar era saw the further rise of industries based on sciences which had been created de novo, such as semiconductors, computers, and modern pharmaceuticals. A disproportionate fraction of public attention went to the magic men of Los Alamos, their weapons, the ensuing space and missile races and their rather unsuccessful atomic power program; a disproportionate share of genuine technological innovation, on the other hand, came from a small number of corporate laboratories, notably Bell Labs, and especially from the cross-fertilization between science and technology whose early beginnings centered around Bell Labs and MIT during the war years and whose products included coherent spectroscopy, semiconductor technology, the laser, cryogenics technology, ferrites, information theory, much of computer science, etc. Scarcely an individual who was at Los Alamos has failed to publish his memoirs; only one major Radlab or Bell Labs figure of the same period has done so.

A major political effect on science which, in the long run, may be more deleterious to our survival as an innovative technological society than any other act, is the legal and economic assault on the entities which were responsible for this progress. The two rather dull-sounding legal and economic concepts of the destruction of vertical integration and the enforcement of deregulation, along with the assault on bigness *per se*, make it essentially impossible for a company to benefit economically from sponsoring basic open-ended research. Research focused on specific devices, chemicals, airframes, or copyrightable software, is not open-ended — and often not very competent — research, unless for instance it considers the total environmental consequences of the chemical, the system in which the device, airframe, or software is situated, etc. In addition, small economic entities exploit small innovations fast but cannot sustain the "critical mass" to make any truly basic discoveries or found new industries.

Forgive me for running on about a pet subject; I think it an important one. Let me get back to more direct science and politics issues. The first postwar political flap, connected with the insecurities brought on by the Russian bomb and the early opportunistic Russian moves in the cold war, was McCarthyism, a code name for the national hysteria which saw treason on all sides and especially among scientists and diplomats. The scientific damage was a few valuable individuals, a heritage of bitterness and division in the advisory apparatus, and perhaps worst of all a lasting disdain of many idealistic young scientists like myself for classified work and for FBI and CIA types.

The scientific consequences of the next flap were even more interesting. I'd like to read you a few paragraphs from Tom Wolfe's book *The Right Stuff* which I think will bring many of us back into the atmosphere of those days right after Sputnik I. Speaking of the famous test pilot Chuck Yeager:

"... By October 4, 1957, he was back in the United States, at George Air Force Base, about fifty miles southeast of Edwards, commanding a squadron of F-100s, when the Soviet Union launched the rocket that put a 184-pound artificial satellite called Sputnik I into orbit around the earth.

Yeager was not terribly impressed. The thing was so goddamned small. The idea of an artificial earth satellite was not novel to anyone who had been involved in the rocket program at Edwards. By now, ten years after Yeager had first flown a rocket faster than Mach I, rocket development had reached the point where the idea of unmanned satellites such as Sputnik I was taken for granted ... The Air Force was interested in a rocket-glider craft, similar to the X-15B, that would be called the X-20 or Dyna-Soar, for "dynamic soaring"; an Air Force rocket, the Titan, which was under development, would provide the 500,000 pounds of thrust that would be required ....

So what was the big deal about Sputnik I? The problem was already on the way to being solved.

That was the way it looked to Yeager and to everybody involved in the X series at Edwards. It was hard to realize how Sputnik I looked to the rest of the country and particularly to politicians and the press ... It was hard to realize that Sputnik I would strike terror in the heart of the West.

After two weeks however, the situation was obvious: a colossal panic was underway, with congressmen and newspapermen leading a huge pack that was baying at the sky where the hundred-pound Soviet satellite kept beeping around the world. In their eyes Sputnik I had become the second momentous event of the Cold War ... The panic reached far beyond the relatively sane concern for tactical weaponry, however; Sputnik I took on a magical dimension — among highly placed persons especially, judging by opinion surveys. It seemed to dredge up primordial superstitions about the influence of heavenly bodies. ... *The New York Times*, in an editorial, said the United States was now in a 'race for survival' ... the House Select Committee on Astronautics, headed by House Speaker John McCormack, said that the United States faced the prospect of 'national extinction' if it did not catch up with the Soviet space program ... The public according to the Gallup poll was not all that alarmed. But McCormack, like a great many powerful people, genuinely believed in the notion of 'controlling the high ground'. He was genuinely convinced that the Soviets would send up space platforms from which they could drop nuclear bombs at will, like rocks from a highway overpass ...."

Even within the terms of the space program itself, recent discussions of the shuttle we actually have and the shuttle we might have had — I again rely on the ubiquitous *New Yorker* of about two years ago — suggest that the successor of those glide planes might have got our scientific instrumentation into space quicker and with a lot less expense, and that a natural evolution might have been a much better way to run a space program. But in any case, for the actual effect of the Apollo program, which was of course totally politically motivated, I quote a man who knows a great deal about science vs. technology, John Bardeen, in a recent statement about the possible effects of Star Wars: "... while the Strategic Defense Initiative is supposed to be a research program, the projected funds are larger than the entire amount spent by the government for all non-military research, including the National Science Foundation (the main supporter of university research) and the National Institutes of Health. At a time when our civilian economy needs all the help it can get to remain competitive in world markets, the best scientific and technical brains in the country may be drawn off to work on a project of dubious feasibility."

Perhaps the country can afford the 30 billion dollar gamble against long odds in monetary terms, but not the diversion of the top talent of scientific manpower to nonproductive ends. The Apollo program in the early sixties was a great technical achievement, but the cost was far greater than the dollars spent. It was during the sixties that the aerospace industry expanded rapidly to meet military and space needs and drew the scarce top technical talent from both the United States and Western Europe. Consumer-based industries could not compete for talent, leaving Japan to establish a lead in the civilian electronics market and in other areas. Since then, the lead has expanded. The aerospace industry grew larger than could be sustained in the long run. Within a decade engineers with training ill—suited for consumer products were being laid off ....

To continue with the space boondoggle, one keeps running into NASA's totally political need to get some kind of non-satellite science or technology program into space. Of course, the communications satellite — another Bell Labs invention — and its relations in terms of intelligence, weather resource etc., surveillance can hardly be seen as other than a beneficient use of space, the only good technological spinoff from the military I know of (though of course fully paid-up in an actual cost sense). But the idea of space factories and space condensed matter experiments is a truly classic example of political meddling with science. I show you a table of a small selection of the goodies of the past 5 years in condensed matter physics, and ask you — what kind of a good condensed matter physicist is going to waste time and dollars on doing these experiments badly in space? (See Table*.)

---

*The Table mentioned here would by now be very out of date. The bonanza continues, however, and I just list some of the recent important discoveries in the quantum physics of matter:

1) Graphene, of course, and its bizarre electronic properties.
2) Topological insulators, spintronics, and the like.
3) "Supersolidity" of solid He.
4) "solid state" experiments on cold atoms — many, many aspects, e.g., the "perfect fluid" in the unitary case.
5) Ong's vortex fluid: vortices in the normal-appearing metal.
6) and many, many others — mysterious phenomena revealed in cuprates by STM spectroscopy, for instance.

Of course, scientists are not immune to the pork barrel and there will be plenty of people willing to spend NASA's money, just not the best and the brightest. NASA itself is the result of an alliance of hot pilots and politicians, and science has rather played second fiddle throughout its history. The marvelous scientific missions reached their apogee just in time to be pushed aside for the Shuttle, the first and one of the worst examples of the Reagan militarization of science. For the overall effect of Apollo, the Shuttle, and the future space station on science, see J. A. van Allen's recent *Scientific American* article.

I was reminded of yet another result of the Sputnik flap when I recently went to a Grand All-Class reunion of my high school. This high school is a fairly extraordinary one, in that the classes of 1935, 1940, and 1948 each produced a scientific Nobel prizewinner. It was calculated that if other high schools in the US had performed as well, there would be roughly 1,300,000 American Nobel prizes. But when it came time to have the obligatory 5 or 6 person panel discussion, all classes after 1955 were represented by lawyers, writers, deans, pediatricians, etc. — all of them bright ladies, as it happened. What had changed at Uni High? I puzzled about this for a while and then suddenly light dawned — the NEW MATH! Uni High had been a center of this Sputnik-inspired aberration, and while the three of us had all mentioned particularly the math teaching in our autobiographies, no class after 1957 had had a good, conventional math education. Fortunately that too has subsided and pupils are no longer being required to think in a straight line — "von Neumann architecture"-like mathematicians instead of in parallel processing — i.e., intuitively — like people.

As the postwar period wore on into the euphoric early '60s it seemed science budgets would continue to grow exponentially, the country could even afford the moon program on top of a healthy science budget and we could defend Europe with a few cute little tactical nukes. But then a few cracks began to appear. Not everything the scientists did turned out perfect.

First a few, and then the entire community, realized that fallout was a real problem, and to everyone's delight we did end up with an atmospheric test ban. Other programs were more controversial: the ABM, the SST, the whole complex of Silent Spring issues, joined Vietnam and a growing uneasiness affecting especially the young with the whole problem of nuclear destruction to produce an anti-science backlash in which the scientists themselves were often the most enlightened participants. Of the whole complex of issues I want to focus on one particular one because it too is an enlightening example of the different kinds of interaction of science and politics, namely the SST (Supersonic Transport) and Concorde.

In this country the SST program was fascinating because it was the first and perhaps only example of the science advisory apparatus going head to head with the administration and winning. The American SST was a uniquely aberrant and ill-timed project. Supersonic flight is a beautiful example of clean, straightforward application of the laws of physics. An object moving faster than the velocity of waves in a medium radiates a bow wave which normally represents the greatest percentage of the dissipation of energy — this is true of particles, of ships, and of planes. The magnitude of the bow wave for an object of a given size is nearly independent of its shape if that is reasonable, and easy calculations show that, first, supersonic flight is much less energy-efficient than subsonic flight, and secondly that all of the excess energy is felt on the ground as a boom; and third, that to make the boom bearable — for a whole year they boomed Oklahoma City believe it or not, to see if people would mind (they do!) — stratospheric flight is necessary so that a calculable and not small percentage of the world's oil supply is going to be transferred permanently and directly into the stratosphere if the aircraft industry goes supersonic. That alone is utterly ridiculous from any environmental standpoint. In the midst of all this came the early beginnings of the oil crisis, making the energy cost totally uneconomic as well.

First the National Academy had to clean up its act. Its comfortable habit of appointing industry advocates to advisory studies produced a classic pro-SST report which even us greybeards couldn't stomach and the Academy's present rather careful reviewing procedures stem from that time. The President's NAA man managed to repress reports from PSAC, from NAA itself, from a specially appointed committee, and even one from the Academy; and at this point a few of the scientists began to go public. This incident in fact is the genesis of a delightful little book by Frank van Hippel and Joel Primack called "Public Interest Science", and may be the point at which a few scientists began to take on that role in the political process which has nowadays become so indispensable. Anyway, the program finally died with a close vote in the Senate and we all breathed easier.

Not so in Europe. Britain and France differ from us in two ways in that they never developed the non-radical, avuncular or Goldberger-Garwin-Bethe type of public interest scientists, because the Official Secrets act and the whole culture enforces the limitation of serious discussion to behind closed doors. Often this leads to enlightened results, as in the excellent British environmental record and the well-engineered and developed French nuclear power program; but it can also lead to disasters as in the case of the Concorde. Even more severe was the fact that the Concorde supersonic transport was initially undertaken as an international project to demonstrate the romance of a high-technology collaboration between the British and the French and to help repair their political difficulties on British Common Market membership. At no time could one partner drop out when its government had come to some sane resolution on the crushing expense and idiotic environmental insult caused by this machine. But neither can they afford to fly more than 10 of the damn things.

We in Cornwall still have to live with the boom. Somewhere in darkest Cornwall there is a beautiful Elizabethan mansion called Trerice, whose finest feature was an enormous Elizabethan window almost all of whose panes were original — before Concorde. I had the unique honor of belonging to organizations fighting supersonic transports on both sides of the Atlantic. Concorde was one of the reasons why I so recoiled when I read the proposal to create an international fusion laboratory to foster US-Russian collaboration, which was the proximate cause of this essay.

It is often said that if we had socialized the horse and buggy it would still be our main means of transportation. Frankly, I'm not sure that would be so bad and in fact sounds like one of the best arguments for socialism I know, it *can* stop progress — but seriously, socialization is no match for internationalization for the preservation of irrelevance.

Unfortunately, although in the end Nixon was driven to sign Salt I and the ABM treaty and to drop the SST and the ABM, the net casualty of those years was the White House science apparatus — partly because of the scientists going public, partly because of the less public institutions such as JASON disagreeing so repeatedly on policy matters and not confining themselves to helpful calculations, so that a deep distrust of scientists arose within the bureaucracy. Even within the Ford and Carter administration, which were more or less friendly to science and had a working advisory apparatus, scientific opposition on such issues as the Bl bomber, the MX, and wiring up Wisconsin for submarine communication seemed to be totally ignorable and to be automatically regarded as a symptom of effete liberalism rather than a genuine regard for the security of the country.

The scientists and technologists themselves have not been entirely blameless for the mistrust in which they were held. The great energy fiasco is perhaps unfairly held against them, even though many of the sensible ones were opposing the extravagances and inefficiencies they could see being propagated around them in the name of solving the energy crisis. It is hard to apportion the blame for the nuclear power failures; it is partly the public interest lobby gotten out of hand, to be sure, and partly the inefficient non-vertically integrated economic and managerial structure of the power industry, where a machine designed by a genius has often to be run by an idiot. But also we failed, most of us at least, to make the sensible economic and technical projections, to recognize the possibilities of even a small application of appropriate technology, and — let's face it — a lot of scientists were "on the take" to take advantage of another Sputnik-like flap. I could tell you some sad stories about photovoltaic technology, and I felt laser inertial fusion was an even more sordid and extravagant mess, not to mention the oil shale game with its absolutely insoluble environmental problems waiting for it at the end of the line. With the true energy crisis, not to mention the CO_2 crisis, still waiting for us around the corner, it is a terrible pity that we have spent our goodwill on that issue already.

With the disarray of the old friendly — in fact often cronyistic — relationship of scientists, their patrons in Congress, the military, and the White House, the physicists find themselves in a new and often disturbing world. The sum total of our mistakes and extravagances have left the US no longer the absolute power in world science. We have spent ourselves up to the point that the next steps always require more expensive technology, at the same time that a military out of control is hogging an ever greater fraction of available funds. It is natural, in this breakdown of the usual channels, that direct appeals to the pork barrel instincts of Congress are going to become commoner. I even was somewhat involved in trying to stop a pork barrel scheme, the so called NCAM in which the President's science adviser himself was allied with the California Congressional delegation to bypass the normal DOE reviewing procedure on a major light source costing about 10^8 dollars.

But let me before closing reassure, to some extent, those of us who are disturbed by the pork barrel. This alone is possibly the least objectionable type of political interference with science — it is capable of contributing to lasting and productive institutions as in the case of many of the National Institutes of Health and their congressional patrons, or the heyday of agricultural experiment stations. We physicists ourselves have repeatedly used it to gain sites and advantages for our large high-energy facilities. What we must do is to have no hesitation in going public with our peer review if necessary, as we did in the case of NCAM, forcing Congress to listen to scientific strategy as well as press and group arguments. If the community doesn't want a facility, often Congress will not fund it.

I'd like to finish by discussing what are conditions for a healthy science and technology. Above all, this involves *stable*, and preferably non-invasive, even if not enlightened, support. The Bell Labs in its heyday hardly grew, but it did not fluctuate downwards ever. Reading the history of the '30s to '50s one is struck by the disproportionate effect of small amounts of Rockefeller money granted with no strings attached, and in the modern day the Max Planck system in Germany stands out. As a horrid example of stop-go funding I see the Thatcher government's savage cuts followed by the so-called "New Blood posts" which were supposed to attract back the best young men. But of course the best young people are all stably funded here or in Europe and wouldn't dream of going back to England, as anyone who knows scientists would assume. Sure enough, famine again followed the infusion of New Blood; in fact the restrictions set around the New Blood jobs were so severe that they were almost guaranteed to fail.

I *have* mentioned NASA and the Explorer missions with the destruction of that magnificent team just at its moment of success. There are even continual threats to kill off the tiny funds for the final stages of the Voyager mission. In a discussion with Harold Furth of Princeton's Plasmalab he told me that what he dreams of is stable, long-term funding, and fears most is a feast-famine regime.

A sound technological base is almost purely a matter of economics: make it profitable to innovate and innovation will happen. But we must *train* and *reward* the innovators. It is not a coincidence that Japanese engineers are the most prestigious and well-paid college graduates, and engineering attracts the best students, while in Russia engineers are paid less than blue-collar workers. The Japanese do not trust manufacturing to businessmen alone — engineers are in charge at all levels from the factory floor up. This too was a unique strength of Bell Labs, that all the managers were technical men, along with the coupling of technology to innovative science. This "technology transfer" problem is the bane of every technical manager, and is the one hardest part of the whole technology cycle. May I repeat again that by far the best solution to it is the legally proscribed (in the USA) device of vertical integration.

Finally, I must mention that another main priority must be the reestablishment of collaboration and trust between the Administration and the scientific community. We are the bemused and flabbergasted witnesses of an era in which government redefines and creates facts rather than responding to what everyone else sees as objective reality. This, more than anything else, leaves scientists isolated and, in many cases, bitter; scientific facts cannot be erased without in effect erasing the scientists' work as well, among other dire consequences. In the long run our leaders will have to learn that valid and impartial scientific advice is a necessity in the modern world for long-term national survival.

The Case Against Star Wars[*]

I am not an expert on strategic weapons. I'm a theoretical physicist who has been involved in almost all of physics except atomic bombs. I have not done classified work since 1945, and that was on radar. My total contribution to the laser — a major technical component of the Strategic Defense Initiative, which is better known as Star Wars — was roughly that when one of the scientists at Bell Laboratories who originated the things asked me to predict whether a certain seminal version of it would work if they built it, I said, "Well, maybe."

Fortunately, most of the scientific issues that come up in discussing Star Wars are very simple ones which require neither specialized nor especially technical — and therefore classifiable — knowledge. One needs to know that it costs everyone about the same amount to put a ton of stuff into a given orbit and that this is a major portion of the cost of any space system; that signals can't travel faster than the speed of light; that it takes roughly as much chemical fuel to burn through a shield with a laser as the shield itself weighs; that Americans are not measurably smarter than Russians; and a few other simple, home truths. Given these, almost everyone comes to much the same conclusions.

If you go through the enormously detailed kinds of calculations on specific configurations which Richard Garwin and his fellow opponents of SDI felt necessary to convince the stubborn, you leave yourself open to the kind of errors of factors of 2 or 4 which Martin Muendel '86 found in his widely publicized junior paper last spring [PAW, May 8] and which then — to the lay person — seem to weaken the whole structure. This is a particularly tough game because Star Wars advocates do not themselves

---

[*]Originally published in *Princeton Alumni Weekly*, 25 September 1985, pp. 10–12.

propose specific configurations and present specific calculations that can be shot down; their arguments are given in terms of emotional hopes and glossy presentations. This is why I think it is good for the argument against SDI to be made by a lazy non-expert person like myself who isn't particularly fascinated by the technical detail.

The reasons for not building Star Wars are essentially identical to those which led both us and the Russians to abandon, for practical purposes, the antiballistic missile in 1972 and to sign a treaty restricting ABMs. It is important to understand that reasoning — and perhaps it is less emotionally charged than Star Wars since it is now history and not even controversial history anymore. Why would anyone feel that a defense against missiles was useless and, in fact, dangerous and destabilizing?

There are three stages, each more certain than the last: (1) It probably wouldn't work, even under ideal conditions. (2) It almost certainly wouldn't work under war conditions. This puts us in the dangerous and unstable situation of the gunfighter who doesn't know if his gun is loaded. (3) Most certain and conclusive of all, *each defensive system costs, inescapably, at least 10 times more than the offensive system it is supposed to shoot down.* Thus it pays the other side to increase its offensive arsenal until the defender is bankrupt, and the net result is an *increase* in armaments and a far more dangerous situation, without any increase in safety.

The offense has, inescapably, enormous advantages: its missiles are sent at will, in any desired sequence and quantity, with any number of decoys and other deceptive countermeasures, preprogrammed at leisure to hit their targets; the defense has to find them, sort them out, get into space at a time not of its own choosing, and then kill the warheads it finds with nearly perfect accuracy. In the case of ABM, there were other problems, such as that the explosions were over the defending side and that the first few explosions probably blacked out the whole shooting match, but that was sufficient argument against.

As far as almost everyone in and out of the Defense Department was concerned until March 1983 this situation was an accepted fact. No technical breakthrough had or has changed those realities. The change has been purely political and emotional, and hence now financial. President Reagan's March 1983 speech, as far as anyone can ascertain, was not preceded by any serious technical review, but quite the opposite: the most recent and urgent internal study of antimissile defenses had come out negative on all possible schemes.

Apparently, the President based his speech and his subsequent program on a collection of rather farfetched suggestions — farfetched but by no means secret and previously unknown — which, to the outside scientific observer, seem to deserve the oblivion that the last pre-Star Wars study consigned them to. These schemes amount to a way for the defense to spend more per missile and still let through a large fraction of the offensive missiles. The defensive hardware that has to be got up into space still has to have roughly the same mass as the offense; in many schemes it has to get there faster; and it still has to be much more sophisticated and therefore vulnerable and delicate. Key components, in most schemes, have to be left in space indefinitely, inviting the enemy to track them with space mines, perhaps the most dangerous trip-wire mechanism for starting a war that one can possibly imagine.

Some Star Wars advocates will protest that I do not mention the one idea which doesn't founder just on the problem of total mass in space. This is the scheme of exploding hydrogen bombs in space and directing the explosive energy of the bombs with lasers to kill very many missiles per bomb — several hundred to several thousand, if one is to kill an equivalent cost in missiles! If I could think of any way such a monstrosity could work, as opposed to the many ways it could not work or be frustrated, I would take it more seriously. Apparently there has been some good and interesting science done on these lasers, but unfortunately it is classified; no one, however, seems to claim that it helps much with the technical problem. I cannot, incidentally, see any way to do meaningful development on such a weapon without exploding H-bombs in space, a terrible pollution as well as violation of what treaties we have.

I think the above would represent reasonably well the views on the technical realities of most trustworthy physicists to whom I have spoken, in or out of academia and in or out of the Star Wars program. In academic physics departments, which receive relatively little support from the DOD, a pledge form has been circulating stating that the signer opposes SDI as unworkable and will not seek SDI funds; this has had a high percentage of signers everywhere it has been circulated and its preliminary circulation in Princeton over the summer encountered only a few holdouts. Those who do not sign feel, primarily, that research in any guise shouldn't be opposed, while agreeing personally that the systems proposed are unworkable and destabilizing.

Perhaps it would be worthwhile, therefore, for me to explain why I feel the large increment of research funds remarked by President Reagan for SDI is a very bad thing for the research community, as well as for the country as a whole. You will note that I said *increment*: every year before Star Wars, we spent $1 billion in ABM research and development. My main reason is that, on the whole, Star Wars will represent a further acceleration of three extremely disturbing trends in the direction of research funding in this country.

First, we are seeing a decrease in basic research relative to mission-oriented, applied research. The basic research agencies — National Science Foundation, Basic Energy Sciences in the DOE, and National Institutes of Health — have been maintained at level funding while their missions have been gently skewed toward applications and engineering by piling more applied responsibilities on them. At the same time, while the Administration has cut back on development in some civilian sectors, it has more compensated by increasing the amount of applied work for the military.

Second, there is a trend away from scientific administration of federal research money — mostly done by the system of "peer review" — to micromanagement either by bureaucrats, or, increasingly, by Congress, with all the logrolling possibilities that entails. The three institutions mentioned above, especially NSF and NIH, operate by subjecting each grant to a jury of other scientists. Like most democratic procedures, this system is worse than everything except the alternatives; its effect has been reviewed repeatedly and there is no serious doubt that it works. Military "research," on the other hand, has always operated on the arbitrary whim of the contracting officers. In the early days after World War II this administration was a benevolent despotism, but the adjective has long since lost its meaning. Most of the in-house DOD laboratories have been rather a scandal in the research community. The dominant motivation in this system seems to be the standard bureaucratic one of "empire building".

Third, from the point of view of the country as a whole, perhaps the most dangerous trend is the shift from civilian to military dominance of our federal research and development spending. Under the Reagan Administration, this has grown to 72 percent military, up from about 50 percent a decade ago. Everyone has been told — the DOD sees to that — of the great economic benefits of "spin-off" from military development, but if they exist (and I have never found an economist who believes in them), they are not evident in our recent economic performance vis-à-vis Japan and Germany. In fact, in a country like ours with a serious shortage of trained engineers and scientists, a shortage which would be crippling if we did not attract great numbers of them from overseas to staff our universities and research laboratories, the waste of our precious technical expertise on military hardware is a serious economic debit.

From Princeton's point of view, all of these trends are disturbing. As a top-flight research university, a heavy percentage of our funding is in individual support of independently functioning basic scientists, mainly peer-reviewed and to a large extent from the agencies mentioned above. We have not had to resort to logrolling political tactics, nor have we had to accept micromanagement, DOD control of publications, or limitations on citizenship of students to keep our research funded. SDI control of funding, and in general the shift of research funding to the military, is a serious danger to the independence of Princeton as a research university.

Of course, this is a narrow and slightly parochial view, but it is nonetheless serious. Certainly it is more important that the naïve emotional appeal of the Star Wars concept is being used to blatantly defuse the country's strong desire for nuclear disarmament, and to turn this emotional pressure into yet another excuse for enriching the arms manufacturers and building up a dangerous and worthless arsenal of nonsensical armaments. To paraphrase Murph Goldberger's testimony on the ABM: Star Wars is "spherically" senseless — that is, silly no matter how you look at it.

Princeton and the Bomb

I suppose Princeton can be given the dubious honor of being the birthplace of the Atomic Age: it was here that Einstein wrote the famous letter to President Roosevelt which, at least in a symbolic sense, set off the development of the bomb. The scientific and political history of the Ultimate Weapon is dotted with Princetonians: Eugene Wigner and Richard Feynman both made major contributions to the development of the bomb; J. Robert Oppenheimer directed the Institute for Advanced Study during and after his period of power and influence; our Henry De Wolf Smyth in 1945 wrote the report which summarized the Manhattan Project; John von Neumann midwifed the calculating machines which made modern weaponry possible; and there are many others.

What is less well known is that Princeton is, in this latter day, a center for the rational — and occasionally irrational — discussion of what to do with the damn things. Our two most famous philosophers of the strategic equation are at the Institute for Advanced Study: George F. Kennan, the patient and hard-headed advocate of living with the Russians instead of dying with them; and Freeman Dyson, whose book *Weapons and Hope* is a very personal but characteristically sane discussion of the entire spectrum of the problem of nuclear armaments.

On two other levels Princetonians concern themselves with these problems. For one, over several decades our scientists have worked within the many advisory bodies to the government which study strategic questions, such as — when it existed — the President's Science Advisory Committee (PSAC), and the still viable organization known as JASON, which brings the talents of the country's most brilliant scientists to give private (in fact classified) advice to the Secretary of Defense. As a member of PSAC, for instance, Marvin L. Goldberger (then a member of Princeton's Physics Department, later its chairman, and now president of Cal Tech) is known to have been an extremely influential advocate of the ABM treaty. Physicists at the University and the Institute continue to work in JASON.

Second, on the level of open study and public discussion of arms, strategy, and arms control, and occasional advocacy of specific measures, we have a very active group in the Center for Energy and Environmental Studies of the Engineering School, with close contacts to the Wilson School. A regular, even busy, program of seminars and studies on arms control is maintained by this group, and the main topic of interest this fall will be the Strategic Defense Initiative (SDI), better known as Star Wars.

Senior personnel of this group are: Frank von Hippel, a Wilson School professor and past chairman of the Federation of American Scientists, the major scientists' lobbying group on arms control issues; Robert Socolow, director of CEES, and Hal Feiveson and Barbara Levi, members of the center; and Richard Ullman of the Wilson School. Many of us from other departments, notably physicist Jeff Kuhn, participate in such functions as the group's regular Thursday lunch seminars.

It was through Kuhn's interest in such issues that SDI first came to PAW's notice: he was the local supervisor and contact for the widely publicized junior paper by Martin Muendel '86, written under the instruction of Major Peter Worder of SDI. Last spring Martin was reported by PAW (and several other publications) to have "refuted" the calculations of Richard Garwin and his fellow opponents of SDI. When I objected that this story, while factually accurate in the narrow sense, did not properly represent the situation or the attitude of Princeton physicists toward Star Wars, I was invited by PAW to represent what I see as the majority attitude, which I attempt in the accompanying article.

— P.W.A.

A Dialogue About Star Wars[*]

Somehow this dialogue caught the august eye of the editor of the prestigious Le Monde Diplomatique, and it was translated and featured on the front page of their December 1986 issue. This led to a rather delectable episode: our annual visit to Cornwall in April '87 ended with a side-trip to France, during which we were feted at a slap-up dinner near the Pantheon by Le Monde; Joyce sat next to a very charming man we later learned was the editor-in-chief of Le Figaro. As I said elsewhere, none of the scientific advice given to the government had any effect, and the missile "defence" system continues to limp along — as unsuccessfully as we predicted, to be sure — as Reagan sainthood becomes ever more politically invincible.

By George P. Shultz[†]

The article by Professor Philip W. Anderson entitled "The Case Against Star Wars" in the PAW of September 25, 1985, has prompted me to respond with my thoughts on this important subject.

Soviet Strategic Defense Program

The first point to make is that the United States is not the first and not the only country to launch a vigorous research effort to explore new technologies for strategic defense. Professor Anderson's article contains

---

[*]Originally published in *Princeton Alumni Weekly*, 26 March 1986, pp. 10–12.
[†]George P. Shultz, an economist who has held several high government posts, received Princeton's Wilson Award in 1971.

no mention of the Soviet strategic defense program, and no discussion of the potential consequences of a Soviet monopoly in this field. Soviet strategic defense efforts have proceeded largely unacknowledged by the Soviet government and certainly free of the public discussion that accompanies every important defense decision undertaken by the U.S.

While the build-up of Soviet strategic offensive forces is well known, less attention has been given to the fact that over the years the Soviets have devoted the same magnitude of effort to strategic defenses — air defense, missile defense, and civil defense. They have taken full advantage of the ballistic missile defense deployments around their capital permitted by the ABM Treaty, which they are modernizing at substantial costs, and have gone beyond this to construct a large radar near Krasnoyarsk which violates provisions of that treaty.

Each of the advanced defense technologies being pursued in the SDI research program is also being pursued by the Soviets. For example, more than 10,000 scientists and engineers are at work on laser weapons at Soviet research and development facilities and test ranges. The Soviets have a laser at the Sary Shagan test range which could be used against U.S. satellites. They also are conducting research on kinetic energy weapons, particle beam weapons, and radio frequency weapons. The Soviet strategic defense program is described in some detail in a report published by the Secretary of Defense Weinberger and myself in October 1985.

These efforts long antedate the President's initiative of March 1983. We do not know whether or when the Soviet strategic defense program might lead to a Soviet decision to abandon the ABM Treaty altogether and deploy a large-scale missile defense system. It seems only prudent that our research keep pace, both as a hedge and as a deterrent to Soviet breakout.

Soviet Strategic Offensive Forces

Since the signing of the ABM Treaty and the SALT I agreement on strategic offensive forces in 1972, the Soviets have deployed a variety of new strategic offensive weapons. They have increased the number of warheads on their missiles by a factor of four and substantially improved their accuracy as well. Through such enhancements, Soviet forces increasingly threaten the survivability of our retaliatory forces, and especially our land-based missile system and command structure. We have under way a program to modernize our strategic retaliatory forces in order to re-establish and maintain the offensive balance in the near term.

However, over the long run, the trends set in motion by the Soviet build-up suggest that continued long-term dependence on offensive forces alone may not provide a stable basis for deterrence.

One objective of the SDI is to provide future options for ensuring deterrence and stability over the long term. We do not seek superiority or unilateral advantage; we are exploring the potential of new technologies to strengthen deterrence by reducing the role of offensive ballistic missiles and placing greater reliance on defenses.

Criteria for Evaluating Defenses

We have publicly stated that we will judge defenses to be desirable only if they are survivable and cost effective at the margin. Survivable defenses would not tempt an adversary to strike first in a crisis. On the contrary, they would contribute to deterrence by sharply reducing the potential gains an adversary might realize from such a strike.

Professor Anderson asserts that defenses inevitably can be overcome by offenses at one tenth the cost. Many talented scientists and engineers believe that this assertion is not correct, and are exploring the possibility that modern technology may permit defenses that meet our criteria and would contribute to stability and security. A defense whose capability can be expanded at a cost which is small compared to the cost of the additional offense necessary to offset it would discourage further expansion of offenses and encourage their reduction.

The ABM Treaty

SDI research is being conducted in strict compliance with the Anti-Ballistic Missile Treaty. When the treaty was negotiated, both sides considered limits on research to be neither desirable nor possible. Former Soviet Defense Minister Marshal Grechko stated at the time the ABM Treaty was signed, "At the same time, it [the ABM Treaty] imposes no limitation on the performance of research and experiment work aimed at resolving the problem of defending the country against nuclear missile attack." Both parties recognized that it would be impossible to devise effective or verifiable constraints on research.

If and when our research criteria are met, and following close consultation with our Allies, we intend to consult and negotiate, as appropriate, with the Soviet Union pursuant to the terms of the ABM Treaty on how deterrence could be enhanced through a greater reliance by both sides on new defensive systems. The ABM Treaty provides for such consultations. In fact, we are already trying to initiate a discussion of the offense-defense relationship in the arms control negotiations in Geneva. It is our intention and our hope that, if new defensive technologies prove feasible, we and the Soviets will jointly manage a transition to a more defense-reliant balance.

SDI and Our Allies

President Reagan has made it clear from the start that SDI research is being conducted with the objective of enhancing Allied security as well as our own. We are working closely with our Allies to ensure that their views and interests are taken into account.

We have offered to our Allies an opportunity to join with us in research program. Several countries are making arrangements under which their scientists, laboratories, and firms can participate in the research effort. We welcome this, and believe it will be in the interest of both the United States and the participating countries.

Conclusion

We do not have any preconceived notions about the defensive options that the SDI research program may generate. We have set in motion a broad research program to explore many different technologies, and have identified criteria which will be used to judge the results. We recognize that our goals for the SDI are ambitious, and success is not certain. Nonetheless, the potential benefits are so great that they justify a substantial investment to see if we may in the future base our security more on defending ourselves and our Allies and less on the threat of retaliation.

Professor Anderson Replies

I am flattered Secretary Shultz has chosen to discuss my article on SDI. His remarks are characteristically reasonable and moderate. I would like to address his points one by one.

Soviet "Defense" Program

We should not confuse the questions of Soviet intentions and integrity with whether the American SDI concept is physically viable or economically possible. Nor should our decisions be governed by our perception of what the Soviets are doing. There are many things the Soviets do which we need not copy; as an example, the Soviet manned space program is even more extravagant and less scientifically productive than our own and there would be little point in imitating it.

Prior to March 1983, we had a program of ABM research costing S1 billion a year, a sum equivalent to the NSF budget for all of civilian science and engineering research. At that time much of the research that is now considered as particular to the SDI effort was already in place. A number of studies judged that this level of expenditure was already a more than adequate hedge against Soviet breakout.

Soviet Offensive Forces

It is of utmost importance to emphasize that two-thirds of the U.S. deterrent forces (assuming that deterrence is meant to be the role of our weaponry) are submarine- or air-launched and invulnerable to the SDI concept of boost-phase intercept as well as to any Soviet first strike. The Soviets, for understandable if complex reasons, have a more vulnerable deterrent force. Under these circumstances, while I am sure Secretary Shultz's assurance that "we do not seek ...unilateral advantage" is sincerely meant, it is unlikely to impress the Soviets, especially as the Secretary cannot speak for future administrations.

As an aside, at this point I would like to say there is a problem in that most laymen do not really understand what is involved in SDI. Certainly just now there is an awareness of human fallibility. For our purposes here, I would like simply to explain that "boost phase" is the main difference from older ABM concepts, and it means that an umbrella of "kill vehicles" in space would cover a hostile country to guard against weapons (missile) being launched. Even assuming such an umbrella could be made to work, it seems likely that not only the Soviets but any other member of the nuclear club — say, France — would not accept this absolute control over its strategic (or deterrent) strength.

Criteria for Evaluating Defenses

I am glad to hear the Secretary restate these. The scientific community has seen sound technical arguments on weapons systems fail in the face of political pressure.

Secretary Shultz says, of the scientific dissent regarding SDI, that "many talented scientists believe otherwise". We should look at what talented scientists are trying to tell us. It is not at all usual to be able to sign up 60 percent of the physics faculties of the nation's top universities; nor to find near-unanimity among American Nobel Prize winners, nor to hear an anti-Administration resolution on a highly controversial matter from the Council of the American Physical Society (including several past presidents). It is truly extraordinary how few scientists of real reputation are willing to support SDI publicly — I can pretty well count them on the fingers of one hand, and of those I can count not a single one is still active in research or has been for at least a decade. Even among scientists who receive SDI funding, support is extraordinary soft; one hears privately, with some dismay, that it is justified because the concept of SDI is a useful bargaining chip, or a stick to frighten the Russians with, or that research money should be accepted from whatever source.

The ABM Treaty

There is a perceptible semantic problem when the Administration's political spokesmen speak of "defense" and the "defense" philosophy, while the actual SDI program, as described by Gerald Yonas of the SDI office or in the Office of Technology Assessment reports recently published by Princeton University Press [PAW, February 26], constitutes essentially a strengthening of our "deterrent" — a very offensive sort of defense. Many aspects of SDI (e.g., "Kill vehicles" over Soviet territory and local "hard target" defense of our ICBM's) may sound more belligerent than defensive to the Soviets.

SDI and Our Allies

Is it not too early to address the problem of what worldwide reaction to a global U.S. policing of space might be? (See "aside" above.)

Conclusion

I am heartened that Secretary Shultz "has no preconceived notions" about SDI. But in speaking of potential benefits, he neglects the many potential costs: the militarization of research, for instance, and, should the system be deployed, a great shift in our entire economy. It seems to me the greatest danger the Soviets can put us in is that of becoming more like them, and this is, in terms of military dominance and obsession with "security", the direction in which Star Wars leads us.

The secret of our extraordinary success in building and maintaining history's most powerful, permanent, and effective defensive alliance has been the perception that our civilian-dominated economy and open society are far more attractive and far less threatening to any nation which has anything to lose than is the secretive and militaristic Soviet system. In militarizing our economy we can lose that most valuable of all advantages.

No Facts, Just the Right Answers[*]

I don't know how it is in Britain, but in the U.S. science and politics are never very comfortable with each other. The message of Chris Mooney in *The Republican War on Science*, substantiated by a number of well-researched case studies, is that under George W. Bush the relationship of science with the Republican hegemony is worse than uncomfortable: it has deteriorated into something like open warfare. Mr Mooney dates the beginnings of this deterioration to the Reagan Revolution of the early eighties, but notes a remarkable acceleration under Bush the younger.

It was Ronald Reagan who formed the alliance of conservative Southern and Midwestern populism with the economic radical Right of "new money" capitalism which dominates the present political scene in the United States. The populist wing of this alliance subscribes to various forms of religious fundamentalism, with its strong bias against the study of evolution and modern biology in general, while the economic right opposes science-based regulation in any form. The two streams coalesced in such figures as James Watt, Reagan's first Secretary of the Interior (responsible inter alia for the national park system), who, believing in the imminence of the Second Coming, opposed all forms of conservation.

Prior to Reagan, science had fared much the same at the hands of either party. President Nixon, for instance, fired his science advisor in a fit of pique at scientific opposition to his pet projects of the SST and the Anti Ballistic Missile system (ABM), yet signed into law several of the key environmental measures now under threat. In response to the

---

[*]Book review of *The Republican War on Science*, by Chris Mooney (Basic Books, New York, 2005). Originally published in the *Times Higher Education Supplement*, 23 December 2005.

Nixon administration's downgrading of the science adviser's office, Congress came both to rely more and more on its own Office of Technology Assessment (OTA), and to contract for reports on controversial subjects from the National Academy of Science or other impartial sources. Also there grew up a competent regulatory civil service in the Food and Drug Administration (FDA), the Environmental Protection Agency (EPA), the Centers for Disease Control (CDC), and other agencies. The industrial giants of the midcentury, such as AT&T, IBM, Dupont, and GE, contributed major figures to the scientific advisory apparatus and strongly supported technical and scientific education, even though, not unwittingly, they were at the same time resisting environmental regulation and leaving behind horrendous pollution problems. Throughout that period public opinion polls gave scientists a high score for trustworthiness relative to other professions.

With the accession of Ronald Reagan the more or less comfortable relationship between science and government began to unravel. Reagan's own attitude toward science seems to have been complete and cheerful ignorance, and a seemingly deliberate determination to ignore any bad news it brought to his attention. Reagan may have been the first presidential candidate since the nineteenth century to question evolution on the campaign trail, and as governor of California he had promoted weakening the teaching of evolution. A number of his appointees were antievolution; even his science adviser, George Keyworth, equivocated on evolution during his confirmation hearings, not his last abandonment of scientific integrity.

An anecdote may serve to illustrate Reagan's attitude toward science. After two years in office he was persuaded to award the ten National Medals of Science, due in 1981, for 1982; but the formal award ceremony was put off until fairly late in 1983. His five-minute speech for the occasion made two points: that he regretted his inattention in science courses in school, which precluded his being himself a medalist; and that science is wonderful at producing better weapons. In attendance myself, as far as I can remember none of us had worked on weapons except for Edward Teller, added as an eleventh recipient at the last minute. Reagan then promptly took off for a photo op in the Rose Garden with a professional hockey team.

The biggest scientific brouhaha of the Reagan administration was undoubtedly SDI, a.k.a. Star Wars, the program for a space-based antiballistic missile defense which emerged in a Reagan speech of April, 1983. The program seems to have originated with a private fringe group called

High Frontier, and to have been firmly quashed within the Pentagon, revived only by Reagan's embrace. A detailed history of the Star Wars episode was published in the book *Way Out There in the Blue* by Frances Fitzgerald (which I was surprised to find was not referenced in the present book). Her conclusion was that the program was finally dropped not because of the highly professional public reports on its impracticality produced by OTA and by organizations outside the government such as the American Physical Society, but because the Pentagon's own internal advisory apparatus worked as intended. This fact is frightening because, as we shall see, under George Bush such internal checks are steadily being dismantled.

It is regrettable that President Clinton felt it politically necessary to reinstate a reduced program of missile defense. Despite this reduction the program continues to be an embarrassment and a waste, although its very incompetence mitigates the threat of the militarization of space, the worst aspect of Star Wars.

An aspect of the Reagan administration's record which Mooney recounts in some detail is the struggle of the surgeon general, Everett Koop, to get any attention focused on the rise of the HIV pandemic, a rise which was almost coincidental with Reagan's terms of office. During the entire first term Koop was prevented from speaking on AIDS publicly — Reagan himself never mentioned it before 1987. As with the present Bush administration, there was strong opposition to preventive measures such as the distribution of condoms.

Mooney does not feel, however, that the Reagan administration and its successor, the first Bush administration, were unequivocally anti-science. Funding for medical science was constantly increased by the Democratic congress, and physical scientists managed to switch to an extent from civilian to military funding in spite of a widely successful self-denying oath against SDI funding which was circulated in 1983–4. A congressional assault on the international agreement on ozone-destroying chemicals was turned back by a Reagan appointee; Everett Kopp was eventually ungagged on the AIDS crisis; President Bush senior's science adviser was the relatively effective and well-respected Allan Bromley.

Actually, the opening battle of the Republican war on science, according to Mooney, was won under Clinton's presidency, by the Congressional majority achieved in 1994 by Newt Gingrich's Contract With America: Gingrich as Speaker abolished the OTA, Congress' only official source of scientific expertise.

As Mooney emphasizes, well before 1980 the right had begun to develop an intellectual-political wing. A variety of conservative "think-tanks" such as the Heritage Foundation, the Hoover Institution, and the American Enterprise Institute, as well as the National Review of William Buckley, grew up in the 1960s in response to what seemed to be the disastrous collapse of conservatism in the 1964 election. The "neoconservative" movement of today, which has such a strong influence in the G. W. Bush administration, also dates from this period. As the influence of this "intellectual" outlier of the conservative revolution grew, the attacks of the right on science began to take on a new and more damaging aspect. In my opinion, this book is strongest in its analysis of the tactics developed by this fusion of Reagan populism with the intellectual sophistication of the neocons and their conservative brethren.

One of these tactics is broadly characterized as the "sound science" approach, that phrase being a Newspeak (Newtspeak?) designation for something which is precisely the opposite. Once Congress had abolished the OTA (whose mandate was to report the scientific consensus on any controversial issue) they were free to assemble their own panel of "expert" witnesses, and to pick and choose among them. Since there are inevitably "outliers" somewhere in the scientific community who are available to present views which contradict the consensus, it is always possible to bias the selection of witnesses toward any given opinion, and to call the fringe opinions "sound science", in contrast to the consensus opinions voiced by such organizations as the National Academy of Science. The outliers can often be identified on the basis of two characteristics: first, that a little digging often reveals a financial interest in the outcome, and second, that the same few names seem to recur as witnesses on issue after issue.

One of the more persistent of these latter, for instance, is Dr S. Fred Singer, a member of the small but very vocal group of contrarians outside the government who supported SDI publicly; he has testified before Congress against the reality of "acid rain" pollution and against the ban on fluorocarbons imposed to save the ozone layer, and now is called on to testify against the reality of global warming. He seems to represent nothing but an organization called the "Environmental Policy Institute" of his own creation, and to have no widely recognized expertise on any of these subjects. But he gets around; I was shocked to find him giving our physics colloquium here at Princeton — on global warming, no less, or call it global non-warming. Since then the data he was using to refute the consensus view has been shown, predictably, to have resulted from a calibration error in one type of study.

It is with the Bush administration that the present condition of virtual open warfare between politics and science comes to a head. The first rumblings came from the handling of appointments to advisory committees within the department of Health and Human Services (HHS) under Bush's new Secretary, Tommy Thompson. These committees have little official but considerable actual power, since the agencies need them to test the scientific consensus on a given subject. Appointments to them have always been considered apolitical, and membership can outlast several administrations. Quite suddenly one began to hear of cases where long-term members were eased out, and where new members seemed to have been vetted on their politics — at least one prospect being asked directly whether he had voted for Bush.

On several issues, and these perhaps the most important ones, the problem has not been one of stacking the advisory committees but of disregarding technical advice altogether. Vice-president Cheney was put in charge of the "task force" developing a national energy policy, and since the very first year of the administration he has absolutely refused to reveal whom the task force consulted; no scientists are known to be in that number. Energy company executives are Cheney's main personal contacts. The only "scientific" initiative on energy has been the futuristic dream of a hydrogen fuel cell economy; here as elsewhere it is a feature of this administration that the financial burden of its policy is timed to mature under its successors. (I believe that there has been a directive to White House staffers to turn off computers at night.)

There also appears to have been no advisory input on the "Moon–Mars" program, the only big scientific initiative to be announced by the president. In the past decade or so there had grown up a highly successful program of science in space carried out by unmanned robot missions, of which the Mars Explorers are the poster children, a shining contrast to the disasters of the manned shuttle and space station programs. With great fanfare Bush has announced that the unmanned program is to be curtailed in favor of an enormously expensive manned program for colonization of the moon and Mars (the major fiscal burden predictably falling on later administrations).

The book's central chapters, however, focus on a series of issues where administration policy has come into greater explicit conflict with scientific advice. Climate change — in Republican Senator Imhofe's words, the "greatest hoax" — is the most important of these, starting with one of Bush's first acts as president, the abrogation of the Kyoto agreements, and followed by virtual abandonment of fuel economy standards.

Again and again, as Mooney describes, the attempt has been to attack the science and the scientists who identified and researched the warming phenomenon, both in the halls of Congress and in the media. Given the unanimity of accepted scientific opinion, these attacks indeed amount to a war on science generally, even when they are transparently disguised as calls for "more research". (Perhaps "sound science" refers to acoustics?)

As Mooney describes in two central chapters of his book, comparable assaults have been mounted on the science behind the Endangered Species Act, the control of power-plant emissions both with regard to mercury emissions and acid rain, and on a United Nations WHO-FAO report on nutrition, obesity and health. This latter case involved an unprecedented tactic, the assertion of a right of the Administration to select the government employees who were to be called upon by UN bodies for technical assistance and advice.

Since before George Bush assumed office, perhaps the most heavily politicized set of issues related to science have been those involving reproduction and sex. Here Bush has been assiduous in following the doctrines of his religious supporters. Already in a campaign speech he misstated the scientific facts in proposing impractical limitations on stem cell research; his administration continues to promote the views of "outliers" on the realism of his regulations. The most recent of many incidents has been about the "Plan B" emergency contraceptive, where political appointees within the FDA bypassed the normal approval procedure and ignored a unanimous advisory committee recommendation.

In concluding chapters Mooney describes what amount to ineffectual attempts to restore scientific integrity in the administrative councils. He perhaps ascribes too much impact to the manifestos organized among scientists to bring the situation to public attention — scientists were, for instance, almost equally unanimous in condemning SDI, and the public ruckus we created had little effect. Certainly Jack Marburger, Mr Bush's nominal science advisor, cut a rather pathetic figure attempting to counter these manifestos, but even the well-informed public hardly noticed. In my opinion we can hope only that the multiplication of disastrous policies resulting from exclusive solicitude for the Republicans' "base", and the determination to weaken the federal government by diverting power to corporate leaders and the churches, will in the end lead to the breakup of this Republican majority.

A good point is that a big part of the problem is a mistaken insistence on "balance" in the media. Science is not about negotiation and compromise, as is the political world that reporters work within; it is about the search for right answers. Of course many scientific issues are not open and shut, but reporters should realize that the outcome can be very one-sided.

While his assessment of the political weight of scientific opinion may be dubious, Mooney's epilogue titled "What We Can Do" is certainly an excellent guide to how the advisory apparatus could be reformed. He is correct that restoring the Office of Technical Assessment is key. He is correct that the legislative and administrative measures of the "sound science" movement must be removed. We must hope that a centrist president can restore the science advisor and his staff organization to the inner circles. The world is global now and fast-changing, and fact-based advice is vital.

This book is a well-researched guide to the recent history, and is much to be praised both in its original analysis of the tactics employed by the New Right to starve the scientific advisory apparatus, and in bringing out the confrontational nature of the attitudes of the Bush administration and its allies. It should be read, certainly by Americans, and will be interesting to scientists everywhere. We should be grateful to Chris Mooney for his diligence.

VIII. Futurology

Introduction

I have only one slightly negative prejudice about science, a somewhat strange one since I started out as an avid reader of science fiction. I really react badly to those who enthuse extravagantly about the future wonders of technology, in fact to anyone who presumes to see in any detail what the future is likely to bring. This chapter contains a few instances where I let my irritation show — there are others.

Futurology

I feel a bit shamefaced publishing this lecture from Volker Heine's, Mike Mulkay's, and my course of 1972–3. But except for the observation about tobacco — and that did take a number of years even in sensible countries, and is still true for most of the world — it is as true now as it was then. Adjust the numbers for inflation, and the personnel for the passage of time, and you will find most of it reads like this year.

Substitute Freeman Dyson for Dennis Gabor; one of several bloggers for Ivan Illich; etc. When I gave the lecture Erio Tosatti donned a robe and ran out on the stage with a fishbowl screaming "Master, Master!" — but my wife was unimpressed.

Futurology is the kind of non-word that gives American English a bad linguistic reputation. Like much such lingo, it has a validity in that it expresses a vaguely understood concept, somewhere short of prophecy, and beyond statistics. Sometimes our thoughts give shape to words, and the quality of the word may affect the quality of the thinker, as those who watched Herman Kahn, the King of the Futurologists, on BBC2 the other night will appreciate.

The purpose of this lecture is primarily to, as it were, help us all to find our way around the future, and for that purpose I'd like to divide my talk among these chief categories (aside from pure science fiction, to which much of this stuff has a close resemblance):

(1) Prophets (including anti-utopians, etc.)
(2) Extended Gadgeteers: Innovation Prediction and Futurology, versus Technology Assessment
(3) Doomwatchers versus Cheermongers

Of which the last really usurps the majority of what I have to say. Then I will end with a vague category, perhaps as yet still empty, which we might call

(4) Synthesis: The Third Watershed?

Before discussing any of these things I'd like to make four remarks which amount to injecting a note of skepticism, where perhaps 20 or 50 such notes belong.

Historians usually are found to disagree, often strongly, about even very recent historical events — say the Chinese revolution — not just about causes but sociological and material *facts*. This leaves one a bit queasy about predicting the future.

Even if we understood the future reasonably well, could we do anything about it? The note of skepticism here comes from watching human responses in such areas as smoking and health. Since the effect of smoking comes really about a 30-year lead time after the cause for most young people, and also, in the case of pregnant mothers, clearly concerns unborn children, I think it a very good test of how effective mass education may be in respect to the quality of life 30 years from now; which is not very, if radical changes are required.

It is very true that engineers and technologists envisage with a certain glee a future in which technology and technologists are king; but one must realize that literary intellectuals are capable of the same thing, and that the futures seen by skinny prophets like Ivan Illich are just as comfortable for prophets as the futures seen by fat technologists like Herman Kahn are for themselves. I remember reading a utopia by the poet and druidophile Robert Graves in which the world was ruled by poets and witches — this is clearly a not uncommon syndrome.

More important still may be to look for the patron. A seller of futurology can make big money only from giant corporations or the Pentagon. Thus not only will his future look very good for fat rich futurologists, it will also contain industrial growth to the nth degree.

1. Prophets

These things having been said, I don't have to say too terribly much about the prophets. I would in this connection like only to mention Ivan Illich. Illich's thesis in his books and his teachings are that science, medicine, education, technology, and in fact the whole industrial complex have passed through two "watersheds", one on the way up and one on the way down (only a prophet could use the word "watershed" for what is clearly a contour line). At the first "watershed" these institutions begin to actually benefit humanity, at the second — because of excessive institutionalization, mechanization, dehumanization — they reach the point of negative marginal utility. One response to that is to try to remember just how high up that first "watershed" is. A few pictures in my mind: any probable damage North Sea oil will do to the crofters of Scotland amounts to a fraction of the total atrocity of the "Clearing of the Glens", when the crofters were often burned right in their cottages. Equally, I remember in villages in the hills of Japan, seeing old peasant women who had clearly lived in the preindustrial community their whole lives. They had lovely faces, but many of them had literally permanently bent backs. Simple return to the old days isn't on, or at least shouldn't be.

2. Gadgeteers vs. Technology Assessors

Extended gadgeteers — these are the futurologists of the Commission on the Year 2000, Dennis Gabor's book, and I class Herman Kahn here — he wrote a brief version of his stuff for the "Year 2000" collection — because he satisfies the two basic criteria for this group. First, their concern is to predict innovations along very much the same line which we have already been following: linear extrapolators, if you like. Their attitude is: we will show you how you can be *even happier and richer* with our new technology. And second, there tends to be a great lack of sensitivity to any of the sociological, ecological, and in some cases even economic, consequences of the innovations and the growth which are predicted. For instance, in Gabor's book slum clearance is mentioned as a fit area for innovation; but one gets the clear impression that he is suggesting the invention of better bulldozers, rather than of new social mechanisms for rehabilitating neighborhoods, and does not acknowledge that past "slum clearance" has often had negative results.

The list of projected inventions, especially in the biomedical and bioengineering field, leaves one with a very queasy feeling indeed and a serious question in one's mind whether better mechanisms for control of technology shouldn't be worked out. The attempt to do so is the concern of the advocates of the idea of technology assessment, an excellent idea which has even passed the U.S. Congress and become enshrined as an official agency of the U.S. Government: the Office of Technology Assessment. The article in the reading list on control of sex of one's children is an excellent example of the kind of thoughtful analysis which one would hope would be the goal of the technology assessors: it points out the disastrou indirect consequence of such control that societies with an excess of males, as would be quite likely under free sex choice, are very much more violent than balanced ones.

But that is a relatively straightforward, and not excessively emotion-ridden, example. To me the real difficulties of technology assessment in the full sense are two-fold: one, political, and the second scientific. An example of political difficulty: imagine for instance that the OTA were to be asked to assess the possible value and effect (a) of a full-scale program such as has been pushed through in the U.S. to find a cure for heart disease; (b) of finding that cure itself. It is not at all obvious what the right answers would be. First, are we using our medical research resources mainly to alleviate the diseases which are most feared by the prosperous middle-aged male who is in the position of appropriating the money, rather than the diseases which are responsible for the most human misery? (One could at least argue that these are old age and schizophrenia, not cancer and heart disease.) Second, until we understand more about how to alleviate the more unbearable miseries of old age, is it certain that it is the right strategy to cure heart disease and as a result condemn greater numbers of our population to those miseries? But it is inconceivable to me that the result of such an assessment could possibly have any effect, one way or the other.

The scientific difficulty is about the problem of the random interactions of technology with extremely little-understood and probably totally irrational sociological and psychological forces. It seems clear that — to go back a few years to something we are not deeply involved in at the moment — worldwide communications, easy and cheap travel, and the pill were all in some way related to the 1968 outbreaks on the campuses of universities all over the world. These also involved, for instance, some kind of generational — and possibly class as well — conflict. It also seems clear that until the social sciences can give us some kind of answers which makes sense about the theory of such widespread phenomena — let these theories be as value-ridden as you like — that we cannot expect to gain enough understanding to assess the consequences of any major scientific or technological development.

Somewhere between this topic and the next belongs the whole question of technological developments which lower the quality of life directly. It is easy to condemn the Concorde and the SST, and indeed after years of battle it may be that sanity has finally won on those issues. Then there are caravans, mobile homes, motor cruisers, scuba sets, and, an enormous growth in the U.S. at this time (and not one predicted by any of the technological forecasts I looked at) the ORV — off-road vehicle. While you consider this picture* of the destruction of a square mile of desert ecology as a result of everyone being as rich as Herman Kahn wants him to be, let me try to confuse you a little about the values here. In the first place, under the Illich value system, I would contend that this exhibition represents a net plus. After all, this is a picture of a lot of people "doing their own thing". Where the emphasis is on the positive values of community, of conviviality in a serious sense, of individual rather than institutional recreation, this is a far more positive picture than one of a hundred thousand screaming fans at a football game, or an equal number watching a T.V. lecture on pulsars.

Deplore ORV's and their ilk as one must, it is very hard not to fall into the kind of trap which leads to the bromide definition of a conservationist as a man who already has bought *his* vacation cottage. There is indeed evidence of an elitist reaction against observable evidence of workers' attainments. The affluent, educated world — the men of culture rare — may to some extent have been trying to create a conserved Garden of Eden, in Western Europe and the U.S., into which the serpent has entered in the form of the oil sheikhs. These utopian ideals would have encountered other snags. There can be a very severe reaction of the working middle class, as in the Nixon-McGovern election, or in Chile, to a serious hint of redistribution of wealth, reduction of its standard of living, or of restriction of its freedom to use its recently acquired wealth. (This was 1973, not the tea-party year 2010!)

3. Doomwatchers vs. Cheermongers

Let me then, trying not to be pejorative in either direction, describe the battle over resource availability and over the disamenities of crowding, and of waste disposal and pollution, which has been taking place over the past few years. The Doomwatchers have produced two major manifestos, Limits to Growth and Blueprint for Survival, and many supporting books, pamphlets and articles. The essential message of all of this can be summarized in two simple graphs. The first is given in BFS showing the result of exponential growth of use running up against finite capacity (in

---

*I have lost this illustration, which was a photo of an ORV convention somewhere in the Western desert, with literally hundreds of powerful motorcycles speeding along, trailing their dustclouds and annihilating anything tenuously living in that fragile environment.

this case, of oil). The second shows the effect on a complex, interrelated world economy of the battle between resources, pollution and population. I don't think I have to repeat all the arguments of the ecologists, because I'm sure you have heard them all. I would like to show one slide on the question of resource availability, because so much has been made of our current problem of energy that this actually far more pressing question has been lost sight of. This graph from BFS shows the resources and maximum (linear) or minimum (exponential) lifetime of a number of universal resources. On the same scale oil would be one of the more available resources, and coal off the graph. In a complex, interrelated world economy, this widespread ceiling on a great variety of important resources can only lead to early difficulties — aggravated by the effect of prices, which I will discuss shortly. Second, the question of waste disposal — as an example which has reached the news media, there seems no doubt that N.Y. city seafront will be drowned in a pool of sludge in a few years. (Wrong guess! But it will happen.) These problems are not by any means distant. Finally, while I agree that the World Dynamics model is fantastically oversimplified, I see no reason to believe that it is oversimplified pessimistically, as seems to be the general objection to it. There is a special feature which is clearly adding to our present serious problems: that a wide variety of underdeveloped countries of very large population, which did not in the past play any role in world resource depletion because, while growing, their consumption was negligible, are suddenly leaping into the world markets: first Japan, then Italy, Spain, USSR, Mexico, Brazil, and soon Egypt, Iran, China above all, and many others.

In view of this, what hope do the cheermongers see? This group is led by the economics establishment, and it focuses on one idea: the price mechanism or, a fancy word for it, "sub-optimization". Let me explain this in a couple of ways. There are three ways of doing economics projections by computer: (1) Kahn and the Futurologists. One simply projects present growth rates of a sequence of economic variables: Population, GNP's, standard of living, etc., and lets them all rip. Naturally they grow exponentially. (2) The Limits to Growth group says that's not true; these are interrelated by negative feedbacks via resources, food, waste disposal and pollution, quality of life. One gets a barrier phenomenon where all run into dead ends together. (3) Economists say even *that's* not right. Every time a resource threatens to run out, it gets pricier. As a result the Invisible Hand reaches in and sub-optimizes that resource, either finding more of it or finding a substitution. This softens the barrier enormously.

Let me, to illustrate this, give you a quotation from one of the most respected of the world's economists, Prof. R. Solow, which will indicate as well both how to be a prophet in spite of oneself, and that this situation is not universal.

> "The most glaring defect of the Forrester-Meadows models is the absence of any sort of functioning price system. The price system is, after all, the main device evolved by capitalist economies for reacting to relative scarcity. One way that a price system might radically alter the behavior predicted is by inducing more active search for resource-saving innovations as resource costs bulk larger in total costs. Other, more pedestrian modes of operation of the market mechanism exist. Rising prices of exhaustible resources will lead to the substitution of cheaper materials. To the extent that it is impossible to design around or substitute away from exhaustible resources the prices will rise and consumers will buy fewer resource-intensive goods. All of these effects will work automatically to reduce resource requirements per unit of GNP.
>
> This is not at all an argument for laissez-faire. Many markets are 'imperfect' for many reasons, and public agencies should intervene, for instance by providing better information. But I don't have the slightest confidence in models that make *no* provision for market forces."

So far so good — we can only agree with him. But what really would be the effects of scarcity: read on!

> "As a matter of fact, the relative prices of natural resources show no tendency to rise over the past half-century."

(This was true, at least of energy, as shortly ago as 1972!)

> "This means that there have been offsets to any progressive depletion. But this situation could change. If the expert participants in the market now believed that resource prices will become high, those prices would already be rising.
>
> Suppose you owned one of the few remaining deposits of X and felt certain that by the year 2000 it would bring the high price of $1000, per ton, because of scarcity. If the interest rate is now 5% per year (sic!) then the value today of $1000 in 2000 is only $250, because that is what you'd have to deposit at compound interest. That means your price now for X should be around $250

per ton. At any lower price, you could do better to leave it in the ground and borrow. Next year you would have to get 5% more, or $262.50; and every year the price will have to rise at least at the rate of interest, in order to induce the owners to mine any."

It was only a few years ago that that was written — but even so it is hard to see how he could have missed the worldwide price inflation which was already underway. Several genuinely nonrenewable resources had been the leaders in this inflation; land, art, gold. Now with a partially artificial shortage of oil, the world as a whole has begun to realize the genuine perils of the Limits to Growth, and to learn that these perils come, for precisely the reasons Solow has described, long *before* the actual limits are reached. We are busily suboptimizing; but in the process someone has got to take a cut in his standard of living; and aside from the poor of the world, who have no say in the matter, the different classes in the advanced nations are determined that someone else will take this cut. The effect of this is to provide a multiplier which drives prices up far more rapidly even than the Solow mechanism would predict, and carries all others with the resource prices. In fact, it is by no means clear that the fabric of our societies is going to hold together even under the very first mild precursors of the Limits.

I'd like, from a position of some ignorance, to continue to speculate on the economics of what is going on at the moment; at least, I gather, I am not impinging on a field where everything is wholly settled and understood. It seems to me that the recent rise in the price of oil is not the inevitable result of exhaustion but rather an interesting, but a little irrelevant, exercise in the imperfect market phenomenon which Solow mentioned in passing in the above passage. In essence, the Arabs suddenly realized that it was open to them to leave the oil in the ground until the year 2000; until then, they had indeed been in an imperfect market in which they were the fall guys. In actual fact, energy is almost the least exhaustible resource we have — far more so than land, waste disposal, or other minerals — and we have a very great range of possible substitutes at the $10/barrel level. (Do not take these numbers seriously; not only are they '73 dollars, the inflation in resource prices has already begun.) Oil's convenience and cleanliness must, of course, come off its price, and we may find ourselves paying a premium for that, and even more for the fact that it is not capital intensive in an inflationary era.

Land and food, on the other hand, are in my opinion a far more serious matter. Again we have here an imperfect market. In Western Europe,

Japan and elsewhere uneconomic producers are being subsidized to conceal the fact that the economic producers are not getting the true marginal price for their food. Note here the recent excursion in the price of sugar. We can expect with some fear the formation of an OFEC and then we *may* have an inflation which will make the current one look tame, because land really *is* a nonrenewable resource, and you need land to grow food.

Let me summarize, then, some of the essential results.

— The Limits to Growth are upon us, in the current inflation of world prices. But the limits are relatively flexible: the severe oscillations of LTG will not occur. The effect is and will continue to be big redistributions of wealth.
— In contradiction to such linear extrapolators as Kahn, as resource substitution occurs, human labor and human ingenuity will become more, not less, necessary: we will increasingly substitute labor and technological understanding for resources. Also, we may, gradually or rapidly, undergo societal changes in this direction.
— We are having and can expect a rising level of social dislocation and disruption; and we can have no idea of whether it will be contained or not.

4. Synthesis: The Third Watershed

At this point in my outline I put down synthesis, which I guess I meant, at least somewhat, in the Hegelian sense. The question is what to do — specifically what physicists should do, and then what educators, engineers, medical, biological, and social scientists should do as well. To borrow Illich's figure of the two "watersheds", what one hopes to do is to get back up past a third "watershed", and for one thing to overcome the legacy of institutional and personal rigidity and insularity with which the recent fat years for science have left us.

I do not however think we should undersell the value and meaningfulness of contributing to our respective sciences. Value-ridden, possibly irrelevant, and enormously frustrating in its haphazard illumination of the world around us, as it may be, of scientific truth we must keep in mind that it is another word for the literal truth, and in fact is the only stock of that we happen to have. A genuine increase in the amount of known literal truths must always be a good thing. Whatever goal or value we choose, in striving toward it we can only use the scientific truth as our guide.

More specifically, one might mention the following points where physics and physicists can play a role — in other words, why physicists?

Physics as a model of successful quantitative science. This has already been attacked in these lectures, and I think rightly so to some extent — to use physics as a model may precondition one to a certain limited view of truth which may not be appropriate for all sciences.

Physics as a source of techniques, of bright people, and of confidence in dealing with the world. There is a component of excessive arrogance here, but also a core of truth. It seems to be as true that having mastered one hard subject helps with another as that climbing one mountain helps with another. Problems have a technological component in which in many cases the physicist may be specially competent to deal. It is important for the technology assessor to have skills which prevent the SST or ABM advocate from hiding behind technological gobbledygook. The fundamental conservation laws of physics have unmasked a great many charlatans and scoundrels. Even this morning I was very unhappy to hear the BBC giving time to a faker who wanted to run a car on cups of water. Snow's point of the two cultures is still worth thinking about.

Aside from the generalist role in technology assessment and overview, there is no lack of actual relevant application of physics in particular across the whole spectrum of the problems of ecology, pollution resource exhaustion, and maintenance of the quality of life. For example:

— Resource substitution;
— Pollution measurement and control;
— Atmospheric and geophysics;
— Energy: catalysis, energy transport, aside from the obvious.

What do we carry with us, then, on the way to that third "watershed"? A valuable bag of tools, indeed; but more important, ideally, certain attributes and values; openness with professionalism, and a kind of skeptical confidence in the solubility of problems.

Bibliographical Note

The above was the last lecture in the 1974 Lent Term Science and Society lectures in Cambridge. The reader will probably guess that it is to a great extent a reply to a number of sources, and looking at those which are available will clarify some of the more obscure references. One is not available: a few days before the talk was given, BBC2 ran a program entitled "Futurology" which featured Herman Kahn globe-hopping and pontificating on a future of endless economic growth. The following books are mentioned.

I. Illich, *Tools for Conviviality*, New York: Calder Press, 1973

D. I. Bell, *Towards the Year 2000*, Boston: Beacon Press, 1969

D. Gabor, *Innovations: Scientific, Technological, Social*, New York: Oxford University Press, 1970

T. W. Forrester, *World Dynamics*, New York: Wright-Allen Press, 1971

D. H. Meadows *et al.*, *Limits to Growth: A Report for the Club of Rome's Project on the Predicament of Mankind*, London: Earth Island Ltd., 1972

There are two useful magazines:

"Blueprint for Survival", *The Ecologist*, January 1972
"Doomwatchers and Cheermongers", *Cambridge Review*, February 2, 1973

Dizzy with Future Schlock[*]

When I composed my first draft of this review, I was unaware either that the author was an attractive woman or that she was a baroness; much less that she had charge of the venerable Royal Institution or that she was in the habit of lecturing in a short, if not mini-skirt. I clearly learned some but not all of the above facts by the time it was finished; but none of them influenced my opinions. She wrote an indignant letter accusing me of antifeminism which, given her eminence, was published in the same issue of the THES as my review. I accept that the remark about cereal boxes was a little harsh, but in the final paragraphs I concede her a certain level of right thinking and I don't think the review as a whole was unfair. It certainly was not motivated by any distaste for attractive women scientists, who usually find me a pushover. I have written similarly acerbic reviews of Freeman Dyson, whom I admire greatly as a scientist but question his judgments of the future for similar reasons.

It is very seldom that reading a book for review is really a chore for me. My curiosity even to hear what slant the author may take on familiar material, not to mention voracity for new facts and ideas, is almost insatiable. I read the backs of cereal boxes at breakfast. But somehow Susan Greenfield's *Tomorrow's People* was an exception to the general rule — I was having to grit my teeth and really work at each chapter. The book is well enough written, with occasional felicities; and the theme, "How

---

[*]Book review of *Tomorrow's People: How the 21st Century Is Changing the Way We Think and Feel*, by Susan Greenfield (Allen Lane, London, 2003). Originally published in the *Times Higher Education Supplement*, 31 October 2003.

21st-century technology can change the way we think and feel", should have intrigued me.

I began to think that I must be the kind of older person well described somewhere in the middle of the book as having a "crystalline" intelligence, capable only of relating the new to the already known, incapable of the flexibility to appreciate the radically innovative effects of new technology.

For about seven of the ten chapters I slogged through a dense forest of what you might call technohype, or even "future schlock", the sort of thing that comes from the pens of writers such as Michio Kaku, Mihalyi Czikszentmihalyi, and the author Ray Kurzweil, all of whom are quoted, Kurzweil extensively.

Since Greenfield has the entire 21st century to work with, she can and does speculate that the wildest dreams not only of the historically overoptimistic artificial intelligence community and other information scientists, but even of a number of sci-fi writers, are likely to come true. On each of a sequence of subjects, such as lifestyle and reproduction, she brings forward in rapid fire the wonders that the future holds for us. For instance, in the second chapter, on lifestyle, within a few pages she had set my head spinning with "lighter-than air megastructures, or oxygenated subocean habitats", the "hyperhouse" in which the floor plan, walls and appliances are all flexibly located, virtual rooms in which the walls and windows present whatever information or aspect one calls for; "inside vs outside might become a thing of the past", and that people may need a "reality room" where they can relax away from information technology. Then "the world itself could become the interface", whatever that means and however one might do it. A few pages further on, at random, "even the bright colors of unmodified carrots, spinach and tomatoes would seem dull and unappetizing compared with their iridescent modern counterparts" — and, turning a couple of more pages, "even your family members can be virtual: they can be of whatever age, gender, sexual orientation, in whatever number you wish ..." And, of course one's health and well-being are automatically monitored.

Only at the end of the chapter does one come back to some kind of serious consideration of what such an environment might mean. Greenfield asks, for example, whether one's relationship to reality has not, by this time, decayed almost to zero — would there remain any sense that there is a hard, unresponsive, *real* reality anywhere? This and many other questions about the effects on the individual have, of course, arisen in one's mind long since — but did one have to go through the

30 pages of hyped futurology to get there?

Then the reader must work through chapters on robots, work, reproduction, education and science, each containing another heterogeneous mass of information, overstuffed but underevaluated. The topic of robots includes the various schemes for coupling "silicon" and "carbon" intelligences, that is to say, enhancing our capabilities by computerizing us (Jaron Lanier, incidentally, has written convincingly about how fundamentally difficult interfacing digital technology with the operating systems of the body and mind will really be). As for the nightmare scenario of being taken over by our own creations, which so exercises Bill Joy of Sun Microsystems, Greenfield sensibly poohpoohs that particular bugaboo, having as a neuroscientist some appreciation of the massive leaps in understanding necessary before we produce really autonomous, self-aware machines. Even so, the possibilities she mentions seemed endless.

Considering work, she details some of the amusing history of past predictions — G. B. Shaw and J. M. Keynes predicted that by now we should be working only about two hours a day. But of course what changes is the nature of work and, massively, the composition as to age, training and flexibility of the workforce. Here, for some reason, she has a brief excursion into energy supply, where her biological and IT background, for once, fails her — she takes seriously President George W. Bush's fuel cells and hydrogen economy, for instance, which simply move the hot potato from one hand to the other.

Reproduction, in the genomics age, is a fertile field for the imagination, and most of the possible changes are rung on cloning, artificial genomes, and life prolongation. A favorite fantasy is the "24th chromosome", on which one hangs all kinds of protective or life-enhancing genes for use as needed. And here too much is made of the fact that with *in vitro* methods, anyone can be the parent of anyone (or anything) so long as the parent is older than the child (… I think).

Education: Here again there is some good sense and just too much information — the good sense being the awareness that just because all information is available at the fingertips one cannot do without education, which is not just, or even mostly, about facts. The overdose of information comes from a heavy injection of psychology and neuroscience of all sorts.

On Science, true to Greenfield's own specialty, there is quite a perceptive and perspicuous discussion of the great problem of consciousness, and a discussion, in passing, of the benefits of interdisciplinarity and the endless frontier of complexity. As with the weakness on energy, there was all too much credibility given to the recent fad for "nanotechnology", which to my mind has been partly evolved to resuscitate a moribund hardware science.

With the first seven chapters under our belts, we come to "Terrorism", "Human Nature", and the "Future". I sense, somewhere in the middle of "Terrorism," that the author had experienced — or intended all along? — a change of heart. After regaling us with all the dreadful possibilities inherent for terrorism in the 20th century which will become much more refined for the 21st, she begins to think about why this should be a problem in the first place — which brings her to the questions that had been on my mind almost from the first page of the book, which are: first, what about the many of the West's population, and most of the rest, who can't afford — and may not be able to tolerate — any of this future? A corollary — are there potentialities in the human race, especially when divided by the effects of technology, more dangerous than anything the robots can do to us? Secondly, what about human nature? Is it not deeply ingrained in the mechanism of our minds and bodies, and dare we replace it with something else based on genomic and information technology? Thirdly, won't further breaking up family and community, and the literate culture, and replacing them with whatever — chat groups, websites — lead to an even greater proliferation of cults and other aberrant, intellectually isolated social groups? Can an evaporating connection to reality support democracy?

To my mind the great conflict of our times is between the Enlightenment culture of tolerance, humanism and individuality, and the coercive and dogmatic fundamentalist cultures, whether in the West or the East. What Greenfield correctly foresees is that IT and its technological comperes simultaneously sap our social coherence, and strengthen fundamentalism, especially among those left behind economically or intellectually. In her final chapters Greenfield completes her turnaround and argues, often with some eloquence, in favor of taking thought while we still can about such measures as using IT and modern technology to improve the lot and to alleviate the ignorance of the masses of humanity. She lays out a number of alternative scenarios for the future, each more desperately pessimistic than the last. And in the end the best she can do is plead for the "ultimate priority" of "not just the preservation, but also the *celebration, of individuality*".

Baroness Greenfield has an important and chilling message to impart. Somehow, out of all this mass of miscellany she has read the trend line of the future and identified a number of tendencies towards disaster which, in fact, one can see in process all around us. I think she is correct to warn us of them. But if at least part of her isn't all in favour of the indiscriminate application of the technological possibilities, why does she dwell on them in such loving detail — and why does she denigrate those who push back against them as "cynics" or "technophobes"? In a preface, Greenfield submits that in her mind the book began as a novel. Frankly, a novel might have made her point in a much cleaner way, and certainly she would have had to be more selective in her choice of technological tricks to present. As it stands, despite my relief at the conclusions reached, this book is flawed.

Einstein and the p-branes[*]

Any theoretical physicist reading Stephen Hawking's books must be consumed by the question: "What is his secret?" How can he reach an audience of millions, when try as we may to write for the public we can reach at most a small subset of the readership of, say, *The New York Review of Books?* The sour grapes answer, involving his spectacular disability and his equally spectacular fortitude in coping with it, won't do, although I am sure that helps. So does the fact that he works in a field naturally fascinating to layman and scientist alike, scientifically attempting to satisfy the natural human questions about origins and the cosmos for which almost every religion has had to have an answer. Also it is a field permeated by the magic name of Einstein. But in the end we have to face the fact that he is an excellent writer. His prose is deft and witty and is a pleasure to read. He has in common with Einstein the characteristic that he seems never to express himself inaptly. (Although to compare him with Einstein as a theoretical physicist, as the blurb on the book jacket does, is an awkward reach.) He also has the gift of saying nothing irrelevant to the line of thought he is pursuing — he does not have the scientists' tic of saying "now, on the other hand" (after all, he has no usable hands at all!) Note that I did not say that he is a superb expositor. I will have more to say about that later. Thus, as prose, and as a compendium of current buzzwords in this part of physics, the book is very much to be recommended; but I think that, as with *A Brief History of Time*, the community of professional theorists may cavil. His exposition is supplemented by elaborate illustrations at the level of more than one per page, designed to help the reader

[*]Book review of *The Universe in a Nutshell*, by Stephen W. Hawking (Bantam, New York, 2001). Originally published in the *Times Higher Education Supplement*, 1 February 2002.

over the hard parts, although in more than a few cases I found them to be misleading or less transparent than the text.

Hawking describes the organization of his book as a "tree structure". Chapters 1 and 2 are intended to tell you literally all you need to know about the part of modern physics underlying what he is going to discuss; the remaining five chapters branch out roughly independently of each other, and follow in any order you like. Chapter 1 is a very concise history of Einstein's career and of the development of special and general relativity. It even contains, at the end, a neat little discussion of the fact that Einstein missed the two most immediately tangible consequences of general relativity, the black hole and the expanding universe, adding an unnecessary constant to the equations to make the universe stand still (though he lived to regret the latter).

The story of Einstein's life and work has been told by others at almost every level of detail and of intellectual rigor, but still Hawking makes it seem fresh and exciting. (And after all who needs irrelevant gossip?)

Chapter 2 is a tougher assignment. In 34 pages, with about 30 figures, several sidebars, and one table, Hawking discusses, and gives a semblance of explaining, the following: the cosmic microwave background; gravitational lensing; his theorem proving the necessity of the big bang; quantum electrodynamics and the infinities of field theory; spin and statistics; string theory, supersymmetry and the compensation of bosonic and fermionic infinities; branes and M-theory; complex (imaginary) numbers and cosmic geometry with imaginary time; and he tops it all off with the Bekenstein limit on the entropy of a black hole (for which he claims sole credit) and the holographic principle. At the end you can have some hope of knowing what all those things are, and that in itself constitutes quite an education.

On the other hand, I hope you will not believe that all this material can be accurately described in such a brief compass. Amazingly Hawking manages to make a coherent-sounding story of it, but he does so by leaving enormous gaps — he writes the way Tarzan travels through a jungle, taking great leaps over the voids and catching a happily placed liana at the last minute. In my opinion, he actually does miss one or two of those handholds, but then I am not the intended lay reader and may require a bit more scientific exposition. Some examples of the hiatuses which bothered me were the jump to the black-body spectrum of the microwave background (the word spectrum appears nowhere else in the book) and from there to "matter opaque to microwaves"; the introduction of spin (the illustration on this matter is gratuitously confusing); what Grassmann numbers have to do with; and both the discussion of "imaginary numbers" — perhaps the least transparent one I have ever encountered — and how they can simplify the geometry of space-time. (I guessed, not being up-to-date in this field, that somehow the continuation into the complex plane turns open geometries into equivalent simpler closed ones, but even a relativist I consulted seemed to think this only complicates things.) Incidentally, in addition to the query about the "Bekenstein limit" referred to above, I think that the "egghead of p-branes" would in most venues outside Cambridge be expected to refer to Ed Witten, not to Paul Townsend as an illustration avers.

The remaining chapters are, according to the author, meant to describe different, branching, areas of research. The third, for instance, has the same title as the book, "The Universe in a Nutshell". It describes our particular cosmos and the possibility of other, perhaps coexisting, alter-natives. He starts by describing well the Slipher and Hubble discovery of the galaxies and the universal expansion, although I have to quibble with his remark that it was the astronomers, not the geologists, who first discovered the antiquity of the cosmos. Dropping in a page or two about quantum mechanics and Feynman's average over histories, he also describes the inflationary hypothesis which describes the early moments of the Big Bang as involving an enormous expansion, so that everything we can see, no matter how distant, initially derives from our immedi-ate neighborhood. So far most of what he says is not outside the box of "ordinary" cosmology. There is in the field at least a minority who discuss many versions of the idea of multiple universes, his next topic. This leads into a maze of speculations about the Anthropic Principle (the idea that the universe is rigged to accommodate us because otherwise there'd be no one there to see it) and the no-boundary boundary condition, a maze into which I was unable or unwilling to follow.

The next chapter, "Predicting the Future", contains a little something for everyone: a typically brief and effective, if perhaps beside the point, condemnation of astrology; a rather nicely done description of black hole physics including the recent idea that what falls in is preserved indefinitely because time slows down to a stop, and will eventually be reemitted as radiation; and a waggish bit about Hawking's fling with the Star Wars TV program. I have a deep scientific problem, however, with this and later chapters, about his general view of time, Laplacian deter-minism and quantum indeterminacy, and the casual way in which he imagines astronauts and even the starship Enterprise tossed into worm-holes, black holes and similar spacetime singularities. As John Wheeler pointed out long ago, there are three kinds of time, each of which has its own "arrow" pointing to the future. There is Einstein's fourth dimension time, thermodynamic time defined by dissipative processes and the Second Law, and biological time which allows animals to remember and plants to have grown in the past and not the future. Probably the third is an obvious derivative of the second, but it was Wheeler's insight that both were controlled by the first, in the form of the cosmic expansion. It is this which allows us to dissipate entropy into the surroundings, and eventually into the cosmos, and hence to irreversibly record what has happened to us — for instance, where an electron landed on a fluorescent screen. Without dissipation, time has no sign, and there is no such thing as memory. By a somewhat more involved argument, we can see that dissipation also fogs our knowledge of what quantum state any system is in and destroys our ability to perfectly predict the future. Any scientist who still believes in Laplacian determinism is doomed to disappointment. And wherever something really singular happens to the cosmic time coordinate, surely the biological (and informational) one will lose its meaning — there is never any problem with going back and fouling up the past, which seems to be the subject of the entire next chapter. This fifth chapter is a rather arch bit of horseplay with Hawking's friend Kip Thorne involving bets about hypothetical time traveling possibilities.

The next chapter quite properly points out that the idealized society depicted in the science fiction series Star Trek is not likely to happen. A few homilies about exponential growth of population, energy use, and information preface a horrifying — to me, at least — picture of a cyborg future, but including a brief reference to the little details of how we avoid the destructive potentials of the human race in the near run — perhaps this is excusable and even reasonable in SH's unique situation. (The only real howler in all those illustrations is that the human population starts its history in 8000 BC at 1 billion instead of 0 billion — a mislabeled axis.)

A final chapter is so completely devoted to the latest fads and speculations about the proliferation of dimensions that I could not develop much interest in it. I am possibly wrong to think thus, but it has seemed to me that really useful science has mostly had either an experimental basis or a reasonably constraining logical necessity. I am not sure that letting the general public into the free-association kind of speculation where our theories of reality are sometimes born is good either for them or for us.

So — science popularization cum science fiction as an art form — or genre? Will the public buy it as they did the Brief History? Tune in next year, folks — I don't know.

Forecaster Fails to Detect Any Clouds[*]

This reviewer is old enough to have seen a lot of futurology come down the pike, pass, and disappear into history. It was 25 years ago that I first wrote up my thoughts on the subject long enough ago that some of the futurologists I then quoted are just now, as myself, becoming what one might call "chronologically disadvantaged". There was a book called *Toward the Year 2000*, an issue of *Daedalus* published in 1965, which had already begun to come seriously unraveled by 1975 and is now completely detached from reality (although, I note, it is being plugged for millennial reading).

At that time I classified the art of futurology into three, possibly four, categories.

These are, first, Prophets — in general, creators of utopias, or anti-utopias, like Huxley, Orwell, Ivan Illich, Samuel Butler, and many science fiction writers from Jules Verne to H. G. Wells. Often they are wildly off the mark — Illich, for instance, saw medicine as going down through a "second watershed" and becoming less and less beneficial; Huxley saw cloning as a way of producing "etas", mindless people who could do the work an upperclass Englishman couldn't imagine himself doing. Such utopias tend to represent either the desires or the fears of their creators, who seldom take the broad view, but there are so many that a few such as Verne couldn't help occasionally hitting the jackpot.

The second category was "Doomwatchers". These were a set of futurologists whose predictions were based on oversimplified computer models

---

[*]Book review of *Visions: How Science Will Revolutionize the 21st Century*, by Michio Kaku (Anchor Books, New York, 1998). Originally published in the *Times Higher Education Supplement*, 7 August 1998.

which showed the imminent exhaustion of all kinds of world resources. These have given such modeling a bad name, since economists (whom I described as "cheermongers") pointed out essentially correctly that price-driven technological innovation could often prevent the collapse. But the present state of the world's fisheries shows that such Malthusian collapse can happen, especially when aided by both technology and government policy. One of these doomwatchers turned out to be a true prophet, Fred Hirsch, with his little book *Social Limits to Growth* which predicted, spot on, the enormous growth in the cost of those goods which are truly not expandable: amenity land (Aspen, Cape Cod, Manhattan), elite education, fine art, antiques, and the like. What appears to be the case is that the consequences of unlimited growth are inevitable and, eventually, disastrous, but not amenable to such simple-minded modeling.

My third category, "Extended Gadgeteers", is where I place Michio Kaku and his new book, *Visions*. These are those who focus on the prediction of technological innovation rather than on modeling or technology assessment. This seems to tempt the futurologist into his most incautious predictions.

Contrary to a claim in the introductory chapter of his book, it is not a new idea to attempt compiling a consensus of the "experts" in each field of endeavor: this is precisely the idea of the "Delphi" scheme which was popular a couple of decades ago. To my knowledge, the results of Delphi have been no more reliable than any others. It is a common failing for such futurologists to ignore either the social and economic consequences of their predicted technologies, or the likely interactions between them. 35 years ago Herman Kahn, for instance, envisioned a world full of rocket ships with no thought of the environmental consequences and the conflicts of amenity they would lead to.

Michio Kaku is, understandably given the passage of time, much less naïve and much more sensitive than Herman Kahn or, for that matter, Denis Gabor, who wrote a book about his selections of the 100 greatest innovations of the future, of which the most memorable was a "technological fix" for slum clearance. Kaku devotes almost a whole chapter to such problems as global warming and overpopulation. But, having claimed awareness of these problems, consciousness that hard choices will have to be made and serious political conflicts will arise disappears beneath the surface.

In the end, what we have here is what is implied by the title: visions of possibilities that lie within the reach of present-day science, as foreseen by consultation with acknowledged "experts" in the various technical fields.

Kaku slices this large cake three ways in subject matter and three ways in time: his three subject areas being the Computer Revolution, the Biomolecular Revolution and what he calls the Quantum Revolution; his three time slices 20, 50 and 100 years.

In the first two revolutions, which occupy well over half of the book, Kaku has done precisely as advertised: he has gone to a great deal of trouble to consult the acknowledged experts and brought himself well up to speed in these active fields. For his 20-year period one feels that his kind of linear extrapolation from what is now going on is pretty good. After all, we are in the earlier years of the Age of the Internet and close to halfway through the Human Genome project, and one can see at least vaguely, or at least imagine, where these are leading us. And for 20 years of the Computer Revolution, we are to benefit from "Moore's law" (one of the few really accurate predictions ever made) of exponential increase of computing power, to have ever faster and more all-pervasive computerization of every facet of everyday life.

For the 50 and 100 years time frames, Kaku has consulted a selection of scientists who are working on the creation of machines that *really* think and really act independently, and the ideas he uncovers — if not the timetable — are fascinating and, indeed, to my knowledge, the best available. From 2050 on he predicts, we shall begin to have to worry about machine consciousness. I, for one, am flabbergasted by the acclaim with which he seems to greet that idea, and do not believe that conscious robots can possibly be programmed to remain under the control of, or even remain benevolent to, their creators. Happily, my personal estimate of the requisite time scale is much more extended than this book's.

Next he deals with biomolecular technology. Again, for the first 20 years he can simply list the expected benefits of the human genome project and other ongoing initiatives involving science which is understood in principle: cures for cancer, yes, almost certainly; mitigation of many hereditary diseases, and a massive increase in life expectancy. But, in the next era, up to 2050 in a passage called "Angels in America" which left me aghast, he envisages the generations of human beings without genetic defects.

It was at this point that I began seriously to question what I was reading. In the passages on the Computer Revolution I had felt a certain doubt as to whether the future that Kaku was projecting referred to the multibillionaire (Bill Gates as a prime example) individual or to the ordinary guy like myself. But in this biological reference frame it became quite clear that only the elite could possibly afford to be genetically scrubbed and to produce only "angelic" offspring. In both his chapters on computers, and his chapters on biology, the implication is that the wonders he envisages are more or less instantly available to all. The world does not work that way, and the serious future problem of allocation will not go away, nor will the political and social chaos that seems its inevitable consequences.

When we got to the "quantum revolution" I stopped and looked back at the list of scientists who were consulted and interviewed. For his 20-year slice, he discussed practically only electrical cars and the possibility of molecular machines, and up to 2050 his list focused primarily on fusion power and on room temperature superconductors, and on all of these subjects the level of discussion was no match for that in *Scientific American* or the news columns of *Science* and *Nature* (not that these are necessarily authoritative sources). It is a useful rule of thumb to judge a book by those subjects with which one is familiar, and on my recent pet subject of superconductivity the book fails rather badly. (On fusion, incidentally, his estimate of the time until practical realization is 35 years, almost exactly the interval that "experts" were projecting 40 years ago. He does not mention, as is also true of many of the spokesmen for fusion, the very relevant problem of 13 MeV neutrons.)

What Kaku gets to his century time scale he comes into his own: antimatter engines, space warps, suspended animation, you name it. He is clearly an avid reader of science fiction, a deep believer that the validation of string theory is just around the corner, and an enthusiast for space travel. Kaku may be the kind of string theorist who sidesteps communication with those of his physicist colleagues who work with actual materials or do seriously applicable work. It is interesting that this dismissal of applications does not extend to biomedicine or to the mechanical engineering of robots, or to computer engineering and software. Physics has become increasingly dichotomous, with one camp, of which string theorists are the epitome, concerned only with the pure subjects of elementary particles, cosmology, and gravitation, which are more and more losing touch with the rest of science and technology. The attitude towards the rest of us was once summed up by Murray Gell-Mann in the immortal phrase, "squalid state physicists". Unfortunately for Kaku, this is the part of physics which underlies not only the Quantum Revolution but his two previous revolutions as well; it has been called "the physics of complexity". In the event, Kaku's list of consultants did not contain a single member of the "other" three-quarters of physics, not a single materials scientist of any kind (chemist, chemical engineer, condensed matter physicist, metallurgist) and no one knowledgeable in the active and fashionable fields of nonlinear mechanics (popularly "chaos"), pattern formation, polymer physics and chemistry, granular materials, and so on.

In fact, what is striking about the list of names of scientists he consulted is the enormous areas of human knowledge which are *not* included in it: no geology (though lots of cosmology!); no history, though one of the most significant lessons for the prospective futurologist is to know how little we really know about the past; no social science except a smattering of economists; no sociobiology (surely a vital area); no serious archaeologists. An indication of the author's attitude is found very early in the book — a remark casually agreeing with John Horgan that the "End of Science" is at hand. To believe this you must restrict your definition of science sufficiently narrowly, as Horgan and Kaku do, to only those subjects it is easy to write popular books about.

There are two vital facts which a study of some of those subjects might have taught Kaku. First, in human affairs, as in many truly complex dynamical systems such as evolution or tectonic motions, it is normal for change to occur very suddenly, in earthquakes, avalanches, backruptcies, extinctions, rather than gradually. This is one of the things which makes the future particularly hard to predict (as first stated not by Yogi Berra, who is quoted by Kaku, but by Odd Dahl, I believe). The second is the well-known law of unintended consequences. The real result, positive or negative, of any given bit of technology is never exactly what you expect, and can be far from expectations. Who could predict, for instance, that one of the major results of the globalization of technology would be a worldwide explosion of fundamentalist religion?

Kaku's book is, then, an entertaining job of setting forth a selection of the goodies with which future technology may bless us. It will have an enthusiastic popular audience and will quite properly help generate public support for some aspects — though not necessarily the most useful ones — of ongoing science. As a reliable guide to the future — oxymoron is the word that comes to mind — it seems to this reviewer to lack depth, skepticism, and, in some areas, the relevant scientific background.

IX. Complexity

Introduction

I do not come even close to being an inventor of "complexity" science in the modern sense of that phrase, although my 1967 lecture "More Is Different", published in 1972, was influential in publicizing the field and in providing a useful slogan. Groups focused on various aspects of what is now called by that name preceded that time: the "Santa Cruz Collective" around Ralph Abraham and a vibrant gang of young people; the Institut de la Vie group including Prigogine's and Eigen's schools; the cluster around Warren McCullough including John Holland and Art Iberall, were only a few of these. Only later, starting in 1978, did I begin to make contact at least with the periphery of these activities; my initial experience was a session at the annual "neuroscience+skiing" meeting in Keystone, Colorado. Other speakers were Abraham, Iberall, and Arnold Mandell. Although I have mostly forgotten the matter of the talks, including my own, it was the incredibly wide-ranging discussions among the five of us which hooked me on the dream that really complex problems such as the origin of life and the organizing principles of the brain could be investigated by serious scientists, using ideas such as chaos and the spin glass theory which were beginning to crop up in the physics of complex systems. Gene next invited me to a meeting in Dubrovnik in 1980 on "self-organization" where I first met such complexity mavens as Harold Morowitz, Brian Goodwin, Leslie Orgel, Peter Schuster, and biological luminaries such as Gunther Stent, Stephen Jay Gould, and Francisco Ayala. By this time I was sufficiently part of the community to be invited, in 1981 I think, to the San Francisco mansion of the EST guru Werner Erhard, whom someone had persuaded to convene a meeting on complexity. Our every need was taken care of by zombiefied EST apprentices; but the program was completely under the scientists' control. There I first met Stuart Kauffman, Doyne Farmer, and Norman Packard, and heard from the latter his ideas on "the edge of chaos."

It cannot have been irrelevant to my growing interest in complexity that it seemed always to happen in pleasant places — which Dubrovnik, then, very much was — and then I was drawn, mostly by David Pines, into the planning workshops for what became the Santa Fe Institute, organized by a collective of Los Alamos alumni. There were other visits to pleasant places: an Aspen workshop organized by John Hopfield where we interacted with the Thinking Machines crowd, for instance. SFI started its life in an ancient ex-convent on Canyon Road, a short walk from the center of the art and antiquities scene of Santa Fe; and there Ken Arrow and I hosted in 1987 what must have been the very first econophysics workshop, by a long shot, supported by Citibank and its then president, John Reed.

By this time I had my own contribution to the mix of simple physical models showing complex behavior, namely the spin glass, which initially was a model proposed by Sam Edwards and myself for an observed phase transition in certain dilute magnetic alloys, but soon, in the hands of numerous collaborators and friends, reappeared as a model for evolutionary landscapes, for neural networks, as a source of highly practical computer algorithms, and as a new way of thinking about classifying computational complexity. It adds two new words to the physics glossary: "frustration" and "non-ergodicity". The pieces reproduced here are only a selection; somewhere I have a full set of lecture notes for a course I gave at Gustavus Adolphus College in Minnesota; and several of the pieces in Chapters III and IX represent other slices through the material.

Physics: The Opening to Complexity[*]

In the minds of the lay public, or even of scientists from unrelated fields, physics is mainly associated with extremes: big bangs and big bucks; the cosmic and the subnucleonic scales; matter in its most rarified form such as single trapped atoms; or measurements of extraordinary precision to detect phenomena — dark matter, proton decay, neutrino masses — which may well not be there at all.

The intellectual basis for this kind of science has been expounded by Victor Weisskopf, Leon Lederman, Stephen Hawking, and particularly Steven Weinberg, in his book *Dreams of a Final Theory*. The buzzword is "reductionism", the idea that the goal of physics is solely or mostly to discover the "fundamental" laws which all phenomena involving matter and energy must obey and that ignorance about these laws persists only on the extreme scales of the very small, the very cosmic, or the very weak and subtle.

The glamorous image of physics that this preoccupation projects is not necessarily all good: with the end of the Cold War and of expansionist public spending, physics is seen by all too many policy makers as too expensive for its practical return or simply too big for its boots. Some pundits have called the past half-century "The Age of Physics" and suggested that this age is coming to an end.

This pessimistic view may or may not be true even of Big Science physics. It seems to me that cosmic physics, at least, is in the midst of a very fertile period, not near collapse. But what it ignores is the fact that most physics and most physicists are not involved in this type of work at

[*]Originally published in PNAS, Vol. 92, July 1995, pp. 6653–6654. © 1995 National Academy of Sciences, USA.

all. Eighty percent or so of research physicists do not classify themselves as cosmic or elementary particle physicists and are not much concerned with testing the fundamental laws. Admittedly, some portion of this 80% are concerned with applications of physics to various practical problems, as for example prospecting geophysicists or electronic device designers. But another large fraction are engaged in an entirely different type of fundamental research: research into phenomena that are too complex to be analyzed straightforwardly by simple application of the fundamental laws. These physicists are working at another frontier between the mysterious and the understood: the frontier of complexity.

At this frontier, the watchword is not reductionism but emergence. Emergent complex phenomena are by no means in violation of the microscopic laws, but they do not appear as logically consequent on these laws. That this is the case can be illustrated by two examples which show that a complex phenomenon can follow laws independently of the detailed substrate in which it is expressed.

(i) The "Bardeen–Cooper–Schrieffer (BCS)" phenomenon of broken gauge symmetry in dense fermion liquids has at least three expressions: electrons in metals, of course, where it is called "superconductivity"; ^{3}He atoms, which become a pair superfluid when liquid ^{3}He is cooled below $1-3 \times 10^{-3}$ K; and nucleons both in ordinary nuclei (the pairing phenomenon explained by Bohr, Mottelson, and Pines) and in neutron stars, on a giant scale, where superfluidity is responsible for the "glitch" phenomenon. All of these different physical embodiments obey the general laws of broken symmetry that are the canonical example of emergence in physics.

(ii) One may make a digital computer using electrical relays, vacuum tubes, transistors, or neurons; the latter are capable of behaviors more complex than simple computation but are certainly capable of that; we do not know whether the other examples are capable of "mental" phenomena or not. But the rules governing computation do not vary depending on the physical substrate in which they are expressed; hence, they are logically independent of the physical laws governing that substrate.

This principle of emergence is as pervasive a philosophical foundation of the viewpoint of modern science as is reductionism. It underlies, for example, all of biology, as emphasized especially by Ernst Mayr, and much of geology. It represents an open frontier for the physicist, a frontier which has no practical barriers in terms of expense or feasibility, merely intellectual ones. It is this frontier that this colloquium was destined to showcase. A typical (but incomplete) selection of the papers given at the colloquium are reproduced in *PNAS*, issue of July 18, 1995.

The subfields of complexity which we chose to represent are only a fraction of the available material. This frontier of complexity is by far the most active growth point of physics. Physicists are also finding themselves, more and more, working side by side with other scientists in interdisciplinary collaborations at this frontier. The flavor of many of the talks was interdisciplinary.

We chose four areas under which to group the talks. These highlighted four kinds of physics. A brief description of each follows.

(i) Conventional solid state physics is really the quantum physics of materials, and as such deals mainly with electronic properties. There has been a resurgence of interest in this area because of several new and startling discoveries of new types of materials: We called this "The New Physics of Crystalline Materials".

(ii) Noncrystalline materials such as glasses and polymers have become increasingly subjects of interest in physics, so we included a section on "The New Physics of Noncrystalline Materials." Some of the deepest problems in theoretical physics still surround the dynamics of glasses. It is slightly unsuitable to include biophysical materials under this heading since biophysics is an important and growing frontier in itself but in order not to subdivide the sessions infinitely this was done.

(iii) Physicists are increasingly moving into realms very far from equilibrium, studying processes which form many of the natural objects we encounter as well as describing the highly nonequilibrium behavior we see all around us in turbulent convective flow (weather) and the nonequilibrium states we see in our geological surroundings. The phenomenology of this sort of phenomenon ("fractals," chaos, pattern formation) far outstrips our theoretical apparatus for dealing with it. The emphasis is on the "search for generalizations".

(iv) Finally, there is a field that has grown up around the new statistical physics developed for some fascinating materials problems of disordered dynamical systems, which has overlapped into problems of computational algorithms for complex problems and into the theory of neural networks. This is a development which promises to allow us to deal with true complexity with physical rigor and needed to be presented: "New Theories of Complexity".

A fifth topic was omitted because of overlap with a recent colloquium: complexity in astrophysics, and I am sure those interested in complexity will have their own candidates. But, in the end, this program is what we had room for and the attendees seemed to have enjoyed the program as presented.

Is Complexity Physics?
Is It Science? What Is It?[*]

This is the first of two Reference Frames for Physics Today about complexity and the Santa Fe Institute. The Reference Frames were initially designed to fit a 900-word page, so tend to be very succinct. SFI was one of the thrilling experiences of my life and I am grateful to George Cowan and David Pines for the opportunity to be an early member.

Questions like the ones posed in the title of this column are being asked nowadays by faculty members in university departments of physics — and many other disciplines as well. They find themselves having to assess graduate work or applicants for jobs or tenure in strange fields like neural networks, self-organized criticality applied to earthquakes or landforms, learning systems, self -organization and even such older fields as nonlinear dynamics or spin glasses. The easy way out in these decisions is to give the nod to more orthodox and traditional work. So it is a matter of some urgency to decide whether "complexity" is a part of physics — or even vice versa.

For most of its history, and for its most well-known practitioners, physics has been the ultimate reductionist subject. Physicists reduce matter first to molecules, then to atoms, then to nuclei and electrons, then to nucleons and so on — always attempting to reduce complexity to simplicity. We found there were, in all of physics, only four forces, which then were reduced to three; now string theorists tell us all three boil down to a single quantum gravitational, supersymmetric, utterly featureless earliest universe. In condensed matter physics, the diversity of the world of solid

[*]Reprinted with permission from *Physics Today*, Vol. 44, July 1991, p. 9. © 1991 American Institute of Physics.

matter seemed to be reducible to a universal band theory of electronic structure plus a few elementary defects such as donors, acceptors, grain boundaries and dislocations.

With the maturation of physics, a new and different set of paradigms began to develop that pointed the other way, toward developing complexity out of simplicity. I happen to have written one of the early manifestos for this infinitely quiet revolution, a Regents' Lecture (given at the University of California, San Diego, in 1967) and magazine article called "More Is Different" (Science 177, 393, 1972). I emphasized the concept of "broken symmetry", the ability of a large collection of simple objects to abandon its own symmetry as well as the symmetries of the forces governing it and to exhibit the "emergent property" of a new symmetry. ("Emergent" is a philosophical term going back to 19th-century debates about evolution, implying properties that do not preexist in a system or substrate. Life and consciousness, in this view, are emergent properties.) The canonical example was the Bardeen-Cooper-Schrieffer theory of superconductivity, which could be described as broken gauge symmetry. The elementary-particle theorists Yoichiro Nambu, Geoffrey Goldstone, Peter Higgs, and eventually John Ward, Abdus Salam and Steven Weinberg used a BCS-like theory to unify particles and forces, but one could just as justifiably look at broken symmetry as a scheme for getting complex behavior from simple origins. (A primer of broken symmetry is to be found in' my book *Basic Notions of Condensed Matter Physics* [Benjamin-Cummings, 1984], in the second chapter and especially among the reprints.) More recently, the study of defects of broken-symmetry systems by Gerard Toulouse, G. E. Volovik and others has opened a whole new realm of complexity.

John Hammersley's theory of percolation and my ideas of localization set a second kind of pattern for developing complex behavior, namely, the idea that randomness and disorder could result in generic properties that are utterly different from those of merely somewhat impure regular materials. For instance, they result in new types of transport theory. The spin glass is in a way the ultimate illustration of this situation: It requires a whole new version of statistical mechanics. Glass itself remains one of the deepest puzzles in all of physics.

More recently, dynamical instabilities and deterministic chaos have been brought to our attention. Yet another group of developments involves spontaneous pattern formation, fractals and the beginning of self-organization. The whole ball of wax has been jumbled together by the remarkable idea of Per Bak, Kurt Wiesenfeld and Chao Tang that many of the phenomena of nature exhibit scaling laws determined by self-organized behavior like that at critical points of phase transitions. (See the Reference Frame column by Leo Kadanoff, March 1991, page 9.)

As one probes deeper into the origin of the universe or the interior of the quark, it will never be questioned that one is doing physics. By contrast, the traditional, reductionist physicist and, for sure, the funding agencies can be left vaguely disturbed or hostile as new fields lead us *up* the hierarchy of complexity toward sciences such as geology, developmental biology, computer science, artificial intelligence or even economics. There can be a somewhat surprising lack of understanding of what those of us working in these new fields are doing. It is still possible, for instance, to find Marvin L. Goldberger and Wolfgang Panofsky, who are relatively enlightened physicists, saying, in an op-ed piece for *The New York Times:* "Other branches of physics [than particle physics] ... are, interesting, challenging, and of great importance .... The objectives are *not* a *search for fundamental laws as such* [my italics], these having been known ... since the 1920s. Rather, they are the application of these laws." If broken symmetry, localization, fractals and strange attractors are not "fundamental", what are they?

A movement is under way toward joining together into a general subject all the various ideas about ways new properties emerge. We call this subject the science of complexity. Within this topic, ideas equal in depth and interest to those in physics come from some of the other sciences. This movement is overdue and healthy. On the other hand, one may well be apprehensive — or at least I am — that such an enterprise might go the way of General Semantics, General Systems Theory and other well-meant but premature and intellectually lightweight attempts at building an overall structure. We complexity enthusiasts (perish the thought that we be called complexity scientists!) are talking, at least for the most part, about specific, testable schemes and specific mechanisms and concepts. Occasionally we find that these schemes and concepts bridge subjects, but if we value our integrity, we do not attempt to force the integration.

A number of institutions have grown up to foster this kind of work, and in a future column I'd like to write about one of them, the Santa Fe Institute. Meanwhile, let me give my own answers to my opening questions: Complexity, as defined in the preceding paragraph, is often physics. It is the leading edge of science. And it is surely exhilarating.

Complexity II
The Santa Fe Institute[*]

It was with a sense of pride that, the other day, I realized I should update my curriculum vitae to include the phrase "External Professor, Santa Fe Institute, 1989".

What in Heaven's name, you may ask, is the Santa Fe Institute? And why am I proud of being a spear-carrier (officially, second Vice Chairman of the science board as well as "external professor") at this unknown place? The answer calls for a brief history and for the explanation of two central ideas: "emerging syntheses" and "complex adaptive systems".

In the early 1980s a group of friends centered around Los Alamos and Santa Fe, led by George Cowan and including Murray Gell-Mann, Herb Anderson, Mike Simmons, Peter Carruthers and Nick Metropolis as well as quite a number of others, began to discuss doing something about one of the sociological peculiarities of science: that most scientific revolutions (contrary to Thomas Kuhn's well-known description) occur *outside* or *between* the established areas of science that are enshrined in our universities' departmental structures and in the funding agencies. Radioastronomy, molecular biology, solid-state physics, quantum optics and the like did not grow up in the university departments devoted to those subjects or even in predecessor departments of astronomy, biology or physics. Rather, these fields emerged at a variety of institutions characterized by flexibility — in the power vacuum of the post-Rutherford Cavendish Laboratory, at the Rockefeller Institute and at Bell Laboratories, among other places. "Chaos", for instance, began with a meteorologist and some mathematicians and population biologists, and it matured in a 1960s-type collective atmosphere at Santa Cruz.

[*]Reprinted with permission from *Physics Today*, Vol. 45, June 1992, p. 9. © 1992 American Institute of Physics.

The Santa Fe group began to dream of an institution focused on the vacuum between established fields and on the "emerging syntheses" that arise therein, as we came to call such new and growing areas. Finally, in 1984 the group sponsored two founding workshops, using the beautiful premises of the School of American Research in Santa Fe as a venue, at which a wide variety of candidate ideas and problems were expounded. The proceedings, published at first as an "advertising brochure" for the Santa Fe Institute, became the first book in the series the SFI eventually issued: *Emerging Syntheses in Science*, edited by Gell-Mann and David Pines, who had by then joined the group.

I remember those workshops not just with delight at meeting a large, diverse and congenial group of like-minded people in fantastically beautiful surroundings, but also with pleasure at discovering what one might call the "SFI secret weapon": almost without exception, the more eminent, the more deeply committed, the more successful within a given field a scientist is, the more eager that scientist is to relate to scholars outside his or her field, and the more open he or she is to the Santa Fe message. It was rare indeed for us to identify a first-rate scientist in any field who we felt could contribute to our work and then to find that person was not happy to join. The *more* mature and self-confident a scientist is, the *less* that scientist feels that his or her narrow field contains the whole of knowledge.

George Cowan was chosen as president — since, in the beginning, it was his idea — and Murray as chairman of the board; along with a tiny bank account from a few private contributors, that was the institute. From the first we had probably the highest ratio of scientific eminence, commitment and sheer competence to physical plant and actual funding since Galileo's Accademia dei Lincei, or Academy of the Lynx-Eyed (that is, the far-seeing).

Gradually, however, through sessions organized with foundation help, proposals to agencies and so on, we have reached an annual budget of 2 to 3 million dollars, coming from at least four disparate sources, as well as private contributions.

We run workshops, summer schools and residential programs, much on the pattern of such institutions as Woods Hole, the International Centre for Theoretical Physics in Trieste and the Institute for Theoretical Physics in Santa Barbara. A few postdocs and a numbers of visiting scientists from our "external faculty" participate in programs whose real core is usually a network of like-minded people nationwide.

In much of our work we have identified a common, theme that we call the "complex adaptive system". We see common behaviors in systems as diverse as biological populations, economics, organisms and the immune system, as well as adaptive computer programs such as those invented by John Holland. This commonality is something we try consciously to retain in order to keep the different parts of our enterprise in intellectual contact with one another.

The most fully developed of our programs began as a result of a conversation, sponsored by economist Carl Kaysen, ex-head of the Institute for Advanced Study, between archaeologist Bob Adams, head of the Smithsonian Institution, and John Reed, CEO of Citicorp. Reed's profound distrust of the conventional economic wisdom that, in the early 1980s, landed Citicorp in such trouble with its Latin American debts, predisposed him to be interested in helping to initiate a radical rethinking of global economics. He and Adams ran a one-day trial workshop at Santa Fe, where he eloquently expressed his concerns, seconded by one of his vice presidents, Eugenia Singer, who later became a regular member of our group. This initiative led to two further workshops, headed by Ken Arrow, an economics Nobelist, and me, that brought together natural scientists of several kinds with economists, and to a continuing on-site program led in the first year (1987–88) by Brian Arthur of Stanford. One activity we have run is the Santa Fe Double Oral Auction Tournament, a test bed for intelligent economics computer programs engineered by Richard Palmer and John Rust.

This is the best-known and best-funded SFI program, but we have a lot of other successful activities. Stu Kauffmann keeps a program going on molecular evolution and is on site this year with several visitors and a postdoc; theoretical immunology is a continuing interest that we share with researchers at Los Alamos National Laboratory along with Alan Perelson; our adaptive computation work has led to an offshoot at the University of Michigan led by Holland; and in some generalized sense the whole problem of adaptive, intelligent computers occupies a variety of our people, among them Kauffmann, Norman Packard, Doyne Farmer and Chris Langton (who has run a sequence of successful "artificial life" workshops with, for instance, 4H Club-type prizes for the most successful artificial animals or plants). We have also run three successful summer schools and one winter school on complexity science, in all of which Dan Stein and Erica Jen played leading roles.

I have slighted for want of space many other activities — for instance, our explorations in archaeology, in linguistics and in quantum measurement theory, all fostered by Gell-Mann, as well as a hopeful program on global survivability, began by Cowan.

George Cowan has now turned the presidency of the Santa Fe Institute over to Ed Knapp, and in new and expanded premises in another part of Santa Fe, we head into our next five years. We continue to believe that we hold the world record for ratio of intellectual effervescence to funding.

Whole Truths False In Part[*]

The following is a little out of date, it reflects the state of the field in 1995 when it was written, but it says many things that are true. Perhaps the biggest development since then has been a lot of emphasis on connectivity of graphs — the "six degrees of separation" phenomenon. The Santa Fe Institute continues to thrive, as of this writing.

The science of "complexity" has reached the attention of the chattering classes, or at least of the writing ones, and the subject is experiencing the predictable backlash following the honeymoon. The well-written but frankly adulatory *Complexity* by Mitchell Waldrop, and the heavily publicized *The Quark and the Jaguar* by Murray Gell-Mann, were among the books which rode the crest of the wave, but now there is a spate of books which seem destined through no fault of their own to hit the trough, of which *Frontiers of Complexity* by Peter Coveney and Roger Highfield is one. The earlier books emphasized the role of a small institute in Santa Fe to which Gell-Mann belongs (as well as a remarkably high percentage of other authors of complexity books), somewhat unfairly for what is a very broad field of endeavor, pursued by groups of scientists worldwide. As a result the Santa Fe Institute (to which the reviewer belongs) has become identified as the defining institution of the field, in this book as elsewhere.

If I had to give my own definition of "complexity science", I would say that it is the search for general concepts, principles and methods for dealing with systems which are so large and intricate that they show autonomous behavior which is not just reducible to the properties of the

---

[*]Book review of *Frontiers of Complexity: The Search for Order in a Chaotic World*, by Peter Coveney and Roger Highfield (Ballantine Books, New York, 1995). Originally published in the *Times Higher Education Supplement*, 22 December 1995.

parts of which they are made. It is the search for the operative principles and concepts, as opposed to the reductionist study of detail, which distinguishes the "complexologist", and as such he is subject to such calumnies as that he is searching for a Theory of Everything. What is true is that he is searching for a theory, not a recipe: to take the example of a successful complexologist, natural selection, not the family tree of the horse.

Complexity is an enormous, rapidly growing and diversifying field. Coveney and Highfield's is arguably the best general book so far on this highly "complex" subject, and would for instance make a useful text for an undergraduate seminar. It covers in some descriptive detail most of the subjects which comprise this emerging science, referring to detailed interviews with many of the major players and citing passages from a wide selection of the important books and articles. There is little or no factually incorrect material, and most controversial material is clearly labeled so. l felt, however, that with more effort and a wider coverage, it could have been better.

First let us sketch what *Frontiers of Complexity* is about. Complexity is a subject which has an annoying propensity to define itself, almost in spite of the efforts of many workers (including myself) to avoid the word as a vague, indefinable generalization, welding together a congeries of different ideas. In fact, I was at first confused by finding "complexity" here defined twice, or perhaps many times, in several incompatible ways. On page seven it is defined in terms of emergent phenomena in macroscopic systems, a concept originating in evolutionary biology and introduced into the physical sciences (perhaps by your reviewer in 1967) without reference to the supposedly essential role of the modern computer, which is added in on pages 9 and 10. I am bemused to find that Darwin and his successors in evolutionary biology and ecology, in the absence of computers, cannot have been dealing with "complexity". I guess we can all go along with "nonlinearity," another stated essential piece, and perhaps with "irreversibility," which the authors also toss into the pot. I am unhappy, nonetheless, with definitions which exclude the essentially equilibrium phenomena of broken symmetry and broken ergodicity, the few-dimensional aspect of (true, technically defined) chaos, and the marvelous analytic treatments of neural networks and of complex optimization by such as Gerard Toulouse, Giorgio Parisi, Marc Mezard, David Thouless, Haim Sompolinsky and many others, as well as leaving out most evolutionary biology.

For my money "complexity" is a state of mind, embracing any study of a realistic system which negates the strong reductionist (what I once called the constructionist) point of view which assumes everything follows from the fundamental laws, and emphasizes the appearance of emergent phenomena of all kinds, at least intellectually independent of the microscopic substrate in which they appear. This attitude is often mistaken for holism or for a rejection of scientific ("weak") reductionism but is neither. Many of its ideas appeared either as parts of computer science or as results of computer investigations, but many did not, and in this reviewer's opinion the greatest problem the field faces is maintaining its sometimes tenuous connection with the non-virtual world, especially considering the seductive nature of computer work and the fascination of the lay public with the images the computer produces. I am not sure that this book will be very helpful in cementing this connection.

Complexity science does in the main result from the creative tension between two intellectual traditions: the creative side of computer science, and the natural science of complex systems; and the book alternates its attentions between these.

Starting with an affecting prologue on the twin founding geniuses of complexity theory and computer science, Alan Turing and John von Neumann, we encounter a couple of introductory chapters on complexity itself and on discrete mathematics and its various overlaps with complexity. Next we are introduced to the computer via a very brief history, from Babbage to the quantum computer. The natural science strain comes in via chapters on computing schemes based vaguely on natural analogies: neural networks, cellular automata, simulated annealing, the genetic algorithms. Nature here appears primarily as a model, not as an intrinsically complex object of study. A final chapter in this sequence juxtaposes three quite incompatible bedfellows: the Brussels "dissipative structure" school, chaos and the concomitant relation of chaos to fractal structures, and self-organized criticality. The major links between these seem to be that they all have a relationship to complex phenomena observed in nature, and they lead to striking pictorial illustrations.

Next come two long chapters on life, first "Life as we know it", and then "Life as it could be", i.e., natural and artificial life. "Natural" covers studies on mechanisms of the origin of life; on pattern formation and morphogenesis; on economics and other forms of behavior; and on ecology, "Artificial" covers the whole gamut of artificially constructed and artificially evolved creatures as well as a number of artificial worlds in which creatures or agents compete, i.e., artificial ecologies. A final substantive chapter treats both aspects of the mind: computational attempts to mimic brain and neurological function, and scientific attempts to understand the mind as an emergent phenomenon.

Perhaps my uneasiness about the coverage can be best justified by discussing their treatment of the field in complexity theory I have dealt with from its inception, "spin glass" and its ramifications. The book's scenario rather oddly places the paper of David Sherrington and Scott Kirkpatrick as the central event in the field. This paper was an obvious next step from Sam Edwards' and my earlier work, and it seemed to all of us that it could not be wrong, but it was — that was the main way in which it was important, because the search for this error led to a still vital part of theoretical physics, as well as to insights into the nature of complex systems and complex problems which still have to be appreciated by most computer scientists. These results follow from the Thouless–Toulouse–Parisi concept of replica symmetry breaking, which stands out in many minds as a truly spectacular breakthrough. Also missing is the tie-in to the ongoing effort to understand complex behavior in glasses, gels, polymers, vortex lattices and the like. This is, to be sure, a very tough subject to popularize, but a perhaps prejudiced view is that its importance as a balance to the computer-oriented slant of much of the rest of the book would have made it worthwhile.

The above criticism could perhaps result from my personal involvement with this area, but there are a few other false notes at several other points. In discussing the Brussels school, the "principle of minimum entropy production" is seriously quoted, despite the fact that it is essentially never obeyed (entropy production is known to be at a maximum in the linear regime, and seems to be irrelevant everywhere else). In a similar vein, reaction-diffusion systems (like the B-Z reaction) are extensively used in morphogenesis, but they are never allowed to form their own patterns: that is under strict genetic control. The significance of all this work to the real study of complex systems' behavior is still questionable.

I was disturbed by the space given (though, admittedly, with carefully footnoted caveats) to problems of non-computability and discrete mathematics *à la* Gödel and Roger Penrose, and to microtubules as the new pineal gland. And surely giving the dubious philosopher Nancy Cartwright space at all is carrying the overview of the field beyond its boundaries.

I believe firmly, with Coveney and Highfield, that complexity, however defined, is the scientific frontier. It takes a lot of flak from critics in the more conventional sciences, some of which is brought on by the more extravagantly stated claims of its proponents. These two authors have been accused of wide-eyed acceptance of these claims, but I felt the skeptical side was often well-represented. This book will make clearer what both sides in this debate are on about, even if it does not provide the most critical possible assessment of where the real "meat" in the field is to be found. Perhaps that would be expecting too much, at this point in history.

X. Popularization Attempts

Introduction

This chapter consists of attempts to explain some technical scientific matters to laymen. I have elsewhere disavowed any ability along these lines, and the reader — especially the true layman — will surely agree. I include these both in the forlorn hope that it will fulfill a kind of pedagogical purpose for some, and because in order to give some account of myself, in the end I have to talk about what I *did*.

Who Or What Is RVB?[*]

Friedrich August Kekulé von Stradonitz (how could I not start with him?), in 1865, solved the structure of benzene, the molecule which would be the poster child for the idea of Resonating Valence Bonds. But he had no such thoughts; he deduced the symmetrical six-membered ring from purely chemical data. It wasn't until 1916 that G.N. Lewis invented the idea of the Valence Bond as caused by a shared pair of electrons, one from each atom of the bond; and the valence bond was first explained quantum-mechanically by Walter Heitler and Fritz London in 1928. They showed that it could be thought of as due to binding of the pair of electrons, one from each atom, by Heisenberg's exchange interaction. If we have two overlapping atomic orbitals φ_1 and φ_2, the exchange integral J in $E_{int} = JS_1 \cdot S_2$ will be antiferromagnetic and will cause a singlet bond between the orbitals to be energetically favorable. Linus Pauling whisked that great idea out of his friends' hands, and in the next few years he ran through most of chemistry showing that the pair bond was a very useful heuristic tool. But it is not only heuristic, it has one property that is overwhelmingly common in chemistry, yet is not shared by other theories — the bond is local, and doesn't depend much on what else is in the molecule or solid. So he could take his pair bond and its component atoms from one substance to another and it would have the same length, strength etc.

But the pair bond alone did not explain benzene. There are three "sigma" bonds (symmetric in the plane of the molecule) for each carbon, one to the H and one to each of its neighbors; but that leaves only one

---

[*]Reprinted with permission from *Physics Today*, Vol. 61, April 2008, pp. 8–9. © 2008 American Institute of Physics.

more ("pi" — odd symmetry) valence electron per C. If it chooses one neighbor to bond to, that will cause an asymmetrical structure of three C_2 pairs, what we would nowadays call a Peierls doubling, and this does not occur. So Pauling introduced the idea of Resonance — the first initial of RVB — which proposes that there is an amplitude for the structure to form a quantum-mechanically coherent mixture between the two different pairing schemes. Actually, there is also some admixture of other pairing schemes, where some pairs form right across the hexagon. Such resonances, like any other form of quantum-mechanical zero-point motion, lower the energy and enhance the strength of the bonding, by an amount in good agreement with the experimental accuracy then available.

In the years 1931–33 Pauling added this concept of resonance to his heuristic toolkit, and using that toolkit produced in 1935 his classic textbook, *The Nature of the Chemical Bond*. This remarkable book not only explains qualitatively the structures of most organic and metal-organic molecules, and most insulating solids, he was able to parameterize the electronic quantities on which the bonding depends, thus quantifying most of chemistry.

There is, of course, an alternative way of understanding the bonding of molecules and solids, called the molecular orbital theory, introduced by F. Hund and R. A. Mulliken a few months after Heitler–London. In Pauling's theory one assumes always that an atom has a specific valency and that its electrons never really run free — in modern terms, it is based on starting with a large-U, atomic, limit and allowing the interactions of the atoms to perturb that. The molecular orbital theory (we physicists call it band theory) takes the opposite tack: one shares the electrons among all the atoms, calculating a set of electronic states or "molecular orbitals", and takes the electron-electron interaction which keeps the electrons on their separate atoms into account only later — and often, in the early days, quite inadequately. But it turned out, in the long run, that while Pauling had enormous success initially, he was the hare and MO was the tortoise, which caught up and passed him half a century later: and now even the chemists tend to believe that the *real* electronic structure is that calculated using LDA bands and pseudopotentials and, if necessary, other refinements on the pure MO scheme, and to forget the idea that there are unchanging atoms with certain fixed valences and bonding electron identities. (Though of course everyone really thinks Pauling's way at the beginning of a problem.) There is in fact a way within band theory of understanding the locality of most chemistry, using Gregory Wannier's wonderful transformation from band wave-functions to atomic-like ones, but that wasn't really understood until the late 1960s (and is still unknown to many band theorists and quantum chemists).

But that really only works — the wave functions can only be transformed into localized forms — if the electrons of interest are split off by a gap in energy from the empty states — that is, if the substance is insulating. Pauling ignored that restriction and in the 1940s, (when I was learning solid state physics) he tried to make a theory of metals using a quantum liquid of valence bonds — an RVB — assigning fractional valences to the metal atoms and using his usual bag of semiempirical tricks with which he could, as the saying goes, "fit an elephant." But of course it was not at all a useful way to go and could not possibly, for instance, give any account of all the properties of metals which depend on the Fermi surface. Also, even its account of the physical sources of the binding energy was less satisfactory than that of band theory. It soon disappeared from view.

In the early 1970s I began to wonder what kind of state might actually exemplify this wrong idea of Pauling's. The reason had to do with a subject I had long been involved in, antiferromagnetism. This again involves rather an excursion into ancient history, and also into parts of condensed matter physics that may be not as widely taught as they should be. The line of thinking stems from another giant of early quantum mechanics, Hans Bethe, who in 1931 solved exactly a model for an infinitely long chain of atoms connected by antiferromagnetic Heisenberg exchanges — an infinitely long benzene ring, if you like — and left us with a dilemma: the result was not a metal, to be sure — but it also was not antiferromagnetic. He had found the first RVB! Next time I will describe how fervently people came to hope it was the only one.

The word "antiferromagnetism" is one of those obvious constructions that seem to define themselves — yet someone must have been the first to use it. In the same year, 1936, one finds it used by Louis Néel as describing the ordered state of the spins of an actual substance — for which insight he received the Nobel prize — and by Lamek Hulthen in reference to the linear chain Hamiltonian for which Bethe had found the ground state. (Hulthen's earlier title for his thesis done under H.A. Kramers did not use the term.)

I indulge myself in one final excursion into ancient history. When Néel attempted to describe his experiments suggesting the existence of an ordered antiferromagnetic state at an international meeting in 1939, he met with such implacable opposition to its reality from the quantum theorists, on the basis that the only exact solution of the antiferromagnetic Hamiltonian was Bethe's disorderly one, that he left the meeting in disgust and retained a prejudice against the quantum theory to the end of his life.

It's hard to believe, but true, that in spite of the neat and beautiful mean field quantum treatment of antiferromagnetism which van Vleck published in 1941, there were still doubters such as the great Lev Landau until, in 1952, I pointed out that the difference between Bethe's 1-dimensional chain and real three-dimensional antiferromagnets (which by then had been observed unequivocally with neutron diffraction by Cliff Shull and colleagues) was that in 1D quantum fluctuations diverge, while in 3D they are finite. The method I used was to calculate the zero-point or thermal amplitude of the spectrum of spin-wave fluctuations — the equivalent of the Debye-Waller factor — and see if it came out finite.

Which observations left partly open the intriguing case of 2 dimensions. My calculations said the ground state would be ordered in 2D, barely, but that without some added three-dimensional help, or some anisotropy, the antiferromagnet would not be ordered at finite T. (All of these predictions turned out, in the end, to be beautifully verified in La_2CuO_4, but it was a while before we realized that!) Since my theory was the result of using a series in powers of $2/ZS$, it didn't really guarantee to be convergent, say for the square lattice of Cu^{++} ions with $S = 1/2$ and $Z = 4$.

In 1973 I had been talking to Denis McWhan and Maurice Rice about some interesting data on the compound TaS_2 (which data, I believe, found some other explanation in the end). This turns out to be reasonably well modeled by a 2-dimensional triangular lattice of $S = 1/2$ spins. The triangular lattice has $Z = 6$, but that is compensated by the fact that you can't set the 3 spins on a triangle antiparallel to each other — it's "frustrated", as we came to describe it — so the series converges as poorly as the square lattice. With TaS_2 as an excuse, I set out[*] to see if the Néel state was the correct ground state of the triangular $S = 1/2$ antiferromagnet, or was it indeed a fluid of resonating pair bonds like the Bethe 1D solution? Not surprisingly, my natural optimism led me to conclude the latter; whether I was right is, I believe, still a matter of some controversy. Probably I was not literally right, but neither is the Néel state; but that is getting far, far ahead of the story. What may be important about that paper is that it first posed the questions: what is the nature of that state?

---

[*]*Materials Research Bulletin* 8, 153 (1973).

Does it have an order parameter? what is its relationship to BCS pairing of electron singlet pairs? how can you tell experimentally if you have one? And, of course, what do you call it? RVB?

The paper had a publishing history. Why is it in the MRB? I have never before nor since used that journal. In fact, a letter had just arrived from the editor, Rustum Roy, soliciting papers for a Linus Pauling Festschrift. Since the nature of the paper was that it was a kind of "in joke" on Pauling, suggesting that his theory of metals was a theory of something else, I thought it belonged in the Festschrift. But I didn't say this in the covering letter, and Rustum treated my paper as an "over the transom" submission — it appeared in a totally miscellaneous issue of the journal, though with a nicely apologetic note.

I'd like to acknowledge that the stimulus for this essay came from discussions with Shivaji Sondhi.

More on RVB

After 1974 I have to confess that, until January 1987, I thought very little more about RVB except on those pleasant occasions when I happened to encounter Patrik Fazekas at meetings (he had gone back to Hungary). That first week of January found most of the community that was interested in the physics of exotic magnetic materials gathered in Bangalore, India, for an international conference on "Valence Fluctuations and Heavy Fermions", but in the interstices of the planned program we all found time to discuss intensively Bednorz and Mueller's exciting discovery of the high T_c cuprate superconductors, which discovery had just been confirmed and extended by a number of groups.

Two things triggered my interest. One was discussions with Baskaran in which he informed me of what was known about the "parent compound" of the superconductors, La_2CuO_4 — that it was an insulator, very anisotropic, with the CuO_6 groups displaced from octahedral symmetry into a square planar arrangement. At that point I realized that we had an almost perfect realization of the simplest possible Hubbard model: a square 2D lattice with a spin $\frac{1}{2}$ ion on each lattice site and a single nondegenerate band; and that the crystal arrangement conspired miraculously to reduce any interplanar coupling.

It was also clear that this compound was a Mott-Hubbard insulator: it was insulating not because of having filled one-electron bands, but because each of the Cu ions insisted on being exactly Cu^{++}, because the electron interaction energy necessary to make free carriers via $Cu^{++} + Cu^{++} \rightarrow Cu^+ + Cu^{+++}$ is much greater than the bandwidth.

The second inspiration came from a paper given by Maurice Rice, an old friend and occasional collaborator. He showed that one could very accurately approximate the fabled Bethe solution of the one-dimensional antiferromagnet (which, you should remember from my last article, is the original model for the RVB state) by a very simple procedure: take the non-interacting, metallic state of exactly one electron per atom on a linear chain, and *Gutzwiller project* out of the wave function every component in which there are two electrons or two holes on the same atom. (Martin Gutzwiller is a Swiss theorist — now at IBM — who, contemporaneously with John Hubbard who invented the model, proposed a clever projection method as a way of computing the properties of strongly-interacting systems such as the Hubbard model, but soon thereafter abandoned that field.) The way Rice applied it was to look in the metallic wave function for components where there was exactly one electron on every atom and keep only those. Thus it is a wave function for *spins*, not for electrons, and of course becomes an insulating state. In particular, we leave the phases of the various components just as they were. This wave function has an energy within a gnat's eyelash of that of the 1D Bethe solution. (Much later, Sriram Shastry as well as Duncan Haldane showed that it is even the exact solution of a slightly modified Heisenberg exchange Hamiltonian, which accounts for its amazing accuracy.)

It immediately struck me that Maurice had discovered the secret of how to write an explicit, simple wave function for an RVB (since Bethe's solution was the original RVB!): namely, Gutzwiller project (which may be done in any dimension) the wave function of a metal.

My mind immediately jumped to what seemed to be a corollary: we know how to make singlet pairs in a metal, that is what BCS did — so why don't we make our RVB by projecting a BCS singlet pair wave function for the metal, instead of a simple Fermi sea? That should certainly improve the energy!

Having thought that, I came to a third conjecture, which practically made itself: if there is an RVB; if the RVB is a BCS wave function projected on the subspace with no charge fluctuations; and we "dope" the system with some free holes as in the cuprate superconductors, so as to restore the charge fluctuations; then couldn't the pairing persist into the metallic state, thus making a superconductor with a transition temperature that is limited by the antiferromagnetic exchange integral J, which can be a thousand or more degrees, not by a Debye temperature which is usually quite a bit lower?

Twenty years on, we know now that this pipe dream is, with various caveats, literally true; but the path to this realization, the formalism which expresses it, and the complicated but fascinating physics which it entails, are all much too tortuous and technical for this article. For one thing, I was one of the first to abandon it, and I published a book which was based on the following stupidly wrong statement: "... it is ... evident that this two-dimensional state (viz, the doped RVB) is not superconducting...". I came back to my senses in 1997-9; the comedy of errors which the theory of high T_c cuprates became, and in much of the field still is, is an utterly fascinating story, but not suitable for even summarizing in the few words I have left.

What I can summarize in the rest of this article is what has happened to the original idea of an insulating liquid-like state of antiferromagnetically interacting $S = 1/2$ spins which does not break translation symmetry. That is what I'd call the "true" RVB, while high T_c is an RVB-derived theory. First I should warn the reader that I myself did not stay involved in this work, and so whatever follows will probably be considered a bit amateurish by the real mavens — but so be it.

In the first couple of years, my group and a number of others stayed involved with the original square lattice problem of $LaCu_2O_4$ and analogues. To make a very long story short, and to neglect the marvelous work of many people, if the square lattice is to make an RVB by the original scheme, it was discovered that the optimal structure is the Gutzwiller projection of a BCS wave function with d-like $(x^2 - y^2)$ symmetry. (A group around Ian Affleck, another around Maurice Rice, and Gabi Kotliar, share credit for this.)

At the same time, a group of whom Bob Laughlin was the most voluble member, although it was Affleck and Marston who first wrote down the result, insisted that the best scheme was a so-called 'flux phase" in which one Gutzwiller projects the free electron wave function for the right number of electrons in a giant magnetic field. After various attempts, Affleck and Marston settled upon a field of exactly $\frac{1}{2}$ a flux quantum per elementary square. It seemed odd that both of these alternatives had four nodes in their spectra at exactly the momenta $\pm k_x = \pm k_y = \pi/2$, until we began to realize that the two wave functions are exactly equivalent — as shown by Shastry, somewhat later — and that in fact, using the enormous local SU(2) symmetry first pointed out to me by Baskaran, and mentioned in my 1987 lecture notes (and first published by Affleck) they are also equivalent to a host of other representations of the state which might seem to be density waves, have staggered flux, etc. (The symmetry comes from the fact that in the projected wave function, an up-spin hole and a down-spin electron create the same state.) The resemblance to the four nodal points of the $d(x^2 - y^2)$ order parameter of the cuprate superconductors is *not* coincidental.

Perhaps the most exciting feature of these states is the nature of this excitation spectrum — what are they nodes *of*? The excitation spectrum is itself necessarily projected on states with no charge fluctuations — so it consists of *spinons*, Fermions with spin $\frac{1}{2}$ but no charge — a so-called fractionalized electron. There is a Fermi surface — if there's no gap — or at the very least Fermi points with a Dirac linear spectrum like that of a neutrino, but there is no charge degree of freedom. This would all be very implausible and speculative if it weren't exactly what happens in the simple Heisenberg chain according to Bethe's solution.

What we don't know for sure about this (or other) pure RVB, spin liquid states is a lot. Is there an order parameter, local or topological? Is there a phase transition from the simple paramagnet? Are there topological defects? One answer to some of these questions was given by Todadri Senthil using complicated gauge theory methods: He thinks there is a Z_2 (Ising-like) order parameter which has vortex-like excitations called "visons". (Which are perhaps projected vortices of the BCS representation.) These were looked for experimentally and not found.

All of this, however, was thought to be futile because it turned out that the ground state of La_2CuO_4 is simply antiferromagnetic, thanks to neutron work by a group around Bob Birgeneau using magnificent crystals from Japanese collaborators, as interpreted by a powerful theory group of Chakravarty, Halperin, and Nelson, and checked by calculations by a number of other groups.

The responses of those who still wanted to pursue the elusive RVB were of two kinds. One way was to invent model systems whose outcome would inevitably (or could possibly) be RVB. After a brief fling with more complicated $SU(N)$ spins, much of this community settled on the "short-range RVB" or "dimer model" in which the spins paired only as nearest neighbors, so that one could summarize the statistics in terms of a single parameter, the resonance amplitude for a pair of singlet pairs to exchange partners — a model invented by Kivelson and explored by Sondhi, Moessner and Oganesyan. If this parameter is sufficiently large there appears to be an ordered state.

More to my taste has been the attempt to find theoretical models which behave like the original triangular lattice of Fazekas and Anderson — i.e. models which are characterized by *frustration*, having competing antiferromagnetic exchange integrals which are unsatisfiable with any ordered arrangement of spins. This was begun by a group around Premi Chandra, and has continued in a lot of capable hands. The exciting thing about this line of research has been that at least three experimental candidates for such spin liquids have been found, as was explained recently by Patrick Lee in a perspective in *Science* magazine. The characteristic that is common to these examples is that, although from independent evidence it is known that the scale of the exchange interactions is of the order of tens to hundreds of degrees, the paramagnetic susceptibility is featureless and remains finite — like the Pauli susceptibility of a metal — down to very low temperature. This seems to mean that they have, as per my original conjecture, an excitation spectrum like the Fermi surface of a metal but composed of Fermionic spinons with spin but no charge.

I am sure that we will not be allowed to forget about these wonderful systems, and we should look forward to the continuation of their experimental study.

Brainwashed by Feynman?[*]

Every morning I walk along the corridor towards my office, past a wall chart advertising the Standard Model of the elementary particles, the model which is a candidate for one of the great scientific achievements of the 20th century. This chart was put out by the Department of Energy, but has the endorsement of the American Association of Physics Teachers. It is very clear and concise, explaining everything beautifully. Unfortunately, there is one (and only one) comment on it which is not true, and, by many, accepted to be not true, although it has also become a piece of folklore: it says that the interactions between nucleons "may be modeled by the exchange of mesons".

It is indeed true that the dominant long-range attraction of nucleons is well modeled by the classic "OPEP" — one pion exchange potential — but after four decades of attempts to fit more and more complicated meson exchange models to the observations on shorter-range interactions nuclear physicists now accept that one really has to understand short-range repulsion in terms of the messy business of overlap of the nucleons as bags containing three quarks apiece. In fact, there is as yet no unique microscopic theory of nucleon-nucleon interactions, there are only empirical model potentials, which, try as they may, never quite fit all the facts (see Hans Bethe's article in the American Physical Society's centennial issue of *Reviews of Modern Physics*[1]).

The reason this meson theory of interactions so quickly became folklore — aside from the fact that Hideki Yukawa did earn a Nobel Prize for predicting the pion from it — is that it fits so well with Richard Feynman's

[*]Reprinted with permission from *Physics Today*, Vol. 53, February 2000, pp. 11–12. © 2000 American Institute of Physics.

diagrammatic, perturbative picture of physics as exemplified by quantum electrodynamics, where everything is calculated using diagrams with electron lines backwards and forward in time, connecting via the emission and absorption of photons.

But the nucleon is not an elementary object, it is a bound state of three elementary quarks, and bound states cannot be described using diagrammatic perturbation theory. The essential step in a derivation of Feynman's diagrams, or in fact of any other form of diagrammatic perturbation theory, is to imagine turning on the interactions gradually and to assume that nothing discontinuous happens as we do; and bound states don't form continuously. When bags of quarks come close together, quarks from one nucleon overlap into the bound state wave function on the other, which gives a repulsive "overlap" force, getting its strength from the strong binding via gluons which holds the nucleon together, since quarks are Fermions and can't occupy the same phase space.

Just as simple and straightforward a case is presented by the well-known "van der Waals" potential between filled shell atoms. The attractive part, which falls off like R^{-6}, is the London dispersion force, coming from photon exchange indeed and calculated from perturbation theory by Fritz London. But there is no diagrammatic way to treat the other half, the short-range overlap repulsion, because this is caused by overlap of electrons from one atom into bound orbitals on the other. It can be calculated with reasonable success, but laboriously, and there is no good perturbative approximation to it. I found a very useful rough approximation using my "chemical pseudopotential" method, but there seems to be no formalism to generalize this.

Yet another case of a force which cannot be calculated by some kind of renormalized diagram is the attractive potential of a covalent bond between atoms. The best way to calculate this is to abandon the separate atoms and let the electrons completely change their wave functions to "bonding orbitals". This is the "Hund-Mulliken" scheme of the 1920s, which lies behind modern energy band theory. But Walter Heitler and London earlier found a useful model by leaving the atoms alone, and calculating a spin-dependent "exchange" interaction between the spins of electrons on them. This was extensively exploited by Pauling in his "valence bond" methods, much valued by chemists for their qualitative insights.

The approach using localized electronic orbits with spins is necessarily the correct way to go in insulating magnetic materials, such as Mott insulators like La_2CuO_4. But the interaction between the localized spins has no diagrammatic perturbative description of the normal kind, again because the electrons are in bound states. I calculated it in 1958 (and called it "superexchange") but my method produces a power series that is upside down relative to the diagram series. T/U is the parameter, where "T" is kinetic energy and "U" is the interaction potential, while Feyman diagrams lead to series in U/T. This is because I started by assuming the localizing interaction potential was infinitely strong and reintroduced the kinetic energy as a perturbation. This means that I assume continuity with a state where part of phase space is separated out: a so-called "projective" transformation.

Physicists for a couple of generations — starting with John Slater and Jun Kondo in the 1950s and certainly including me, at various times — have been trying unsuccessfully to find a way to deal with Mott insulators and superexchange antiferromagnestism with conventional diagrams, but not with any notable success, as may be clearly seen from the perennial appearance, decade after decade, of the mantra "strongly interacting electrons" or its equivalent in the programs of our national and international centers of theoretical physics. (Solved problems don't require such workshops!)

Can You Sum All the Diagrams?

Maurice Rice pointed out a number of years ago in connection with the theory of mixed valence ("heavy electron") rare earth metallic compounds that, sometimes, it is possible to introduce "superexchange" in a metal using again projective methods, leading to the infamous "t-J" model, and this idea has been one of the central themes of several of the approaches to high T_c superconductivitiy in the cuprates. But, as in the past, there has been a great deal of confusion caused by the automatic assumption of field theoretically trained young theorists that the methods and ideas which are appropriate to the theory of ordinary metals and to conventional phase transitions can be applied without caution to these borderline materials. The obvious assumption is that if one is able, by dint of very hard work, to "sum up all the Feynman diagrams," one must arrive at the right answer. The problem is that no one has ever been able, over four decades, even to arrive at the right interaction that way; and there are very good formal mathematical arguments that this cannot be done in principle. But the waters are muddied by the fact that there are substances (like the metal Cr) that look like antiferromagnets but in which the spins which order are not the original atomic spins but are a "spin density wave" of quasi-free metallic electrons, and in which diagrammatic approximations more or less work. So no one can really be dogmatically certain that using these arguments absolutely precludes success.

Thus, the theorist is required to abandon tried and true procedures and to exercise taste and judgment; he is driven out of his comfortable, well trodden ways, and is forced to be creative, a skill which is neither taught in courses, nor rewarded, very reliably, by the NSF and other funding agencies.

Actually, the lineal descendants of Feynman, the particle theorists, have long abandoned straight-forward diagrams, for a much more varied toolkit of concepts and techniques. Two examples are cited by Roman Jackiw in his beautifully clear Dirac Prize Lecture[2] given at the International Centre for Theoretical Physics in Trieste: the fractionalization of quantum numbers, and the "chiral anomaly" which we shall discuss shortly. The former is a very fashionable subject this year, after the Nobel prize, and I need merely to say that the theory of Robert Laughlin is antithetic to diagrammatic perturbation theory, as were the earlier ideas of Jackiw and Claudio Rebbi and of Robert Schrieffer on fractionalization of charge in charge density waves.

The chiral anomaly (and similar phenomena) of quantum field theory is an even subtler effect which Jackiw explains neatly in his lecture: where one expects the currents of right-handed and left-handed massless Dirac fermions to be separately conserved, in fact it is only their sum, the so-called vector current, which is perfectly conserved when there are interactions, as a result of dynamics happening at high energies within the supposedly inert sea of negative energy states. The two chiralities of Dirac fermions are not independent, which leads to large effects like the mass of the π meson. This poses, again, a serious problem for condensed matter theorists: when can we ignore the dynamics of our Fermi seas of filled states? When do separate left-spinning and right-spinning (i.e., chiral) quasiparticles suffice, and when do they decay into something else? The notorious "linear T" resistivity observed in the cuprates is very likely to be a manifestation of just this kind of decay. (This is the observed scattering rate whose frequency and temperature dependence is incompatible with quasiparticle theory.) "Proofs" from perturbation theory which purport to show otherwise are not very useful since they ignore the possibility of anomalies, which our field theoretical colleagues have long since shown cannot be consistently treated using perturbation methods alone.

References

1. H. Bethe, *Rev. Mod. Phys.* **71**, no. 2, S1 (1999).
2. R. Jackiw, *Helv. Phys. Acta* **59**, 835 (1986).

Just Exactly What Do You Do, Dr. Anderson?

This is the text of the Abigail and John Van Vleck lecture at the University of Minnesota, given also at a number of other universities. The intent is to give a general description of the nature and significance of theoretical condensed matter physics.

I was recently (1983!) interviewed by Jeremy Bernstein for a series of articles about Bell Labs he turned into a book. He chose to start his report with my description of the recurring nightmare-like feeling I have when a layman asks me "Just exactly what did you do to earn the Nobel Prize, Dr. Anderson?" I know that nine times out of ten I can't tell him anything that won't sound like either putting him off or putting him down, since so few people — even few scientifically literate people — have any idea of what a solid state (or, as we now call ourselves, condensed matter) theoretical physicist does. Since John Van Vleck may have been the very first of our breed — the type specimen, as the biologists say — and as he served specifically as the guide and model for my career, it seemed more than appropriate to devote this lecture to trying to tell you something about our mutual field of science. Incidentally, I don't intend to go into what specifically I did, because that was only one of many fields I've worked in and one of the hardest to communicate.

A second recent conversation may serve to start out with. A group of us at the lunch table were discussing the fact that when a university hires one of my experimental colleagues nowadays it is expected to ante up well over a million dollars for lab equipment, while it gets its theorists absolutely free. My collaborator, Elihu Abrahams, asked, "Is this fair? Which of us unlocks the secrets of the universe?" My own opinion on that was there are actually two functions of the safecracker: one is carefully turning the dial on the lock, making things happen; that is the experimentalist. The other is listening for the "click" which tells you when you have the answer. The latter is what we do.

What the story doesn't tell you is that there are at least three kinds of secrets of the universe, even if one sticks strictly to scientific ones and doesn't wander off into philosophy, aesthetics, or religion. One might classify these types of knowledge into (1) the microscopic, (2) the phenomenological, and (3) the bridges or connections. Van Vleck was a famous builder of intellectual and actual bridges, so he would have liked this view of his science. In most areas of human knowledge, one man's microscopic laws can be another man's phenomenology (see Table 1). For instance, consider the situation in molecular biology and evolution. One may think of Darwin's theory of evolution as a great phenomenological generalization, and the existence of genes, mutations, and Mendelian inheritance as the microscopic laws from which the facts of evolution are to be explained by building the bridges and connections. These are called various things, among them evolutionary theory and population biology. Alternatively, the facts of genetics on the cellular level are the phenomenological generalizations to be explained from the microscopic truths of organic chemistry by means of the connections between the two furnished by modern molecular biology.

I think this latter example is the best analogy which I can give you to condensed matter physics. The molecular biologists have contributed very little to the fundamental laws of organic chemistry: they tell us little or nothing about the strength of the peptide bond, the relative stability of alkanes or alkenes, the configurations of carbon-sulfur bonds,

Table 1. Types of Science

Examples:

Microscopic	Bridge	Phenomenology
Facts of genetics	Population biology	Darwinian evolution
Mendelian synthesis	+ evolution theory	
Organic chemistry	Molecular biology	Facts of genetics
Atomic facts:	Condensed matter	Continuum physics of materials
Quantum theory +	physics	
statistical mechanics		

etc. Even the rates of organic reactions and the nature of enzyme catalysis are not yet considered to be part of the field, but rather of the emerging, new 'bridge" field of molecular biophysics. All of this is taken as given; yet molecular biology has been the center of enormous intellectual growth and of brilliant discoveries, both in terms of fundamental knowledge, of new phenomena, and of unexpected practical applications.

Condensed matter physics is just such a 'bridge" field. It takes as given the fundamental laws of quantum and statistical physics, and takes as its "phenomenological" end the everyday, macroscopic continuum physics of ordinary matter (or in some cases pretty extraordinary matter, such as the ultra low temperature superfluid helium three, or the outer solid crust of a neutron star).

When John Van Vleck started the field, back in the late 1920s, along with a few colleagues, like Bloch, Pauli, and Sommerfeld, there was a real need to know whether what he then called the "New" quantum mechanics was really right and really gave better explanations for the properties of matter than did the classical or the old quantum theories. I don't think that after the early 1930s, 50 years ago, when Van calculated accurately the magnetic properties of the iron group and rare earth crystals, and Wigner and Seitz did a roughly accurate job on the chemical binding energy of sodium metal, there was any serious doubt left, that the underlying laws which explain the behavior of all matter are known, at least at the kind of atomic scale which is adequate for everyday matter. What then is the motivation for keeping on with it? Are we just pursuing more and more decimal places in cut-and-dried, crank-turning kinds of calculations which proceed straightforwardly from known theories to known experiments? Admittedly, a few of us do calculate, but I assure you that the process of doing so is a fairly subtle art, not a cut-and-dried turning of cranks.

There is one clear motivation, which is surely a major one for many condensed matter physicists, and which is the most common. reason why laymen have heard of "solid state" physics. This is that the application of quantum mechanics to condensed matter has totally transformed technology by giving us the technical fields of semiconductor electronics and quantum optics (as well as a number of others of great, but relatively minor, importance, such as superconducting circuitry, superconducting magnets, important new materials, etc.). I can imagine a historian looking back from the 25th century and asking: what did the quantum theory do in its first half-century that transformed the world? If he can do this, it will be clear that atomic weaponry did not, fortunately; fission energy will have had a minor, temporary effect on world energy supplies, probably negligible compared to photovoltaic and other solar devices, most of which are solid-state based; but the computer revolution, the guidance systems which made it possible to go into space, and the myriad uses of lasers and quantum optics, will have had a quantitatively overwhelming effect comparable to that of the industrial revolution or the invention of metallurgy. Perhaps only molecular biology has anything like the same potential for causing qualitative change in our lives. So one can argue that the major impact of modern physics on the world has been through condensed matter, even though, somehow, the public consciousness of physics is as the field responsible only for bombs and nuclear energy. Even among physicists one often finds this is so — I just happened to glance, the other day, at a book called, simply, *The Physicists* by a physicist turned historian, which purported to be a history of twentieth-century American physics. The tone of this book can be epitomized from the index. John Van Vleck, for instance, was indexed only for his early 1925 work on the old quantum theory, as was John Slater. The only two-time Nobel prize winner in physics, John Bardeen, was not mentioned at all in the index, nor were the words "semiconductor" or "superconductor" — the two fields of his as well as many other Nobel prizes. Today's popular scientific literature is still dominated by the glamour, the sheer frightfulness, and the expense of the nuclear and postnuclear era. But actually, condensed matter physics is still growing and is by far the largest field of physics, with 30–40% of all physicists, and at least that fraction of papers published.

Actually, however, to drag myself back from this diatribe on my favorite subject, the real reason many of us — and most theorists, in particular — pursue the subject is its sheer intellectual depth and fascination: that, as we attempted to explain the behavior of matter in detail, and with some intellectual depth and rigor, we found ourselves more and more confronted with intellectual puzzles of great subtlety and complexity.

This explanation is almost never straightforward, and is often conceptually as well as mathematically complicated. One finds more and more ways in which these apparently simple laws can lead to very unexpected, and often spectacular, consequences. Often whole new types of substances show up. Some of them are discovered by the experimentalists and only then understood and "explained" by the theorists. That such things are still happening is instanced by last year's (1984) discovery, by Tsui, Störmer, and Gossard, of the "fractional quantized Hall effect", which appears to be a totally unexpected and mysterious new type of condensed two-dimensional electron liquid, explained by Bob Laughlin. (As of 2010 its ramifications are still creating unexpected phenomena.) The theorist's pride, on the other hand, is the anisotropic superfluid phase of the mass 3 isotope of helium 3, first predicted in 1960 by Morel and myself, and not discovered until 12 years later by Osheroff, Richardson and Lee.

Almost all of the difficulties and puzzles have to do with the fact that the laws of quantum physics are couched in terms of individual, isolated atoms and electrons, interacting with each other a few at a time, while what we are trying to understand is the macroscopic object which, for us to see it at all, contains over 10^{20} electrons or atoms. Thus, to a first approximation, the only sensible way to start is by thinking in terms of what is called the "thermodynamic" or $N \to \infty$ limit, the limit in which the number of particles is so large that we think only in terms of averages, limits, and distributions, not of the individual electrons. But in order to calculate these we have to follow the behavior of the individual electrons in the fields of all the others. Thus in every case what we see involves at least one stage of mathematical abstraction of a particularly difficult and fundamental kind. One consequence is that there are no exact solutions: every calculation involves a judicious choice of approximations, usually not just one but many, and the whole process is much more of an art than a science. In fact, we always have to guess the nature of the answer first before solving the problem. Admittedly, there is enormous computational complexity even in doing a precise theory of a small molecule or large atom, but the real problem of size and complexity is an intellectual, not a computational one. The fact is that quantum mechanics, as you learn it in beginning courses, doesn't seem to have any relationship at all to macroscopic properties such as rigidity, conductivity, friction, color, magnetism, etc., and you have somehow to get from the quantum mechanical behavior of all those little electrons to these everyday properties.

There is one other proviso which makes everything much harder: you mustn't cheat by inventing as many hypotheses and approximations as you have experimental facts. Our science is a seamless web: we really believe that everything we see in ordinary matter **does** follow from the few simple equations from which we start, so that — for example — it is essential that the same ideas which you use to understand the total binding energy of a metal like sodium should also explain why it is a metal, roughly what its conductivity is, its magnetic susceptibility, optical spectrum, etc. This is to a great extent the difference between the physicist and the chemist; until recently, at least, the chemists have been much more concerned to make great empirical generalizations such as Pauling's ideas about electronegativity, or tables of parameters like the "atomic radii" and the energies and lengths of covalent bonds between atoms, in order to summarize in a simple way great masses of empirical data. On the other hand, they usually make no attempt to relate the different types of parameter to each other in a consistent way. This kind of chemical intuition is of enormous value and I have long since learned to defer to my chemical friends when it comes to guessing what compounds can be made and what their properties will be; but it doesn't, I hope, satisfy theoretical physicists, who are always searching for a deeper, more fundamental understanding. For instance, try asking a chemist why his most fundamental concept, the chemical bond, is transferable from one compound to another: why is chemistry local? I have worked on the theory which explains that and, I assure you, it doesn't interest chemists.

I imagine you are, by now, tired of airy generalities and bombastic claims, and would like me to try to get down to some cases, preferably from my own career and without any equations at all, to give you some flavor of how condensed matter theory really works. There is such a bewilderment of choice, even from my own limited career. Questions like — what makes a metal? How does a lodestone work? Why is glass insulating? What makes a neutron star glitch? What makes a superconductor? What really is a superconductor? All of these got answered in the course of my career, and I participated to some small extent in each; and most of them illustrate well the second beauty of condensed matter physics: the intimate interdependence between theory and experiment. In most other fields of physics, especially elementary particle but also, for instance, astrophysics, the theorist's role is like that of a soloist in an orchestral concert: he tweedles away by himself for a time, and then shuts up while an enormous organization grinds away to work out the consequences. Theory and experiment in condensed matter physics, on the other band, ideally works like a string quartet, with each instrument taking up the melody in turn, finishing each others' sentences, so to speak, in intricate collaboration.

Let me finish, then, with one example, a particularly wide-ranging one in its overall implications, because it connects at least three apparently almost totally unrelated phenomena:

(1) The decay of supercurrents in a ring: are supercurrents forever?
(2) Catastrophic breakdowns in superconducting magnets: "flux jumps".
(3) The phenomenon of sudden period changes, or "glitches", in neutron stars (pulsars).

Most of you will have heard of superconductivity, probably because the advance of technology has now brought us superconducting magnets as an important component of giant particle accelerators and of NMR tomography for medical purposes. Even before that, I imagine many people had heard that if you start an electric current going in a superconducting ring, it will go on flowing essentially forever. This delightful property, for reasons we now very well understand, while present in almost all metals, visits us only at temperatures at most a couple of dozen degrees above the absolute zero. Shortly after this phenomenon was fully explained, in 1957, a number of us began to worry seriously about whether superconductivity really was forever — under what circumstances could the state with current flow break down and make a transition to the obviously lower energy state, which is that without current flow?

At that time we at Bell Labs had already discovered the gigantic currents and the large magnetic fields which could be sustained by certain superconducting alloys, mostly of the element niobium (Nb_3Sn and NbZr, particularly) discovered by B. T. Matthias and tested by Gene Kunzler. These are the materials in the magnets you have heard about. That in turn was a case where the experimental-theoretical harmonies had been a bit discordant. In the early days of experimental superconductivity C. T. Gorter and H. B. G. Casimir had remarked rather casually that the critical magnetic field should be low, while the Russian theorist, Alyosha Abrikosov, had in 1956 produced the theory of high critical fields — the so-called type II superconductivity — without bothering to publicize to experimentalists that all bets were off and that he had showed Gorter and Casimir to be wrong. Meanwhile it was pure empiricism that led Bernd and Gene to try Nb_3Sn.

A young Korean experimentalist at Bell Labs, Young Kim, working under Ted Geballe, took over the problem and began to test the physics of very high fields and currents. He discovered that there was a "critical state", depending on the two parameters current and field, above which the current in a ring was no longer stable. In trying to understand this "critical state" theoretically, one day I suddenly realized that, actually, the whole phenomenon was not discontinuous: that really the current in the ring is always slowly decaying, but that the rate simply becomes catastrophically slow at the so-called critical state. The next day I asked Young to take a look, and sure enough he found that there was indeed a very slow process going on all the time which we called "flux creep" because it had the characteristic time dependence of a certain kind of plastic flow

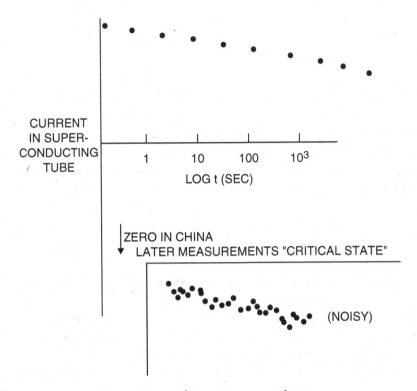

Figure 1. Decay of current in superconductors.

of metals which is called "creep". I remember — incidentally — that when Young Kim and I were invited to give talks about this at the next APS meeting, Bernd Matthias objected bitterly because it made it sound as if his magnet materials were not perfect — even though we calculated the length of time it would take the current to actually decay to zero, and it was something like 10^{99} years (see Figure 1).

To understand the rest of the story we have to have a picture of what was going on. The first key came from Abrikosov, who used an idea which dates back to several very prescient early theorists of superconductivity and superfluidity, namely, Fritz London, Dick Feynman, and Lars Onsager: the quantization of flux and circulation in superconductors and superfluids. The magnetic flux in or through a superconductor can only exist in units $hc/2e$, about 10^{-7} gauss-cm^2.

The cause of this quantization is the condensation of the superconducting electrons into a macroscopically coherent state of Cooper pairs. You don't need to know what all these words mean, but basically, they say that there is a new material parameter, the superconducting wave function, which can be defined at every point at which the material is superconducting. This parameter has a "phase angle" which we can think of as a little clock hand defined at every point in the superconductor, and this clock hand likes to point the same direction everywhere (Figure 2a). In any region where the clock hand is turning as you go through space, there is an electric current given by the rate at which that hand turns: $v_s = h/m$ (rate of turning) (Figure 2b).

Now when a magnetic field passes through the superconductor, it induces currents which act in such a way as to cancel out the magnetic field — to screen it out. It turns out that if exactly one field quantum runs through the superconductor, the current around the quantum can vanish if the little clocks turn through exactly 360° around it.

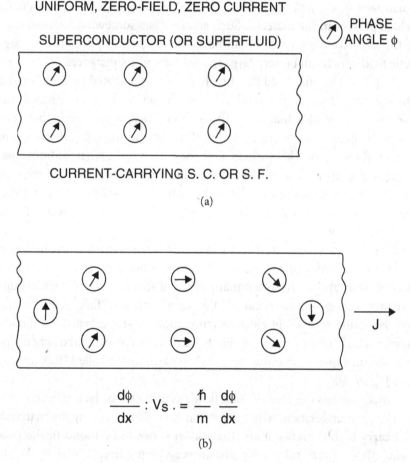

Figure 2. Current in superconductors and superfluids and the "phase angle".

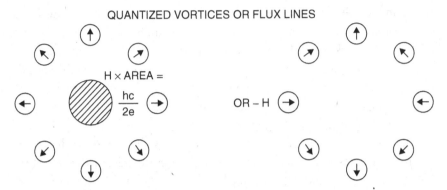

Figure 3. Quantized vortices and flux lines.

The picture, then, of a single quantized flux line is shown in Figure 3. Then Abrikosov's state for a "type II superconductor" is an array of these "quantized flux lines", their density corresponding to whatever magnetic field is present in the material. Such an array is a somewhat strained state of the superconductor, but it can be stable, up to an upper critical magnetic field which can be very large indeed for some substances.

As I said, the quantized flux line had been anticipated by London and Onsager, and derived by Alyosha. What my new insight contributed was the very simple idea that these things aren't necessarily static, but may move. If there is an overall current through the material, that current pushes sidewise on the vortices and they may slide. That sliding causing a voltage to appear (which happens, for not quite so simple reasons, to be given by Faraday's law of induction $V = 1/c \cdot d(\text{flux})/dt$) and this has two consequences: the flux decays, and resistance appears in the superconductor.

From our first insight, that a material could remain superconducting and yet have a very tiny resistance, Young Kim went on to create a whole field of study and to show that many kinds of superconductors show quite large resistances — that is called "flux flow" and not "flux creep" — and we also came to a fairly clear interpretation of the catastrophic breakdowns which often, especially in those days, were exhibited when one tries to run superconducting magnets too hard: the so-called "flux jumps" (see Figure 4).

Once one has the idea of "flux flow", as a matter of fact, it becomes at first hard to understand why the motion is so slow, and why the materials can carry high currents at all. The reason is not to be found in the pure materials — these really don't sustain very high currents at all — but in the presence of all kinds of dirt and disorder in the technical materials

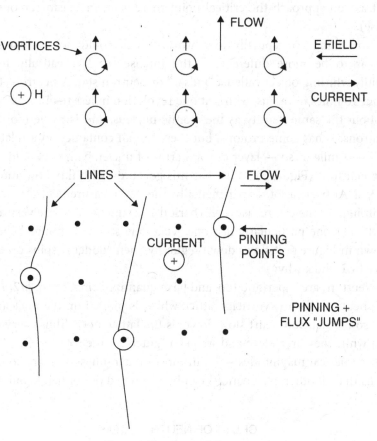

Figure 4. Flux flow and flux creep.

which are actually used. As, actually, all hard metals are so only because they are dirty and messy atomically, this is also true of the good "hard superconductors" which are used technically. This was another phrase we invented at the time. How the same complex of ideas was the foundation for many other fascinating branchings of science — particularly for the discovery of the Josephson effect, which shortly ensued — is another story. It is also another story that this is the first example of a subject which is becoming fashionable in the 1980s: sliding friction between interpenetrating lattices, the atomic underlying lattice of the superconductors and the quasi-rigid, 2-dimensional lattice of the quantized flux lines (Figure 3).

Keep with you this picture: a hard superconductor, carrying nearly too much current, so the flux lines are barely hanging on, every once in a while letting loose and having a 'flux jump" (we showed, in those days, that as you approach the critical point there is an increasing amount of noise).

Now jump conceptually to a world which could not at first sight appear to be more different — the intensely hot, dreadfully heavy, rapidly whirling object called a "pulsar" or neutron star. A neutron star is about 10 miles in radius, whirls at one revolution in less than one second, is about the same density as the atomic nucleus, and is made mostly of neutrons. It has some protons, but these are all contained in a relatively thin — a mile or so — layer of solid crust of nuclei, bathed in a liquid of free neutrons (Figure 5). What has this got to do with flux flow and flux jumps? Answer: a lot. Experimentally, the key measurement for pulsars is timing, so one can measure the period as a function of time very accurately. For one particularly fast one, Vela, one sees the strange behavior shown in Figure 6. It slows down regularly, then suddenly spins up every once in a while: why?

Neutrons are superfluid too, and have quantized *circulation: h/2m*. The free neutrons form a vorticity lattice which is pinned on the nuclear lattice so the neutrons can't slow down as the lattice does. Thus every once in a while they break free and we get a "glitch" in the rotation frequency. The numerical magnitudes — as in most of astrophysics — are so enormous that all other explanations can be dismissed out of hand, and while

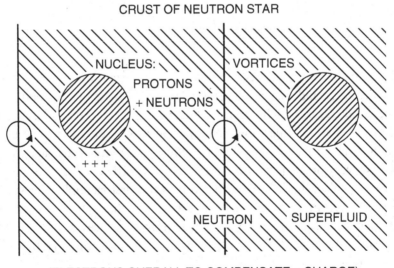

Figure 5. Neutron star physics.

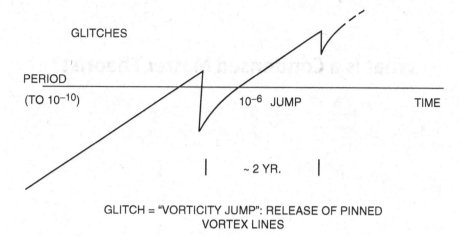

GLITCH = "VORTICITY JUMP": RELEASE OF PINNED
VORTEX LINES

Figure 6. Typical period history of the Vela pulsar (not to scale).

the details are still under adjudication the essential physics can hardly be in doubt. For instance; one of our predictions was that all pulsars will eventually glitch, not just the fast ones, and as observations accumulate this turns out to be the case.

I am particularly pleased because this is a good example of the unification of three branches of physics: nuclear physics, astrophysics and condensed matter physics. Van was fortunate enough to do most of his work before physics had split itself up into its various subdisciplines. What I have been talking about would to him have been simply "physics" and he hardly would have felt it necessary to explain what he did. But now it is a rare and happy chance that one idea can be successful in several fields, and I am very pleased to have been able to share one such instance with you.

What Is a Condensed Matter Theorist?

Most Nobel Prize winners are identified with a single clear-cut discovery or invention — you can say of us, "Oh, he's the Big Bang" or "he's the Josephson Effect" or "he's PCR" or "the transistor" or "quarks" or "the laser". I am not so fortunate; and in fact occasionally lay people (for instance, but not only, autograph hounds) send me snippets of descriptions of the things they have read about me which they expect to be complimentary, but in which I am hard put to it to recognize myself. We condensed matter theorists get a bad press even among professional historians of science and science writers because nobody seems to know what we do. So what does a condensed matter theoretical physicist do? In what does he differ from a computational chemist? Will his field die off as he is replaced by ever more sophisticated computing, as in fact has been confidently predicted for decades now? How can he discover anything unexpected or unusual, when the fundamental laws with which he works, namely electro-magnetic theory, statistical mechanics, and the quantum theory, have been known for 70 years? Can he do anything but just be the computational adjunct to his useful if humdrum experimental colleague, who produces new materials or new hardware devices? The demand for these, mostly military, has after all slacked off since the Cold War ended.

It is true that we condensed matter people have ambiguous goals: we are both curious and useful. Our useful research can have great intellectual spinoffs, just as curiosity driven research can often lead to useful devices — I will give you examples of both.

It has only recently become intellectually fashionable to give a general title to the kind of thing we do: we are engaged in the search for "emergent properties". The idea of emergence began in biology, where it was realized that the course of evolution is one long story of novel prop-

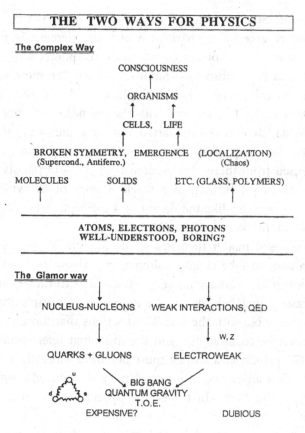

Figure 1

erties emerging out of the synergy of large collections of simpler systems: cells from biomolecules, organisms from cells, societies from organisms. But that emergence could occur in physics was a novel idea. Perhaps not totally novel: I heard the great evolutionist Ernst Mayr claiming that 30 or 40 years ago, when he described emergence to Niels Bohr, Bohr said: "but we have that in physics as well! — physics is all emergent", but at the time, as usual, only Bohr knew what he meant.

In fact, the story of physics in the last half of the 20th century has been one of emergence — Bohr was also, as usual, basically right. But there are two kinds of physicists, who come at emergence from two separate points of view. (See Figure 1.)

One group, the particle physicists and, more recently, the cosmologists, start from the properties of the elementary constituents of the universe as we see them and try to deduce the simpler, smaller or more primitive laws and entities from which they arose. They have a great success in the "Standard Model" of the elementary particles, which is basically an "emergence" scenario in which the elementary particles as we know them — neutrons, protons, electrons, photons — have come through at least two transmogrifications from simpler, more general and more symmetrical entities, the quark-gluon field and the electroweak unified interactions. This fits well with the cosmologists' "hot big bang" scenario in which everything started out as a uniform, symmetrical, inexpressibly hot soup of all the possible elementary fields, and cooled and condensed from there. But modern theoretical physicists are trying to reduce things yet further, to a single Theory of Everything, which sounds to the outsider like the dream of a medieval scholastic. In these senses, Bohr's remark is seen as valid.

The condensed matter theorist's intellectual challenge is to turn the process around: to take as given simple, underlying theories on which the behavior of all ordinary matter is based, and to understand, organize and classify that behavior in all its observed complexity in a way which is compatible with the severe restrictions that those laws enforce. Particle theorists, cosmologists and the like often refer jocularly to the "tooth fairy" principle: a theory must not have more than one "tooth fairy". The T.F. is an ad hoc, arbitrary element introduced simply to make the theory fit the facts. (In respectable speech, this principle is called "Ockham's razor".) But the condensed matter theorist is much more severely restrained: in principle, he is *not allowed any tooth fairies at all*, because the laws of physics at his level are *really exactly known*. It is this aspect of his work that I think layman — or even other scientists — least often grasp. In principle, we should be able to understand the mechanism for everything in the physical world around us.

It sounds easy, but it isn't. The reason is again the phenomenon of emergence: what is possible or characteristic for an atom relevant to what may be possible for a large collection of them.

The phenomenon of "localization", which was one of the ideas cited in my Nobel prize, involves actually a very neat example of this concept of emergence. One of the most obviously emergent properties of large collections of atoms is that of being metals. One can't in any way accuse a single atom of copper or zinc or aluminum of being a metal: a metal occurs when the atoms are condensed into a solid and the electrons free themselves from their individual atoms and form a free "Fermi liquid" in the "energy band" formed from states on all of the atoms. The metal is shiny and ductile, but its defining property is that it conducts electricity by a mechanism which is entirely quantum-mechanical, depending on the fact that because the electrons obey Fermi statistics, they are forced to have a very large "zero-point" kinetic energy, even at the absolute zero of temperature. Therefore in a metal the electrical conductivity not only does not vanish at low temperature, it improves — even in a non-superconductor — as you lower the temperature.

An insulator, such as the plastic which covers electric wires or the glass and ceramic which are used for power lines, is quite different. These materials only conduct electricity when their electrons are freed from their local homes by thermal energy. The defining characteristic of an insulator is that its conductivity is *exactly zero* in the limit as $T \to 0$ and usually is small even at normal temperature, and decreases very fast (exponentially) at low T. This means that its electrons are in locally bound states: they stay at home. "Exactly" zero? Yes. This illustrates the philosophical point I wanted to make: that in condensed matter physics there are precise dichotomies, not compromises. Because we are dealing with materials in the "thermodynamic limit" where the number of atoms is taken to be infinite — $N \to \infty$ limit — and only in that limit — we can say that there is a precise mathematical distinction between a metal and an insulator. My concept of localization was an attempt to understand what that distinction is, and what the "metal-insulator transition" consists of.

There had long been a supposed answer, not a wrong one but partial, imprecise, and incomplete, furnished by Bloch's band theory of metals as interpreted by Mott and Jones: that insulators had an energy gap in their electronic spectrum of energy levels, which had to be overcome to conduct electricity. This is imprecise and incomplete because no real substance is perfect or has a perfect gap, and in many cases the energy levels of the electrons in such a gap caused by imperfections do act metallically.

What I showed is that in fact what is responsible is a very general property of the spectra of waves propagating in a large, disordered system: that the states or "eigen-functions" have two possible natures depending on the parameters of the system, either extended throughout, or localized: and at any energy they are all either one, or the other. It is not important for our purposes here why this is so, or how to show it; it is important to say that while it was utterly unexpected and new, it follows perfectly from laws of physics that were well-known at the time.

It was also what Merton has called a "premature" discovery: so new that the world of science was unable to absorb it for about a decade, and in fact it was almost exactly two decades — after the Nobel Prize — before it began to develop into a really active field of physics, partly due to my own efforts in the two or three years after 1977. (I must not ignore, also, great contributions by Sir N.F. Mott.) What our group (called, appropriate to the times, the Gang of Four) did was in essence mainly a sugar coating which allowed the phenomenon to be seen and studied more in context by conventionally-trained physicists. Nonetheless these are some of my most heavily cited papers.

This, then, has been a fairly typical example of my work as a condensed matter theorist during more than four decades. It is striking how often certain general features of the various discoveries repeat, in different degrees, of course, in every case:

(1) The emphasis on new emergent properties of matter, on singularities or new behaviors occurring in the $N \to \infty$ limit.

(2) This follows often from the recognition of a question, a paradox, or a dichotomy, often one that is quite simple, obvious or evident but is not noticed by the "normal" scientist (in the Kuhnian sense.)

(3) The phenomenon of the "premature discovery": either strong opposition, controversy or skepticism from the community as a whole, or, more characteristic and almost more frustrating, the total lack of response which comes from non-recognition that the *question* exists. Both have to do with the idea that such discoveries tend to lie outside of the normal practice of sciences; the bulk of scientists don't recognize the method or even the question as part of science as they know it; or the discovery may lie so far outside what is assumed to be the accepted wisdom as to appear an irrelevant fluke.

I recently found a wonderful illustrative quotation in a biographical memoir of the late great geneticist Barbara McClintock, who received the Nobel Prize in 1983 for a discovery completed and announced in 1951. Of her early papers on the discovery a very eminent colleague said publicly, "I didn't understand one word she said, but if she said it, it must be so!" This was the attitude of many towards my idea of localization.

Let me continue with one more example from my own history, the example of the phenomenon of broken symmetry. It's actually quite timely, because it starts from a discovery made by the winner of a recent Nobel Prize, Cliff Shull. It isn't emphasized in the publicity around his work, but the big thing he did, which had us all excited in 1951, was the experimental proof of the existence of antiferromagnetism: an alternating spin structure with no net magnetic moment, which he found

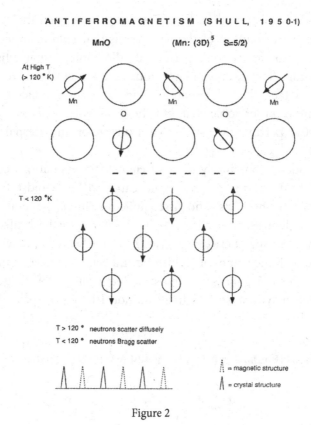

ANTIFERROMAGNETISM (SHULL, 1950-1)

MnO (Mn: (3D)^5 S=5/2)

Figure 2

in, first, MnO. (See Figure 2.) Although this phenomenon had already been postulated by Neèl, Landau and Van Vleck, there was considerable skepticism about it because the only exact solution of a system with anti-ferromagnetic exchange was the one-dimensional chain, solved by Bethe in 1931, and that did *not* exhibit antiferromagnetism. Such great theorists as Landau and Kramers were very much involved in this question. In fact, the experimentalists had no idea they would see it, and the phenomenon came as a great surprise: They saw strong, sharp new Bragg peaks in the magnetic scattering, in the transition metal compounds MnF_2 and MnO.

Some of my very first theoretical papers were aimed at explaining this phenomenon. The objections of Landau and Kramers were perfectly reasonable: it is possible to show that the most stable quantum state of a perfect antiferromagnet should be a singlet, zero-angular momentum state, perfectly isotropic in its properties — yet the antiferromagnetic state is clearly, by Cliff's measurements, very *anisotropic*: the moments

point in a specific direction. What I seem to have been the first to realize is that the resolution of this paradox is that *both* answers are right:
if the antiferromagnet were perfect, its direction of magnetism would
have quantum fluctuations of enormous magnitude which would restore
isotropy. In fact, these fluctuations have a residual effect in reducing
the amount of order somewhat, but the divergent part is easily removed
by any weak perturbation such as anisotropy or an external field. (See
Figure 3.)

These fluctuations imply a set of elementary excitations called "spin
waves" for which the general name came to be "Goldstone Bosons",
for one of the physicists who first applied a similar general scheme to
theories of elementary particles. (I lay no claim to the brilliant idea of
building this analogy between the two fields of theoretical physics. This is
associated with the names of Goldstone and Nambu.) (See Figure 4.)

The antiferromagnet is an almost perfect example of the general theory of broken symmetry — which did not, I may say, spring full-blown

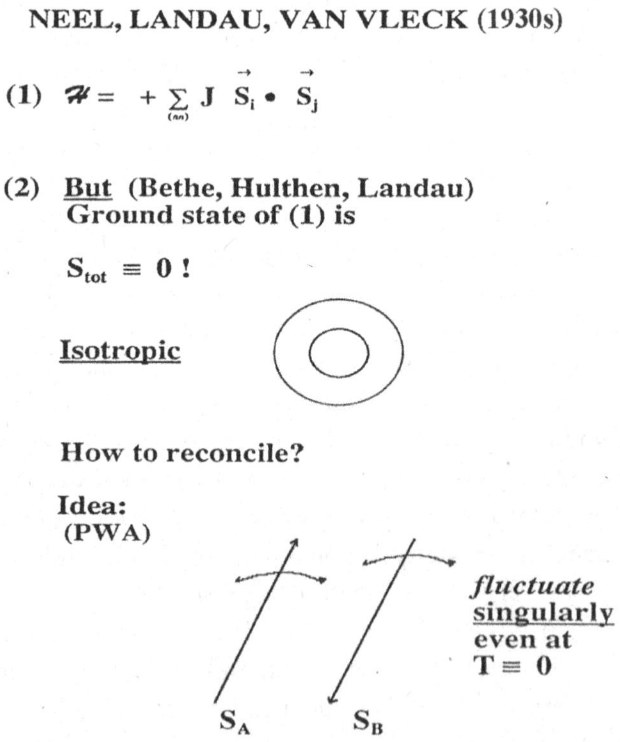

NEEL, LANDAU, VAN VLECK (1930s)

(1) $\mathcal{H} = + \sum_{(nn)} J \, \vec{S}_i \cdot \vec{S}_j$

(2) **But (Bethe, Hulthen, Landau)
Ground state of (1) is**

$S_{tot} \equiv 0$!

Isotropic

How to reconcile?

**Idea:
(PWA)**

fluctuate
<u>*singularly*</u>
even at
$T \equiv 0$

S_A S_B

Figure 3

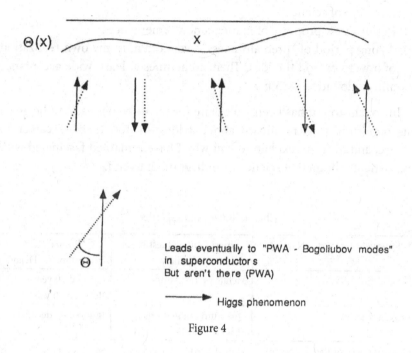

FLUCTUATIONS ARE "ANTIFERROMAGNETIC SPIN WAVES'

(MUCH LATER) - "GOLDSTONE BOSONS"

Leads eventually to "PWA - Bogoliubov modes"
in superconductor s
But aren't there (PWA)

Higgs phenomenon

Figure 4

from my mind at that time in 1952, but was built up piece by piece over many years but particularly in response to the *next* discovery, in 1957, of a hidden broken symmetry, which was the BCS theory of superconductivity. With a mind prepared by the example of antiferromagnetism, I was able to analyze the quantum fluctuations in this case and to see both the necessity for Goldstone modes and how to eliminate them, which is the mechanism now called after Peter Higgs, who reinvented it in a form more palatable to particle theorists.

There is yet more about this general theory which was gradually realized over the years by many theorists: the wonderful topological theory of defects finally codified by Toulouse and Volovik, the general theory of dissipation in condensed matter systems, etc.; but I have no time here to try to explain all these things.

What I would like to do is to take the few minutes remaining and to run you through a partial list of these "emergent properties" with which I have been involved in the course of my career, and reminisce a bit about how well or badly each of them fits this general template which I have been discussing:

(1) Recognition of an anomaly, a paradox, a problem, *outside* of the main direction of science.
(2) Result a new property of matter, a new "emergence".
(3) A long period of "prematureness", even, often, in my own realization of how to exploit the idea. Then, sometimes at least, wide acceptance and exploitation by others.

In conclusion, it has been a pleasure for me to be here and to be given this opportunity to reminisce about a long — too long — career in science, and to try to explain to you why I have remained fascinated with my particular branch of science from beginning to end.

Chart of Emergent Properties

Emergent Property (Names)	Experimental Question of Anomaly	"Premature"? Resistance + Delay
Localization of Waves	Metals vs. Insulators	Ignored 10 years Resisted 10 years
Broken Symmetry	Quantum fluctuations in ordered states	Ignored — decades
Defect Theory of Breakdown of B.S. (Toulouse, Volovik)	How do ordered states dissipate	~2–3 years
Non-Ergodicity (Spin Glass) (Toulouse, Edwards, Palmer, Kirkpatrick)	Phase Transition without Symmetry Change	5 years (ignored)
Superexchange: Magnetic Signal of "Mott Transition" (Mott)	Insulator=Antiferromagnet and vice versa	Mott Transition 30 years
Non-Fermi Liquids in > 1D	High T_c	Bitter resistance: >10 years (now 25)

Global Economy II

Or, How Do You Follow a Great Act?

The following was my summary talk for a second meeting of the central figures in the first (1987) Santa Fe workshop, which took place in October 1988. It was notable for the presence of a number of participants who show up in the national news nowadays — Paul Krugman, Andrei Shleifer are two I remember. Ken Arrow contributed a somewhat disorganized but fascinating discussion — and, of course, many of the participants in these workshops were his students. I cannot help but feel that he and, under his influence, the Santa Fe program, was an important but, in the end, impotent, counterweight to the baleful influence of Milton Friedman and the Chicago school.

Global Economy II, the September 1988 workshop on the G.E. as a C.A.S.,* seemed to those of us participating to be the appropriate sequel to last year's opening workshop. The sense one had was of a consolidation and deepening of a number of lines of thought which originated at the first one but had needed a year's time to develop into a programmatic approach to coherent research directions. The meeting was slightly shorter and considerably less formal than last year's, and leaned much less on formal presentations than on subgroup meetings; one even sometimes had the feeling that almost the entire workshop would meet as a working group on one subject or another.

I would like, more for the sake of opening a discussion than closing one, to identify three basic themes which seemed to continue through the two meetings and, eventually, to lead to certain groupings of people who will follow them up; and at least two new ideas — perhaps "tools" in a

---

*C.A.S. is the acronym of complex adaptive system.

general sense — which surfaced and which seem to be important to the future developments which I hope will happen.

The three themes were:

(I) further developments essentially on prediction and analysis of the economy as a dynamic system. While we are still keeping in mind the whole spectrum of learning algorithms, I felt that a major effort should be to establish a network as Doyne Farmer is attempting to do, sharing data so the different methodology may be used on the same data as well as allowing a maximum input of parallel time series. Doyne seems to be making progress with the troublesome question of how to introduce exogenous data.

Geoff Grinstein presented a model of a large system which did not exhibit deterministic chaos; I felt it was not a good descriptive model for physical systems, where often only one or a few modes exhibit non-linear behavior (the rest being "slaved" in Haken's sense), where Geoff's was a collection of intrinsically non-linear oscillators. This is however possibly a better model for economic systems where the individual agents themselves are quite complex and non-linear.

One question we physical scientists have asked from the beginning, but has yet to be answered by our colleagues in economics: has anyone attempted to fit a "power-law noise" spectrum such as "$1/f$" to economic data, as opposed to the conventional regression analysis? Many possible dynamic models might show such a fluctuation spectrum.

(II) A second theme is a continuation of the "economic web" ideas of the earlier workshop. The grand project which seemed to emerge from this grouping was the possibility of modeling the process of economic and technological growth as a complex web of interdependent agents, each of which can grow in productivity with an "entry cost" which is decreased by the growth of neighboring entities: automobile production encouraging steel plants to modernize as well as the construction of roads and fast food outlets; this kind of dependency infinitely proliferating was the picture we had in our minds. We felt such a web would exhibit "self-organized criticality" (see later) and hence noisy behavior on all scales. Simpler and/or somewhat different models such as the Scheinkman-Boldrin one were explored and many will also deserve further work.

(III) A third theme I would designate as general studies of game-playing and markets. Such special topics as Wicksell triangles and the development of money, double oral auctions, and so forth belong in here: my sense is that coherent themes in this area have not yet emerged but that some quite massive and important issues specifically in the global economy — debt problems, trade barriers, exchange markets, industrial strategies — have their appropriate forum in this area. Personally, I found fascinating the questions about the origin and meaning of money which one can only approach this way. Perhaps the most valuable aspect of this kind of activity is that it offers a wide variety of examples of non-classical economic behavior contradicting conventional equilibrium ideas, and as such is a relatively painless way of liberating our minds from the conventional wisdom.

I was disappointed that the group firmly rejected the possibility of expanding into the area of political or psychological feedback effects (but see below).

One of the new sets of ideas which impacted in several places was the examination of the limits of rationality which Ken Arrow specifically talked about as well as Andrei Shleifer, and which was a major area of discussion. To my mind, rational expectation was shown to be an exceptional phenomenon, which I felt was associated more with economists' hopes that a general equilibrium exists in principle, than with observations of actual economic behavior. It may be that poker players act on rational expectations as they play, but if their behavior were really rational would they be playing poker at all?

The world, possibly, and certainly most peoples' perceptions of it, is much less foreseeable than economists have assumed. One is left wondering whether the common sense which, for instance, ignores prior probabilities, is not rooted in a possibly healthy skepticism about any received wisdom or apparently inevitable outcome.

In any case, the relative weakness of anticipation — or a high discount rate for the future — makes the economy much more similar to ordinary dynamical systems and thus much more likely to have behavior physicists can understand, as well as to have unstable, chaotic, or noisy behavior. I felt that this encouraged me, at least, to look more seriously on the possibility of low dimensional chaos or of more complicated behavior such as self-organized criticality.

This latter topic was a new "tool" which I brought to the community: an idea pioneered by Per Bak for physical systems, it is an attempt to explain the widespread occurrence of "scale-free" behavior such as fractal shapes of beaches, landscapes, aggregates, etc.; and to relate this to scale-free dynamic behavior such as ubiquitous "$1/f$ noise" which is exhibited by such different systems as the river Nile and neuron firing as well as by carbon telephone microphones.

The idea refers to systems for which there is a wide difference in scale between the macroscopics of a system and the rate at which it is driven (the canonical example is sand flowing through an hourglass and sliding down the pile, where the individual grain motion is much faster than changes in the pile as a whole). It is argued that for many models of such systems the system is driven to a new so-called "critical state" where large areas are nearly unstable, and a single additional "grain of sand" can cause an avalanche of almost any scale.

I, at least, was eager to test this idea on models of economic "webs" to see if they can exhibit self-organized critical instability, and I hope that the webs group will eventually get some such system formulated. This is also the motivation behind the question whether economic time-series exhibit "$1/f$" or other power-law fluctuations, to test whether there could be an approximately scale-free internal mechanism.

In general, I sensed a growing willingness among the economists to go a bit beyond the normal boundaries of their subject, and possibly a greater attempt to approach economic professionalism on the part of the physical scientists. I still see the program as an adventure in more or less gently turning aside some channels in economics — hardly the mainstream! — into new and more promising directions. For those new directions non-linearity, positive feedback effects, and lack of foresight dominate economic behavior, in contrast to mainstream economics. If we do not attempt this task, we have no reason for being. We can do no less than try to find a set of paradigms for the future, even if their first versions look very primitive, arbitrary and/or artificial.